AI-Based Data Mobility and Intelligent Modeling for Smart Cities

Sultan Ahmad
https://orcid.org/0000-0002-3198-7974
Prince Sattam Bin Abdulaziz University, Saudi Arabia

Sudan Jha
https://orcid.org/0000-0003-0074-2584
Kathmandu University, Nepal

Md Alimul Haque
https://orcid.org/0000-0002-0744-0784
Veer Kunwar Singh University, India

IGI Global
Scientific Publishing
Publishing Tomorrow's Research Today

Vice President of Editorial Melissa Wagner
Director of Acquisitions Mikaela Felty
Director of Book Development Jocelynn Hessler
Production Manager Mike Brehm
Cover Design Jose Rosado

Published in the United States of America by
IGI Global Scientific Publishing
701 East Chocolate Avenue
Hershey, PA, 17033, USA
Tel: 717-533-8845 | Fax: 717-533-7115
Website: https://www.igi-global.com E-mail: cust@igi-global.com

Library of Congress Cataloging-in-Publication Data

Names: Ahmad, Sultan (Computer scientist) editor | Jha, Sudan, 1975- editor | Haque, Md Alimul, 1981- editor
Title: AI-based data mobility and intelligent modeling for smart cities / edited by Sultan Ahmad, Sudan Jha, Md Alimul Haque.
Description: Hershey, PA : IGI Global Scientific Publishing, [2026] | Includes bibliographical references and index. | Summary: "This book aims to review the role of artificial intelligence (AI), machine learning (ML), and deep reinforcement learning (DRL) in the evolution of smart cities"-- Provided by publisher.
Identifiers: LCCN 2025030706 (print) | LCCN 2025030707 (ebook) | ISBN 9798337342023 hardcover | ISBN 9798337342030 paperback | ISBN 9798337342047 ebook
Subjects: LCSH: Smart cities | Artificial intelligence
Classification: LCC TD159.4 .A434 2026 (print) | LCC TD159.4 (ebook)
LC record available at https://lccn.loc.gov/2025030706
LC ebook record available at https://lccn.loc.gov/2025030707

British Cataloguing in Publication Data
A Cataloguing in Publication record for this book is available from the British Library.

Table of Contents

Detailed Table of Contents

A key component in modern technology, artificial intelligence (AI) is changing industries, promoting scientific research, and influencing social interactions. Fundamentally, artificial intelligence (AI) is the development of systems that simulate human intelligence, allowing machines to carry out operations like learning, reasoning, problem-solving, and decision-making. These capabilities are powered by AI-driven models, which process large datasets to find patterns, forecast outcomes, or carry out actions with little assistance from humans. These models, which have an impact on industries like healthcare, finance, education, transportation, and urban planning, range from straightforward algorithms like linear regression to intricate deep learning architectures. AI models, for instance, help with traffic flow optimization in smart cities and disease diagnosis from medical images. By examining their historical development, fundamental paradigms, important algorithms, mathematical foundations, and development processes, this chapter offers a fundamental understanding of AI-driven model

Abstract Intelligent and contextual decisions at the network edge are opening up exciting new possibilities in real-time data movement as well as urban IoT network safe. Diverse machine learning techniques are deployed in these models which are constantly evolving. They enable the live processing and analysis of data streams from a wide variety of IoT sensors located in an urban environment: this also facilitates adaptation to the rapidly changing conditions surrounding us as well as adapting itself quickly to widely different application needs. The collaborative deployment of large AI models (LAMs) throughout edge devices allows the system to learn across multiple modalities and scenarios. In this way, it increases scalability for smart urban services. Thus, adaptive AI frameworks not only fuse in stringent security measures such as anomaly detection and federated learning but also satisfy the requirements for real-time threat identification as well as data privacy protection under resource-constrained conditions in dynamic networks.

To build smart cities, we need to completely change how we design, run, and manage transportation systems in cities. Artificial Intelligence (AI) is a game-changing tool that enables us to model human movement, enhance transportation networks, and predict travel patterns with unprecedented accuracy. This chapter gives a thorough overview of the most recent developments in AI-based mobility modeling, with a focus on how transportation has changed from traditional methods to smart, dynamic, and data-driven infrastructures. It examines how AI methods, such as machine learning, deep learning, and reinforcement learning, are applied in transportation systems to enhance real-time decision-making, traffic forecasting, demand-responsive services, and autonomous driving. The chapter also discusses the most pressing problems, moral issues, and future directions for integrating AI into urban transportation systems to create smart, sustainable, inclusive, and resilient cities.

The development of many aspects of a smart city heavily relies on artificial intelligence (AI) technology. The most recent AI technologies for building smart cities are examined in detail in this article. According to the analysis, artificial intelligence (AI) is crucial for smart cities both now and in the future. Numerous issues have arisen as a result of the fast population growth, particularly in urban areas. These include garbage accumulation, water scarcity, pollution, transportation congestion, and health issues. The smart city plan is one of the initiatives that governments have launched to address these problems. This plan aims to use contemporary technology to deliver dependable, environmentally friendly services while also enhancing quality of life. In pursuit of this objective, scholars and professionals have developed numerous frameworks for the establishment of smart cities. One of the most important technologies for creating the components of a smart city is AI. Environmental considerations are crucial while creating solutions.

overhead, and data imbalance are critically analyzed, with solutions and future policy directions proposed to ensure ethical, secure, and sustainable adoption of FL in citizen-centric smart cities.

This chapter explores techniques for detecting abnormal behavior in surveillance systems using deep learning. Modern surveillance relies on advanced models to identify suspicious activities crucial for safety and monitoring. The discussion traces the evolution from traditional rule-based methods to deep models such as convolutional and recurrent neural networks, autoencoders, and transformers. It highlights challenges in defining normality, managing temporal variations, handling large-scale data, and ensuring privacy. The chapter examines temporal behavior models, including LSTM encoders, 3D CNNs, and attention-based architectures. Real-world applications in video surveillance, crowd monitoring, and human activity analysis are reviewed. Issues like data imbalance, real-time inference, model explainability, and ethics are discussed. Finally, performance metrics, evaluation protocols, and model comparisons are outlined to guide readers in designing effective anomaly detection frameworks for practical use.

We present an automated machine learning model management system, which provides feedback loops, at-rest and in-flight observability, and continuous monitoring for sustaining and improving model performance in the long term. The system is built to support automated retraining and redeployment to mitigate challenges associated with data drift or the evolution of operational norms. Our models achieved 99.14% accuracy, 99.19% precision, 99.17% recall, and 99.18% F1 scores in Random Forest and Gradient Boosting. Our system demonstrated superior on-line observability and adaptability to AutoWeka and more conventional Gradient Boosting. Continuous feedback with on-going observations allows users to leverage our system in scalable, decision making applications, in non-stationary conditions while ensuring ongoing performance.

Image captioning, which is the generation of descriptive word text summaries from image content, has drawn considerable interest in computer vision and natural language processing (NLP). This research proposes a Python application that combines Vision Transformers (ViT) and GPT-2 for automatic image captioning. The system employs a pre-trained NLP connect/vit-gpt2-image-captioning model from Hugging Face, coupled with a graphical user interface (GUI) designed using Tkinter. The model efficiently extracts features from images and produces coherent, contextually appropriate captions, showing improvement over conventional Convolutional Neural Network-Long Short Term Memory(CNN-LSTM) based models. This study emphasises the architecture, methodology, and comparison of the system, highlighting its applicability in real-world applications such as visually impaired accessibility, content management, and image retrieval. Performance measurement suggests the model's capacity to produce high-quality captions in an efficient manner.

Kerala's waste management practices, this research shedding light on significant rural-urban disparities and emphasizing the importance of targeted interventions in rural regions, particularly addressing kitchen waste disposal. Drawing from a robust dataset of 46,334 respondents in Kerala, sourced from NFHS 5, the study uncovers a compelling commitment to responsible waste management. Both rural and urban areas exhibit a shared dedication to sustainability, with a consistent preference for closed drain disposal and reusing kitchen waste for gardening or farming. Result: Encouragingly, urban regions show higher levels of awareness and engagement in

Preface

INTRODUCTION

The rapid acceleration of urbanization across the globe has created unprecedented challenges for city planners, policymakers, technologists, and citizens alike. As urban populations expand and the demand for sustainable infrastructure intensifies, cities must evolve into intelligent ecosystems capable of efficient governance, adaptive service delivery, and data-driven decision-making. It is within this transformative context that we present *AI-Based Data Mobility and Intelligent Modeling for Smart Cities*.

This edited volume explores Artificial Intelligence (AI)-driven data mobility and intelligent modeling approaches that are shaping the development of next-generation smart cities. Smart cities are envisioned not merely as technologically advanced urban spaces, but as sustainable, inclusive, and resilient environments that optimize energy consumption, promote environmental stewardship, enhance economic growth, and improve overall quality of life. Central to this vision is the effective integration of Information and Communication Technologies (ICT) into urban policy design, implementation strategies, and public service delivery systems.

The primary objective of this book is to examine the evolving role of Artificial Intelligence (AI), Machine Learning (ML), and Deep Reinforcement Learning (DRL) in addressing complex smart city challenges. These computational paradigms enable predictive analytics, autonomous optimization, intelligent resource allocation, and adaptive governance mechanisms that are essential for modern urban systems. By leveraging data-centric intelligence, cities can transition from reactive management models to proactive and self-optimizing frameworks.

This volume presents a comprehensive functional framework and workflow for digital transformation toward smart city ecosystems. Emerging technologies such as AI and the Internet of Things (IoT) are positioned as key enablers of data mobility, intelligent modeling, and interconnected urban services. The chapters included in this book cover both foundational principles and advanced methodologies in data modeling, system integration, and security considerations. Particular emphasis is placed on secure, scalable, and sustainable architectures capable of supporting future-ready infrastructures, especially in the context of the evolving 6G era.

One of the distinguishing features of this book is its interdisciplinary perspective. It brings together leading academic scientists, researchers, and scholars to share their insights, research findings, and practical experiences. The contributions collectively address theoretical advances, implementation strategies, performance optimization, and real-world applications of AI-enabled technologies in smart city development.

We envision this book as a valuable resource for researchers, industry professionals, policymakers, and graduate students. For practitioners, it serves as a guide to designing and deploying AI-powered smart city solutions. For newcomers and early-career researchers, it provides a structured foundation for understanding the principles, tools, and applications of intelligent urban technologies.

As editors, we extend our sincere gratitude to all contributing authors for their scholarly efforts and dedication. We also acknowledge the reviewers and publishing team for their support in maintaining the academic rigor and quality of this volume.

We hope that this book will stimulate further research, foster interdisciplinary collaboration, and contribute meaningfully to the global pursuit of sustainable, intelligent, and citizen-centric smart cities.

ORGANIZATION OF THE BOOK

As editors of *AI-Based Data Mobility and Intelligent Modeling for Smart Cities*, we are pleased to present a collection of chapters that collectively reflect the depth, diversity, and transformative potential of artificial intelligence in urban ecosystems. Each contribution advances our understanding of how intelligent data mobility, adaptive modeling, and secure infrastructures are shaping the cities of the future.

Chapter 1 explores a key component in modern technology as artificial intelligence (AI) is changing industries, promoting scientific research, and influencing social interactions. Fundamentally, artificial intelligence (AI) is the development of systems that simulate human intelligence, allowing machines to carry out operations like learning, reasoning, problem-solving, and decision-making. These capabilities are powered by AI-driven models, which process large datasets to find patterns, forecast outcomes, or carry out actions with little assistance from humans. These models, which have an impact on industries like healthcare, finance, education, transportation, and urban planning, range from straightforward algorithms like linear regression to intricate deep learning architectures.

In **Chapter 2**, the authors introduce adaptive AI models designed for real-time data mobility and enhanced security within urban IoT networks. We see how edge intelligence, collaborative large AI models, and federated learning frameworks enable scalable, multimodal learning across distributed devices. The chapter underscores the importance of anomaly detection and privacy-preserving mechanisms in resource-constrained and dynamic environments, offering a strong foundation for secure and responsive smart city networks.

Chapter 3 shifts attention to intelligent transportation infrastructure, exploring advances in AI-based mobility modeling. The contributors trace the evolution from conventional transport systems to data-driven, adaptive frameworks powered by machine learning, deep learning, and reinforcement learning. We find a compelling discussion of predictive traffic management, autonomous mobility, and demand-responsive services, alongside ethical and governance considerations critical to sustainable urban transport transformation.

In **Chapter 4**, the focus broadens to AI-based data modeling as a cornerstone of smart city development. The authors highlight how AI technologies address pressing urban challenges—ranging from waste accumulation and pollution to water scarcity and healthcare demands. By presenting contemporary frameworks and environmentally conscious solutions, the chapter reinforces AI's pivotal role in enabling resilient and sustainable urban services.

Chapter 5 explores predictive analytics in smart cities through the lens of the Random Forest paradigm. The discussion demonstrates how this robust machine learning technique supports classification and regression tasks across traffic optimization, pollution forecasting, healthcare prediction, and energy management. Particularly noteworthy is the emphasis on model transparency and feature importance, which fosters citizen trust and policy alignment in data-driven governance.

Chapter 6 emphasizes the integration of AI to enhance data flow and urban intelligence. The authors articulate how seamless data mobility across traffic, energy, waste, and safety systems supports predictive modeling and automated decision-making. Through case studies and architectural frameworks, the chapter presents a vision of adaptive, resilient, and citizen-centric digital ecosystems.

Chapter 7 deepens the discussion on data security and privacy in IoT-enabled smart cities. The authors examine vulnerabilities inherent in connected environments and propose advanced cryptographic methods, distributed authentication mechanisms, and privacy-preserving computation models. The integration of AI-driven threat detection and edge computing solutions highlights proactive and latency-aware protection strategies.

Building on privacy concerns, **Chapter 8** explores federated learning frameworks for secure and scalable AI in smart city ecosystems. We observe how decentralized learning models, combined with blockchain, differential privacy, and edge AI, support citizen-centric intelligence without compromising data sovereignty. The chapter critically evaluates practical challenges and policy implications, charting a responsible path forward.

Chapter 9 addresses anomaly detection in behavioral surveillance using deep learning techniques. The authors trace the evolution from rule-based systems to advanced architectures such as CNNs, LSTMs, autoencoders, and transformers. By discussing evaluation metrics, ethical considerations, and real-time inference challenges, the chapter provides a rigorous guide for designing effective surveillance intelligence systems.

In **Chapter 10**, the emphasis turns to automated machine learning model management and observability. The proposed framework integrates continuous monitoring, feedback loops, and automated retraining to mitigate data drift and sustain performance. The reported empirical results demonstrate high accuracy and adaptability, reinforcing the importance of lifecycle management in AI-driven urban applications.

Chapter 11 presents an innovative application of vision-language modeling for image captioning using Vision Transformers and GPT-2. The chapter illustrates how multimodal AI systems can generate coherent textual descriptions from images, surpassing traditional CNN-LSTM approaches. Beyond technical architecture, the work highlights practical applications in accessibility, content management, and intelligent retrieval systems.

In **Chapter 12**, the discussion moves into the socio-environmental dimension of smart cities, examining rural–urban differences in kitchen waste disposal practices in Kerala. Drawing on large-scale empirical data, the authors reveal patterns of sustainability awareness and behavioral variation. The chapter reminds us that intelligent urban transformation must be grounded in community engagement and context-sensitive policy interventions.

Chapter 13 explores voltage control in smart city low-voltage distribution networks using AI-enhanced optimization and power electronics. Through application of the Exhaustive Search Method under worst-case renewable and load conditions, the authors demonstrate significant improvements in voltage stability. This chapter exemplifies how AI-driven control systems strengthen critical infrastructure resilience.

Finally, **Chapter 14** presents real-world case studies illustrating the application of AI and IoT in smart cities. By examining initiatives in Singapore, Barcelona, and Amsterdam, the authors demonstrate tangible outcomes in energy efficiency, carbon reduction, public safety, and service delivery. The chapter also anticipates future trends involving 5G, edge computing, and blockchain, situating smart city evolution within an emerging technological continuum.

Together, these fourteen chapters form a cohesive narrative that bridges theory, architecture, implementation, and societal impact. As editors, we believe this volume not only captures the current state of AI-enabled data mobility and intelligent modeling but also provides a forward-looking blueprint for building secure, sustainable, and citizen-centric smart cities in the digital era.

CONCLUSION

As we bring this volume to a close, we reflect on the remarkable breadth and depth of scholarship presented in *AI-Based Data Mobility and Intelligent Modeling for Smart Cities*. The chapters collectively illustrate that the evolution of smart cities is not merely a technological transition, but a systemic transformation—one that integrates intelligence, sustainability, governance, infrastructure, and human-centered design into a unified digital ecosystem.

Throughout this book, we have seen how Artificial Intelligence, Machine Learning, Deep Reinforcement Learning, IoT integration, federated architectures, predictive analytics, spatial intelligence, automated model management, and AI-driven infrastructure control are redefining the way cities operate. From transportation and energy systems to surveillance, environmental monitoring, waste management, and citizen services, AI-enabled data mobility serves as the backbone of responsive and adaptive urban environments.

A recurring theme across the chapters is the shift from reactive to proactive governance. Predictive modeling, real-time analytics, and edge intelligence enable cities not only to respond to emerging challenges but to anticipate them. Equally important is the emphasis on security, privacy, explainability, and ethical responsibility. As smart cities become increasingly data-driven, safeguarding citizen trust becomes foundational to sustainable digital transformation.

This volume also underscores the importance of interdisciplinary collaboration. The convergence of computer science, electrical engineering, urban planning, data science, environmental studies, and public policy is essential for building inclusive and resilient smart city ecosystems. Technological innovation must align with regulatory frameworks, societal values, and long-term sustainability goals.

Looking ahead, the integration of AI with next-generation communication networks, edge computing, blockchain, and autonomous systems will continue to expand the possibilities for intelligent urban living. However, the success of these innovations will depend not only on computational sophistication but also on transparency, scalability, affordability, and equitable access.

As editors, we hope this book serves as both a reference and a catalyst—supporting researchers in advancing new theoretical models, guiding practitioners in implementing robust solutions, and assisting policymakers in designing informed and responsible digital strategies. We believe the insights presented here will contribute meaningfully to shaping future-ready cities that are secure, adaptive, environmentally conscious, and citizen-centric.

We extend our sincere appreciation to all contributing authors, reviewers, and collaborators whose expertise and dedication made this work possible. It is our aspiration that this volume inspires continued research, dialogue, and innovation in the pursuit of smarter, safer, and more sustainable urban futures.

Chapter 1
Introduction to AI-Driven Data Models:
Foundations of AI-Driven Models

Mohammad Shuaib Khan
http://orcid.org/0009-0006-4774-2572
Chitkara University, Punjab, India

Mohammad Mazhar Afzal
http://orcid.org/0000-0001-8814-2772
Glocal University, India

ABSTRACT

A key component in modern technology, artificial intelligence (AI) is changing industries, promoting scientific research, and influencing social interactions. Fundamentally, artificial intelligence (AI) is the development of systems that simulate human intelligence, allowing machines to carry out operations like learning, reasoning, problem-solving, and decision-making. These capabilities are powered by AI-driven models, which process large datasets to find patterns, forecast outcomes, or carry out actions with little assistance from humans. These models, which have an impact on industries like healthcare, finance, education, transportation, and urban planning, range from straightforward algorithms like linear regression to intricate deep learning architectures. AI models, for instance, help with traffic flow optimization in smart cities and disease diagnosis from medical images. By examining their historical development, fundamental paradigms, important algorithms, mathematical foundations, and development processes, this chapter offers a fundamental understanding of AI-driven model

DOI: 10.4018/979-8-3373-4202-3.ch001

1. FOUNDATIONS OF AI-DRIVEN MODELS

1.1 Introduction to AI and AI-Driven Models

A key component in modern technology, artificial intelligence (AI) is changing industries, promoting scientific research, and influencing social interactions. Fundamentally, artificial intelligence (AI) is the mechanism to evaluate, allowing machines to carry out operations. These capabilities are powered by AI-driven models, which process large datasets to find patterns, forecast outcomes, or carry out actions with little assistance from humans. These models, which have an impact on industries like healthcare, finance, education, transportation, and urban planning, range from straightforward algorithms like linear regression to intricate deep learning architectures. AI models, for instance, help with traffic flow optimization in smart cities and disease diagnosis from medical images. By examining their historical development, fundamental paradigms, important algorithms, mathematical foundations, and development processes, this chapter offers a fundamental understanding of AI-driven models.

1.2 Historical Evolution of AI

Over the course of seven decades, AI development has undergone several notable shifts that have influenced its course. Alan Turing developed the Turing Test in 1950 as a standard to check whether a machine could demonstrate human-like intelligence by having a conversation that was identical to that of a human. The field of artificial intelligence was founded on this theoretical framework.

The term "artificial intelligence" was first used in 1956 at the Dartmouth Conference, which was organized by John McCarthy and others and formally established AI as an academic field. Rule-based and expert systems, which made decisions using preset logic, were the foundation of early AI systems. However, these systems had trouble being flexible and dealing with ambiguity.

Models can now adapt to data patterns thanks to the development of neural networks and the backpropagation algorithm. When IBM's Deep Blue defeated world chess champion Garry Kasparov in 1997, it was a historic moment that demonstrated AI's capacity to handle challenging strategic tasks. Ten significant moments in the brief history of AI.

Geoffrey Hinton's 2006 work on deep learning, which took advantage of more computing power and data availability, sparked a resurgence of interest in neural networks in the 2000s. Significant advancements were made in the 2010s, which are sometimes referred to as the "deep learning revolution." IBM Watson demonstrated sophisticated natural language processing when it won the Jeopardy! game show in 2011. The Whole AI Timeline: The History of Artificial Intelligence. The convolutional neural network Alex Net won the ImageNet competition in 2012, lowering image classification errors and making deep learning one of the top ten AI achievements of the past decade. AI's ability to play complex, intuitive games was demonstrated in 2015 when Google DeepMind's AlphaGo defeated professional Go player Lee Sedol (Wang et al., 2020).

Large-scale models and developments in generative AI have characterized the 2020s. AI Timeline: Key Events in Artificial Intelligence from 1950-2025 OpenAI released GPT-3 in 2020, a 175-billion-parameter language model that can produce text that is human-like. Timeline of artificial intelligence in 2022, ChatGPT, which was based on GPT-3.5, became widely used and reached 100 million users in just two months. GPT-4's multimodal capabilities were further enhanced in 2023, as evidenced by its superior performance on tasks such as standardized tests. By 2025, advancements like Amazon's humanoid package delivery robots and Mistral AI's Le Chat, which can process up to 1,000 words per second, highlight how AI is being incorporated into a wide range of applications. Artificial intelligence timeline.

1.3 Core Paradigms in AI

AI-driven models operate within three primary paradigms; each suited to different types of problems.

1.3.1 Supervised Learning

Supervised learning refers to a class of machine learning techniques in which models are trained using datasets that contain both input variables and their corresponding target outputs. The objective of this approach is to learn a functional relationship that maps inputs to known outputs, enabling the model to generalize and make accurate predictions when exposed to previously unseen data. In practice, this paradigm is widely applied in classification and regression problems. For example, in image recognition tasks, a model trained on annotated images can learn distinguishing visual features and subsequently classify new images into predefined categories. Similarly, predictive modeling in domains such as real estate valuation or email filtering relies on supervised algorithms to estimate continuous values or assign categorical labels. Common techniques employed in supervised learning include linear and logistic regression models, decision trees, support vector machines, and multilayer neural networks. The effectiveness of supervised learning largely depends on the quality, diversity, and representativeness of the labelled training data.

1.3.2 Unsupervised Learning

In contrast to supervised approaches, unsupervised learning operates on datasets that do not contain explicit output labels. The primary aim is to uncover hidden structures, relationships, or patterns inherent within the data itself. Rather than predicting predefined outcomes, unsupervised models attempt to organize or transform data in meaningful ways. This paradigm is particularly useful in exploratory data analysis and knowledge discovery. Clustering techniques, for instance, group data points based on similarity measures, which can support applications such as customer segmentation or anomaly detection. Another important application is dimensionality reduction, where high-dimensional datasets are transformed into lower-dimensional representations while preserving essential information. This is valuable for visualization, noise reduction, and computational efficiency. Algorithms commonly associated with unsupervised learning include k-means clustering, hierarchical clustering, principal component analysis (PCA), and autoencoders. By learning internal data representations without external supervision, these models facilitate deeper insight into complex datasets.

1.3.3 Reinforcement Learning

Reinforcement learning (RL) represents a distinct paradigm within machine learning in which an autonomous agent acquires decision-making capabilities through interaction with its environment. Instead of relying on labelled datasets, the agent learns by performing actions and receiving feedback in the form of rewards or penalties. The objective is to develop a policy that maximizes cumulative reward over time. This learning process is typically modeled as a sequential decision problem, where the agent continuously evaluates the consequences of its actions and adapts its strategy accordingly.

A classical illustration of reinforcement learning can be found in strategic board games such as chess, where an AI system improves its gameplay by associating positive rewards with successful outcomes and negative signals with suboptimal moves. Beyond gaming, RL has demonstrated significant impact in robotics, autonomous systems, resource optimization, and control engineering, where adaptive decision-making in dynamic environments is essential.

1.4 Key Algorithms in AI

Several algorithms form the backbone of AI-driven models, each suited to specific tasks.

1.4.1 Linear Regression

To predict continuous outcomes based on input features, linear regression models the relationship as a linear equation:

$$y = \beta_0 + \beta_1 x_1 + \ldots + \beta_p x_p + \epsilon$$

where y is the output, x_i are inputs, β_i are coefficients, and ϵ is the error term. It is frequently used for jobs like estimating home values according to location and size.

Key Differences:

- **Interpretability:** Decision trees offer high transparency ("if-then" rules), while neural networks are generally considered uninterpretable "black boxes".
- **Data Type & Volume:** Decision trees perform well on smaller, tabular datasets. Neural networks require large datasets to outperform other methods.
- **Performance:** Neural networks often achieve higher accuracy on complex tasks (e.g., image recognition), whereas decision trees can be faster and more efficient for simpler, structured tasks.
- **Training Speed:** Decision trees train quickly; neural networks require significantly more time for training.

When to Use Which?

- **Decision Trees:** Use for tabular data, when transparency is required, or with limited data/computational resources.
- **Neural Networks:** Use for complex, non-linear problems, large, unstructured datasets, and when maximum predictive power is needed over interpretability

1.5 Mathematical Foundations

The effectiveness of AI-driven models relies on mathematical principles that enable learning and optimization.

1.5.1 Gradient Descent

Gradient descent is a fundamental optimization algorithm widely used in machine learning to minimize a model's loss function. The loss function quantifies the discrepancy between predicted outputs and actual target values, thereby serving as a measure of model error. The central objective of gradient descent is to iteratively adjust model parameters in a manner that reduces this error.

$$theta = theta - alpha * grad(J(theta))$$

At each iteration, the algorithm computes the gradient of the loss function with respect to the model parameters. This gradient indicates the direction of the steepest increase in error; therefore, parameter updates are performed in the opposite direction to achieve the greatest reduction. The learning rate (α) controls the step size of each update and plays a critical role in determining convergence speed and stability.

Conceptually, the process can be compared to descending a multidimensional surface in search of a minimum point. Through repeated incremental adjustments, the algorithm gradually converges toward a local or global minimum of the loss function. Gradient descent forms the computational backbone of training procedures for neural networks and numerous other predictive models, particularly in large-scale deep learning systems where efficient parameter optimization is essential.

1.5.2 Backpropagation

Backpropagation, short for "backward propagation of errors," is a core computational procedure used to train artificial neural networks. Its primary function is to efficiently compute the gradient of the loss function with respect to each network parameter, enabling systematic weight updates during optimization.

$$dL/dw = (dL/da)*(da/dz)*(dz/dw)$$

The algorithm relies on the chain rule of calculus to decompose complex derivative calculations into a sequence of simpler partial derivatives. Starting from the output layer, where prediction errors are first observed, the gradients are propagated backward through each hidden layer. At every stage, the contribution of individual weights to the overall error is quantified. This structured backward flow of gradient information ensures computational efficiency, particularly in deep architectures with many layers.

By combining backpropagation with gradient-based optimization methods such as gradient descent, neural networks iteratively adjust their parameters to minimize prediction error. This mechanism enables the network to progressively capture hierarchical and non-linear relationships within data, thereby supporting the learning of complex patterns across domains such as image recognition, natural language processing, and predictive analytics.

1.6 Model Training, Validation, and Testing

Building AI-driven models involves a structured process to ensure accuracy and generalizability.

1.6.1 Data Preparation

Data preparation constitutes a critical preliminary phase in the development of reliable machine learning models. This process involves the systematic collection of relevant datasets, followed by structured partitioning to support model training, tuning, and performance evaluation. Typically, the dataset is divided into three subsets:

- **Training set (approximately 70–80%)** – used to fit the model parameters.
- **Validation set (approximately 10–15%)** – used to fine-tune hyperparameters and prevent overfitting.
- **Test set (approximately 10–15%)** – reserved for final performance assessment on unseen data.

Beyond partitioning, data preparation also includes cleaning and preprocessing tasks such as handling missing values, removing noise and outliers, correcting inconsistencies, encoding categorical variables, and normalizing numerical features when necessary. These steps ensure that the learning algorithm is exposed to accurate and representative information.

Properly prepared data enhances model generalization capability, reduces bias, and improves predictive stability. Inadequate preprocessing, by contrast, can lead to distorted learning patterns and unreliable evaluation results. Consequently, robust data preparation is foundational to achieving high-performing and trustworthy AI-driven models.

1.6.2 Training Process

In order to minimize the loss function, training entails feeding data into the model, calculating predictions, and modifying parameters. Performance is optimized through hyperparameter tuning, such as determining the learning rate. This process is made easier by frameworks like PyTorch and TensorFlow.

1.6.3 Validation and Testing

Model validation plays a crucial role in monitoring learning behavior and optimizing performance. After training on the primary dataset, the model is evaluated on a separate validation set to assess its ability to generalize beyond the training data. This stage is particularly important for hyperparameter tuning, where parameters such as learning rate, regularization strength, or network depth are adjusted to improve performance. Validation also helps detect overfitting, a condition in which the model captures noise or overly specific patterns from the training data rather than learning broadly applicable relationships.

Following validation, final model assessment is conducted using an independent test dataset that remains unseen during both training and tuning phases. The objective at this stage is to obtain an unbiased estimate of predictive performance.

For classification tasks, several quantitative metrics are commonly employed. These include accuracy, precision, recall, and the F1-score, which balances precision and recall:

$$F1 = 2*(Precision * Recall)/(Precision + Recall)$$

For regression problems, performance is often evaluated using error-based metrics such as Mean Squared Error (MSE), defined as:

$$MSE = (1/n)*sum((y_i - y_h a t_i)^2)$$

These evaluation measures provide insight into how well the model performs on unseen data, thereby reflecting its generalization capability. A rigorous validation and testing framework is essential for ensuring that AI-driven data models are reliable, robust, and suitable for deployment in real-world applications.

1.7 Conclusion

The historical development of AI-driven models from the 1950s to 2025, fundamental frameworks (supervised, unsupervised, and reinforcement learning), important algorithms (linear regression, decision trees, and neural networks), mathematical underpinnings (gradient descent, backpropagation), and the training procedure have all been covered in this chapter. These ideas lay the groundwork for later chapters' exploration of more complex approaches and applications.

2. CORE APPROACHES AND MODEL ARCHITECTURES

2.1 Introduction

Artificial intelligence systems rely on a diverse spectrum of computational models to extract information from data, recognize underlying structures, and support automated decision-making. These models range from foundational machine learning techniques to advanced deep learning architectures that enable high-level abstraction and representation learning. Understanding the theoretical principles and structural design of these models is essential for appreciating how AI-driven systems address tasks such as classification, prediction, sequence modeling, and generative content production.

The chapter begins by examining classical machine learning approaches that form the conceptual backbone of predictive analytics. Algorithms such as decision trees, support vector machines, and k-nearest neighbours illustrate how structured decision boundaries and similarity-based reasoning can be applied to classification and regression problems. In addition, ensemble strategies—including random forests and gradient boosting—demonstrate how combining multiple learners can enhance robustness, reduce variance, and improve predictive accuracy through aggregated decision-making.

Building upon these foundations, the discussion transitions to deep learning frameworks that have transformed modern AI applications. Architectures such as convolutional neural networks are particularly effective for spatial data processing, including image analysis, whereas recurrent neural networks are designed to model sequential dependencies in time-series and language data. More recent innovations, including transformer architectures and generative adversarial networks, extend these capabilities by enabling large-scale contextual modeling and high-quality synthetic data generation.

Beyond architectural design, the chapter also addresses the mathematical principles that govern model behavior. Core components such as activation functions introduce non-linearity into neural systems, while loss functions quantify prediction error and guide optimization procedures. Model performance is evaluated using established quantitative metrics to ensure reliability and generalizability.

By integrating algorithmic foundations, architectural insights, and mathematical underpinnings, this chapter equips readers with a structured understanding of AI-driven modeling approaches. Such a framework enables informed engagement with both theoretical developments and practical implementations within the rapidly evolving landscape of artificial intelligence.

2.2 Traditional Machine Learning Models

AI is based on traditional machine learning models, which provide efficient and comprehensible solutions for structured data. These models are especially useful for tasks that prioritize explainability and simplicity (Mumuni & Mumuni, 2024).

2.2.1 K-Nearest Neighbours (KNN)

K-Nearest Neighbours is a non-parametric, instance-based learning algorithm that uses a distance metric to identify the k closest training instances for classification or regression.

How They Operate: KNN determines the distances (such as the Euclidean distance) to every training point for a new data point, chooses the k nearest neighbours, and assigns the average value (for regression) or the majority class (for classification). Performance is greatly impacted by the choice of k and distance metric.

Use Cases: KNN is utilized in image recognition, anomaly detection, and recommendation systems (e.g., making movie recommendations based on user preferences), especially when decision boundaries are irregular.

Benefits include ease of implementation, no need for a training phase, and efficient handling of multi-class issues.

Cons: Affected by irrelevant features, sensitive to k and distance metric selections, and computationally costly for large datasets.

For instance, in image recognition, KNN may identify a creature in an image as either a dog or a cat by comparing its features to those of the k closest labelled images, or K-Nearest Neighbours.

2.2.2 Ensemble Methods: Random Forests and Gradient Boosting

Ensemble Methods Overview: Ensemble methods combine multiple models to improve predictive performance, reducing variance and bias compared to single models.

- Random Forests

Random Forests are an ensemble technique that builds several decision trees and combines their predictions using either averaging (regression) or majority voting (classification).

How They Operate: To lessen correlation between trees, each tree is trained on a random subset of features at each split and a random subset of data (via bootstrap sampling). The mean or mode of each tree's individual predictions is the final forecast.

Use Cases: Because of their accuracy and resilience, Random Forest algorithms are utilized in customer churn prediction, fraud detection, and predictive maintenance.

Benefits include handling high-dimensional data, reducing overfitting, and offering feature importance metrics.

Cons: Less interpretable than single decision trees and slower for real-time predictions.

For instance, a Random Forest may use transaction data analysis to forecast fraudulent activity in fraud detection, utilizing multiple trees to increase accuracy. A Comprehensive Guide to Random Forest.

2.3 Deep Learning Architectures

Deep learning architectures have transformed AI by enabling the processing of unstructured data, such as images, text, and audio, through hierarchical feature learning.

2.3.1 Convolutional Neural Networks (CNNs)

Convolutional Neural Networks (CNNs) constitute a specialized class of deep learning architectures designed to process structured grid-like data, particularly images. Their strength in computer vision applications arises from their ability to capture spatial hierarchies of features through layered transformations. Rather than treating input pixels independently, CNNs exploit local connectivity patterns, enabling the model to learn increasingly abstract representations—from simple edges and textures in early layers to complex object components in deeper layers.

The fundamental computational operation within a CNN is convolution, in which a learnable filter (kernel) slides across the input to generate feature maps:

$$FeatureMap(i,j) = sum_m sum_n Input(i+m,j+n)*Kernel(m,n)$$

This operation allows the network to detect local patterns while preserving spatial relationships. Convolutional layers are typically followed by non-linear activation functions and pooling layers. Pooling operations—such as max pooling—reduce the spatial dimensionality of intermediate representations, thereby lowering computational complexity and enhancing translational invariance. An important architectural characteristic of CNNs is weight sharing, where the same filter parameters are applied across different regions of the input. This significantly reduces the total number of trainable parameters compared to fully connected architectures, improving both efficiency and generalization.

After several stages of convolution and pooling, the extracted high-level features are passed to fully connected layers, which perform classification or regression tasks depending on the problem domain.

CNNs have demonstrated remarkable effectiveness across a wide range of practical applications. In medical imaging, they assist in detecting abnormalities such as tumors in radiographic scans. In computer vision more broadly, they are widely employed for object detection, facial recognition, and automated image categorization. Their ability to learn hierarchical feature representations has made CNNs foundational to modern vision-based AI systems

Benefits include learning hierarchical features, reducing computational complexity through parameter sharing, and being effective with visual data (Hinton & Salakhutdinov, 2006).

Cons: Needs big datasets, is computationally demanding, and is architecture-sensitive.

For instance, in medical imaging, a CNN might analyse X-ray images to detect pneumonia, identifying patterns like lung opacity What are Convolutional Neural Networks?

2.3.2 Recurrent Neural Networks (RNNs)

Definition: Recurrent neural networks are perfect for tasks where order is important because they are made for sequential data and maintain a hidden state that records information from earlier steps.

RNNs operate by processing sequences one element at a time, updating the hidden state before passing it on to the following stage. The formula for the hidden state:

$$a = g_1(W_{aa}a + W_{ax}x + b_a)$$

where (g_1) is an activation function, such as *tan*(*h*). Issues with vanishing gradients are addressed by variants such as LSTMs and GRUs.

Use Cases: RNNs are used in natural language processing (e.g., sentiment analysis), speech recognition, time series forecasting, and machine translation Introduction to Recurrent Neural Networks.

Benefits include the ability to handle sequential data, remember prior inputs, and perform well on tasks like language modelling.

Cons: computationally demanding, prone to vanishing and exploding gradients, and challenging to train for lengthy sequences.

For instance, in speech recognition, an RNN might process audio sequences to transcribe spoken words, leveraging temporal dependencies What is a Recurrent Neural Network (RNN)?

2.3.3 Transformers

Transformers are deep learning models that excel at natural language processing and other tasks by processing sequential data in parallel using self-attention mechanisms.

How They Operate: Transformers are made up of an encoder and a decoder, both of which have feed-forward neural networks and several layers of self-attention. Self-attention uses queries, keys, and values to calculate each input element's relative importance. Sequence order is maintained by positional encodings.

Use cases include text summarization, question answering, language translation (e.g., Google Translate), and vision transformers for image tasks (deep learning architecture).

Benefits: Achieves cutting-edge NLP performance, supports parallel processing, and manages long-range dependencies.

Cons: difficult to implement, computationally demanding, and requiring big datasets.

For instance, in machine translation, a Transformer might translate English to French by attending to relevant words in the input sentence What is a Transformer?

2.3.4 Generative Adversarial Networks (GANs)

Generative Adversarial Networks are frameworks where two neural networks—a generator and a discriminator—compete to generate realistic data resembling the training set.

How They Operate: The generator takes random noise as input and produces fake data, while the discriminator evaluates whether the data is real or fake. They are trained simultaneously, with the generator aiming to fool the discriminator and the discriminator improving its accuracy. The loss function balances their objectives (Seenivasan, D. 2024).

Use Cases include, GANs are used in image generation (e.g., creating realistic faces), style transfer, data augmentation, and deepfake creation Generative Adversarial Network (GAN).

The advantages include generating high-quality synthetic data, help in data augmentation, and learning complex distributions.

Cons: Training can be unstable, requires careful hyperparameter tuning, and is computationally intensive.

For instance, in art generation, a GAN might create new paintings by learning from a dataset of existing artworks What is a Generative Adversarial Network (GAN)?

2.4 Conclusion

This chapter has provided a structured examination of the principal methodologies and architectural frameworks that underpin contemporary artificial intelligence systems. Beginning with foundational machine learning techniques—such as decision trees and support vector machines—and extending to advanced deep learning paradigms including convolutional neural networks and transformer-based architectures, the discussion has highlighted both conceptual principles and practical applications.

By exploring how these models process data, extract patterns, and optimize predictive performance, readers gain insight into the computational mechanisms that enable AI systems to address diverse tasks ranging from structured data analysis to large-scale vision and language modeling. The mathematical foundations discussed—particularly those related to optimization, activation functions, and loss minimization—offer a theoretical lens through which model behavior can be interpreted and refined.

Equally important, the evaluation metrics presented throughout the chapter provide a systematic framework for assessing model effectiveness. Through rigorous validation and performance measurement, practitioners can ensure that AI systems exhibit robustness, generalizability, and reliability when deployed in real-world environments.

Collectively, this integrated perspective equips readers with both the conceptual understanding and analytical tools necessary to engage critically with AI-driven models and to apply them responsibly across a wide range of domains.

3. BUILDING AND TRAINING AI MODELS

3.1 Introduction

The development of AI-driven models represents a central stage in the creation of intelligent systems capable of addressing complex analytical and predictive challenges. Whether applied to computer vision, natural language processing, or forecasting tasks, model construction requires a structured and iterative workflow that integrates design, training, evaluation, and refinement. This chapter outlines the methodological foundations of that workflow, emphasizing the cyclical nature of experimentation and performance optimization.

Model development typically begins with careful problem formulation and architectural selection, followed by systematic training using representative datasets. Modern machine learning ecosystems provide robust computational support for this process. Frameworks such as TensorFlow, PyTorch, and Scikit-learn offer flexible tools for implementing algorithms that range from traditional statistical classifiers to large-scale deep neural networks. These platforms facilitate efficient computation, automatic differentiation, and scalable deployment.

A substantial portion of model performance depends on effective data preprocessing and feature engineering. Transforming raw data into structured, informative representations enhances learning efficiency and reduces noise. Hyperparameter tuning further refines model behavior by adjusting variables such as learning rate, batch size, network depth, or regularization strength. Cross-validation techniques are commonly employed to ensure that parameter adjustments generalize beyond a single dataset split.

Preventing overfitting is another critical consideration during training. Regularization strategies introduce constraints that discourage overly complex models, often by penalizing large parameter values:

$$L_total = L_data + lambda * \mid \mid w \mid \mid^2$$

In addition, dropout mechanisms randomly deactivate subsets of neurons during training, promoting redundancy and improving model robustness. Together, these techniques enhance generalization and stability.

Through illustrative examples and applied demonstrations, the chapter provides practical guidance for implementing and refining AI models across varied data contexts. By integrating theoretical principles with hands-on strategies, readers are prepared to navigate the technical and conceptual challenges associated with real-world AI deployment.

3.2 Popular Machine Learning Frameworks

Selecting an appropriate framework is crucial for efficient model development. Below are three widely used frameworks as of June 2025, each tailored to specific needs in AI development:

- **TensorFlow**: Developed by Google, TensorFlow is an open-source framework renowned for its scalability and production-ready capabilities. It supports distributed computing and deployment across platforms, including cloud, mobile, and edge devices. TensorFlow 2.15, the latest version as of March 2025, offers enhanced performance for large-scale models and improved support for generative AI tasks TensorFlow Releases.
- **PyTorch**: Created by Meta AI, PyTorch is favoured in research for its dynamic computation graph and Pythonic syntax, which facilitate rapid prototyping. PyTorch 2.3, released by April 2025, includes advanced features for distribut-

ed training and generative AI, making it ideal for experimental and academic projects PyTorch Foundation.

- **Scikit-learn**: A leading library for traditional machine learning, Scikit-learn provides simple, efficient tools for tasks like classification, regression, and clustering. Version 1.5, available as of June 2025, introduces new algorithms and optimizations for large datasets, integrating seamlessly with Python's scientific computing ecosystem Scikit-learn Documentation.

These frameworks cater to different use cases: TensorFlow excels in production environments, PyTorch in research, and Scikit-learn in traditional machine learning tasks.

3.3 The Machine Learning Pipeline

Building an AI model involves a structured, iterative pipeline to ensure robust performance. The key steps are:

1. **Data Preparation**: Collecting high-quality data, cleaning it to remove errors or outliers, and splitting it into training (70-80%), validation (10-15%), and test (10-15%) sets. Data quality directly impacts model accuracy.
2. **Feature Engineering**: Selecting and transforming features to enhance model learning. Techniques include normalization (scaling features to a standard range), encoding categorical variables (e.g., one-hot encoding), and creating new features (e.g., combining age and income for financial models).
3. **Model Selection**: Choosing an architecture based on the problem type (e.g., classification, regression) and data nature (e.g., structured or unstructured). For example, convolutional neural networks (CNNs) suit image data, while transformers are ideal for text.
4. **Model Training**: Feeding data through the model, computing predictions, and optimizing parameters using algorithms like gradient descent to minimize the loss function.
5. **Hyperparameter Tuning**: Adjusting settings like learning rate or number of layers to optimize performance.
6. **Cross-Validation**: Evaluating the model on multiple data subsets to ensure generalization.
7. **Model Evaluation**: Assessing performance on the test set using metrics like accuracy, precision, or mean squared error.

3.4 Data Preprocessing and Feature Engineering

Data preprocessing ensures data is clean and suitable for training. Common tasks include:

- **Handling Missing Values**: Imputing missing data using mean, median, or advanced methods like k-nearest neighbour imputation.
- **Normalization and Scaling**: Scaling features to a range (e.g., [0, 1]) or standardizing to zero mean and unit variance to ensure consistent model behaviour.
- **Encoding Categorical Variables**: Converting categories (e.g., "red," "blue") into numerical formats using one-hot encoding or label encoding.

Feature engineering enhances model performance by creating meaningful features. For example, in a retail dataset, combining purchase frequency and average spend into a "customer value" feature can improve predictive accuracy. Scikit-learn provides tools like StandardScaler for normalization and OneHotEncoder for categorical data (Pang et al., 2021)

3.5 Model Selection and Training

Model selection depends on the task and data type. For structured data, traditional models like decision trees or support vector machines (SVMs) may suffice, while unstructured data (e.g., images, text) often requires deep learning models like CNNs or Transformers. Training involves:

- **Forward Propagation**: Computing predictions by passing input data through the model.
- **Loss Calculation**: Measuring prediction errors using a loss function (e.g., mean squared error for regression, cross-entropy loss for classification).
- **Backpropagation**: Adjusting model parameters to minimize the loss using gradient descent.

Frameworks like TensorFlow and PyTorch simplify training with high-level APIs. For example, TensorFlow's Keras API allows rapid model prototyping, while PyTorch's dynamic graphs support flexible experimentation.

3.7 Cross-Validation

Cross-validation is a systematic resampling technique designed to evaluate a model's ability to generalize to unseen data. Unlike a single train–test split, which may produce performance estimates sensitive to a particular data partition, cross-validation reduces evaluation bias by repeatedly training and validating the model on different subsets of the dataset.

One widely adopted strategy is *k*-fold cross-validation. In this approach, the dataset is partitioned into *k* approximately equal subsets, referred to as folds. During each iteration, the model is trained on $k - 1$ folds and evaluated on the remaining fold. This process is repeated *k* times so that every subset serves as a validation set exactly once. The final performance estimate is obtained by averaging the evaluation scores across all folds:

$$CV_score = (1/k)*sum_{i=1}^{k} Score_i$$

For example, in 5-fold cross-validation, the dataset is divided into five segments. The model undergoes five training cycles, each time reserving a different segment for validation while using the remaining four for learning. This methodology provides a more stable and reliable estimate of predictive performance, particularly when working with limited datasets where a single split may lead to high variance in results.

Modern machine learning libraries, including Scikit-learn, offer automated utilities such as the cross_val_score function, which streamlines the implementation of cross-validation procedures and facilitates reproducible evaluation workflows.

By mitigating the limitations of isolated train–test evaluation, cross-validation contributes to more dependable model selection and hyperparameter tuning processes.

3.8 Techniques to Prevent Overfitting

Overfitting occurs when a model learns noise in the training data, reducing its ability to generalize. Techniques to mitigate overfitting include:

Regularization introduces additional constraints into the optimization objective by penalizing large parameter values. This discourages overly complex models and promotes smoother decision boundaries. In L1 regularization (Lasso), the penalty is proportional to the absolute magnitude of model weights, encouraging sparsity and potentially driving some coefficients to zero. In contrast, L2 regularization (Ridge) penalizes the squared magnitude of parameters, leading to smaller but non-zero weights and improved numerical stability.

In neural networks, L2 regularization modifies the objective function as follows:

$L_total = L_data + lambda * sum_i w_i^2$

Here, λrepresents the regularization coefficient that controls the strength of the penalty term. Larger values impose stronger constraints on parameter growth.

Another widely adopted technique is dropout, which enhances model robustness by randomly deactivating a specified fraction of neurons during each training iteration. For instance, a dropout rate of 20% temporarily removes one-fifth of the neurons in a layer during forward propagation. This prevents excessive co-adaptation among neurons and encourages the network to learn redundant, distributed representations that generalize more effectively.

Early stopping provides an additional safeguard against overfitting. During training, performance on a validation dataset is continuously monitored. When validation loss ceases to improve—or begins to increase—training is halted to prevent the model from memorizing noise in the training set. This approach balances optimization progress with generalization control.

Together, regularization, dropout, and early stopping form a complementary set of techniques that promote stable, reliable model learning in both traditional machine learning and deep neural network architectures.

- **Data Augmentation**: For image data, applying transformations like rotation or flipping increases dataset diversity, improving generalization.

3.9 Practical Examples

To illustrate the process of building and training an AI model, consider the task of creating a neural network to classify handwritten digits using the MNIST dataset, a widely used benchmark in machine learning. The MNIST dataset contains 60,000 training images and 10,000 test images of handwritten digits (0–9), each a 28x28 pixel grayscale image. This example demonstrates the key steps of the machine learning pipeline—data preprocessing, model design, training, and validation—while addressing challenges like overfitting (Sugureddy, 2023).

Step 1: Data Preprocessing

The initial stage in training a neural network involves systematic preprocessing of the MNIST dataset. Each grayscale image in MNIST consists of a 28×28 pixel grid. For compatibility with fully connected neural network architectures, these two-dimensional images are reshaped into one-dimensional feature vectors. Consequently, each image is represented as a vector of 784 numerical values: 28*28=784.

Pixel intensities originally range from 0 to 255, corresponding to grayscale values. To enhance numerical stability and accelerate convergence during training, these values are normalized to a standardized interval between 0 and 1. This transformation can be expressed as:

$$x_n ormalized = x/255$$

Normalization ensures that input features share a consistent scale, which improves gradient-based optimization and prevents disproportionate influence from large-valued inputs.

Following preprocessing, the dataset is partitioned into distinct subsets to support robust model development. Approximately 70% of the data (around 42,000 images) is allocated for training, enabling the model to learn parameter representations. A validation subset comprising roughly 15% (9,000 images) is used for hyperparameter tuning and performance monitoring during training. The remaining 15% (9,000 images) forms the test set, reserved exclusively for final evaluation to provide an unbiased estimate of generalization performance.

Each image is associated with a categorical label indicating the digit it represents (e.g., "5" for an image of the handwritten digit five). These labels serve as supervisory signals, guiding the model during the learning process by linking input patterns to their corresponding target outputs.

Through structured preprocessing, normalization, and careful dataset partitioning, the MNIST dataset becomes suitable for effective neural network training and performance assessment.

Step 2: Model Design

To perform handwritten digit classification, a feedforward neural network architecture is constructed with a layered structure tailored to multiclass prediction. The model comprises an input layer, two hidden layers, and an output layer, forming a fully connected network suitable for vectorized image representations.

The input layer receives a 784-dimensional feature vector corresponding to the flattened 28×28 pixel image. These input features are propagated to the first hidden layer, which contains 128 neurons. At this stage, the network begins extracting intermediate representations by combining weighted pixel intensities to identify fundamental visual patterns such as edges, contours, and simple stroke structures. The second hidden layer, consisting of 64 neurons, further refines these learned representations, enabling the abstraction of more complex digit-specific features.

Non-linearity is introduced through the Rectified Linear Unit (ReLU) activation function, defined as:

$$ReLU(x) = max(0, x)$$

The use of ReLU enhances computational efficiency and allows the network to model non-linear relationships between input features and class labels.

The output layer contains 10 neurons, each corresponding to one digit class (0–9). This layer produces a score for each class, typically transformed into probabilities using a softmax function to enable categorical prediction.

To improve generalization and mitigate overfitting, dropout regularization is applied to the first hidden layer. During training, 20% of the neurons are randomly deactivated in each iteration, preventing the network from becoming overly dependent on specific units. This stochastic regularization strategy encourages distributed feature learning and enhances robustness.

Through this layered architecture, the model progressively transforms raw pixel intensities into discriminative feature representations, ultimately enabling accurate handwritten digit classification

Step 3: Training Setup

To train the model, a loss function is defined to measure how far the model's predictions are from the true labels. For this classification task, the cross-entropy loss is used, which penalizes incorrect predictions more heavily. The model is optimized using a method called Stochastic Gradient Descent (SGD), which adjusts the model's internal parameters (weights) to minimize the loss. SGD is configured with a learning rate of 0.01, determining the size of each adjustment, and a momentum of 0.9, which helps the model converge faster by considering past updates. The training data is processed in small batches of 64 images at a time, balancing computational efficiency and learning stability.

Step 4: Training Process

Model training is conducted iteratively over multiple complete passes through the training dataset, known as epochs. During each epoch, the dataset is processed in smaller subsets called batches. For every batch, the network performs forward propagation to generate predictions, computes the loss by comparing predicted outputs with true labels, and then updates its parameters using gradient-based optimization to minimize prediction error.

The training objective is typically defined through a loss function, which is iteratively reduced across epochs:

$$L = (1/N) * sum_{i=1}^{N} Loss(y_i, y_{h}at_i)$$

Here, Nrepresents the number of training samples, $y_i$denotes the true label, and $y_h a t_i$represents the predicted output. The optimization process aims to adjust model weights so that this average loss decreases progressively.

To evaluate generalization performance during training, the model is assessed on a separate validation dataset at the end of each epoch. The validation loss provides an external measure of learning quality and helps detect divergence between training accuracy and real-world predictive capability. If training loss continues to decrease while validation loss increases, the model may be overfitting.

In this implementation, training is scheduled for a maximum of ten epochs. However, an early stopping mechanism is incorporated to prevent unnecessary training once performance plateaus. Specifically, if the validation loss fails to improve for three consecutive epochs, the training process is terminated automatically. This strategy prevents the network from memorizing noise in the training data and promotes stronger generalization to unseen images.

Step 5: Evaluation and Results

After training, the model is evaluated on the test set to measure its accuracy. For the MNIST dataset, a well-designed neural network typically achieves an accuracy of 95–98%, meaning it correctly identifies the digit in 95–98 out of 100 images. For instance, if the model predicts “7” for an image of a handwritten 7, it is correct; if it predicts “2” instead, it is incorrect. The validation loss and test accuracy provide insights into the model’s performance, while techniques like dropout and early stopping ensure it remains robust. This process demonstrates how data preprocessing, model design, and training strategies work together to create an effective AI model for real-world tasks.

This example highlights the practical application of the machine learning pipeline. By preparing the MNIST data, designing a neural network with dropout, training with SGD, and using early stopping, the model achieves high accuracy while avoiding overfitting. This approach can be adapted to other tasks, such as classifying medical images or predicting customer behaviour, by adjusting the data, model architecture, and training parameters accordingly (Brahma & Gore, 2024).

3.10 Conclusion

This chapter has provided a detailed guide to building and training AI models, covering frameworks like TensorFlow, PyTorch, and Scikit-learn, and the iterative pipeline of data preparation, feature engineering, model selection, training, and validation. Techniques like hyperparameter tuning, cross-validation, and overfitting prevention ensure robust models. Practical examples, such as the MNIST classification task, illustrate real-world application, preparing readers for advanced AI development challenges.

4. AI IN INDUSTRY USE CASES

4.1 Introduction

Artificial Intelligence (AI) has become a transformative force across industries, driving innovation, enhancing efficiency, and creating significant business value. AI-driven models are deployed in diverse sectors, including finance, e-commerce, manufacturing, and beyond, to optimize processes, reduce costs, and improve user experiences. This chapter explores key industry use cases, focusing on applications such as algorithmic trading in finance, fraud detection, recommendation systems in e-commerce, and predictive maintenance in manufacturing. Through real-world examples and recent data, it illustrates how AI models are tailored to address specific challenges, delivering measurable benefits like increased revenue, improved accuracy, and enhanced customer satisfaction. By examining these applications, the chapter highlights the practical impact of AI and its role in shaping modern industries.

4.2 AI in Finance

The financial sector has embraced AI to enhance decision-making, automate processes, and mitigate risks. Two prominent applications are algorithmic trading and fraud detection.

4.2.1 Algorithmic Trading

Overview: Algorithmic trading uses AI models to analyse market data and execute trades at high speeds, often outperforming human traders. These models process vast amounts of historical and real-time data, such as stock prices, trading volumes, and economic indicators, to identify patterns and predict market movements.

How It Works: AI-driven models, such as recurrent neural networks (RNNs) or reinforcement learning algorithms, analyse time-series data to forecast price trends. For instance, an RNN might process sequential stock price data to predict short-term price movements, while reinforcement learning optimizes trading strategies by maximizing returns. These models consider factors like market volatility and news sentiment, enabling rapid, data-driven trades.

Impact: According to a 2025 Acropolium report, AI in finance is expected to contribute to a global AI software market reaching $896.32 billion by 2029, with a compound annual growth rate (CAGR) of 32.4% from 2025 to 2029 9 AI Use Cases in Major Industries. Algorithmic trading accounts for over 70% of equity trades in major markets, improving efficiency and reducing transaction costs.

Example: A hedge fund uses a deep learning model to analyse historical stock data and real-time news feeds, predicting price movements for tech stocks. By executing trades within milliseconds, the model capitalizes on small price fluctuations, generating significant returns.

Challenges: High-frequency trading requires robust computational infrastructure and low-latency systems. Regulatory compliance, such as adhering to market manipulation laws, is also critical.

4.2.2 Fraud Detection

Overview: AI models detect fraudulent activities, such as unauthorized transactions or identity theft, by identifying anomalies in financial data.

How It Works: Machine learning models, such as gradient boosting or neural networks, analyse transaction patterns, including amount, location, and frequency, to flag suspicious activities. Anomaly detection algorithms, like isolation forests, identify outliers deviating from normal behaviour. For example, a sudden large transaction from an unusual location might trigger an alert.

Impact: AI-driven fraud detection systems reduce financial losses and enhance customer trust. A 2025 study estimates that AI-based fraud detection saves banks billions annually by preventing unauthorized transactions AI Statistics 2025.

Example: A bank employs a gradient boosting model to monitor credit card transactions. The model flags a transaction made in a foreign country shortly after a local purchase, prompting a verification call to the cardholder, preventing potential fraud.

Challenges: False positives can inconvenience customers, requiring careful tuning of model sensitivity. Ensuring data privacy, especially under regulations like GDPR, is also essential.

4.3 AI in E-Commerce

E-commerce leverages AI to personalize customer experiences and optimize operations, with recommendation systems being a flagship application.

4.3.1 Recommendation Systems

Overview: Recommendation systems suggest products or services based on user behaviour, preferences, and purchase history, driving sales and engagement.

How It Works: Collaborative filtering and content-based filtering are common approaches. Collaborative filtering uses user-item interactions (e.g., purchase or click data) to recommend items liked by similar users, often implemented with matrix factorization techniques. Content-based filtering recommends items based on their features (e.g., product descriptions) and user preferences, using models like neural networks. Hybrid systems combine both for better accuracy.

Impact: Recommendation systems significantly boost revenue. For example, Netflix attributes $1 billion annually to its AI-driven recommendation engine, which accounts for 80% of content viewed on the platform AI Statistics 2025. Amazon's recommendation system drives 35% of its sales by suggesting relevant products.

Example: An e-commerce platform like Amazon analyses a user's browsing history, such as frequent views of running shoes, and uses a neural network to recommend similar products, like sports apparel or fitness trackers. The system also considers purchases by users with similar profiles, suggesting trending items in the same category.

Challenges: Recommendation systems require large datasets to avoid the "cold start" problem, where new users or products lack sufficient data. Balancing personalization with privacy, especially under data protection laws, is also critical.

4.4 AI in Manufacturing

Manufacturing benefits from AI through predictive maintenance, which minimizes downtime and optimizes equipment performance.

4.4.1 Predictive Maintenance

Overview: Predictive maintenance is an advanced industrial application of artificial intelligence that focuses on forecasting equipment failures before they occur. Rather than relying on reactive repairs (after breakdown) or scheduled preventive maintenance (at fixed intervals), predictive maintenance leverages data-driven models to anticipate faults based on equipment condition and operational patterns.

How It Works: AI models, such as time-series forecasting with RNNs or anomaly detection with autoencoders, analyse sensor data from machinery, including temperature, vibration, and pressure. These models identify patterns indicating potential failures, such as unusual vibrations signalling bearing wear. By predicting when maintenance is needed, companies avoid unexpected breakdowns.

Impact: Predictive maintenance saves manufacturers 8–12% in maintenance costs and reduces downtime by up to 50%, according to a 2025 Acropolium report 9 AI Use Cases in Major Industries. This enhances operational efficiency and extends equipment lifespan.

Example: A car manufacturing plant uses sensors on assembly line robots to collect vibration and temperature data. An RNN model analyses this data to predict when a robot's motor might fail, scheduling maintenance during planned downtime, thus avoiding costly production halts.

Challenges: Implementing predictive maintenance requires integrating sensors and data pipelines, which can be costly. Ensuring model accuracy across diverse equipment types is also challenging.

4.5 AI in Other Industries

Beyond finance, e-commerce, and manufacturing, AI impacts other sectors:

- **Customer Support**: AI-powered chatbots handle customer inquiries, resolving issues like order tracking or refunds. For instance, a retail company uses a transformer-based chatbot to answer 80% of customer queries, reducing support staff workload.
- **Content Creation**: AI generates marketing content, such as product descriptions or social media posts. Tools like Jasper AI create tailored content, improving marketing efficiency AI Statistics 2025.
- **Healthcare**: AI models assist in drug discovery by predicting molecular interactions, accelerating development timelines.
- **Logistics**: AI optimizes supply chain routes, reducing delivery times and fuel costs.

4.6 Practical Example

To illustrate AI's impact, consider an e-commerce company implementing a recommendation system to boost sales. The company collects user data, including browsing history, purchase records, and product ratings, from its online platform. The goal is to recommend products that maximize customer engagement and revenue.

Step 1: Data Preparation

The company gathers a dataset of user interactions, such as clicks on product pages and completed purchases. Data is cleaned to remove incomplete records, and features like product categories (e.g., electronics, clothing) and user demographics (e.g., age, location) are extracted. The dataset is split into training and test sets to evaluate the system's performance.

Step 2: Model Selection and Design

A hybrid recommendation system is chosen, combining collaborative filtering and content-based filtering. Collaborative filtering identifies users with similar purchase patterns, such as those buying similar electronics, and suggests items they liked. Content-based filtering uses product descriptions to recommend items with similar attributes, like suggesting a laptop with similar specifications to one viewed. A neural network processes these features, learning complex patterns in user behaviour and product attributes (Zott & Amit, 2012).

Step 3: Training and Optimization

The model is trained on historical user data, optimizing its parameters to predict user preferences accurately. Techniques like regularization prevent overfitting, ensuring recommendations generalize to new users. Hyperparameter tuning adjusts settings like the number of neural network layers to maximize prediction accuracy (Mitchell & Coles, 2003).

Step 4: Deployment and Evaluation

The recommendation system is deployed on the e-commerce platform, suggesting products in real-time as users browse. For example, a user viewing a smartphone might receive recommendations for phone cases or wireless chargers. The system's performance is evaluated using metrics like click-through rate and conversion rate. After deployment, the company observes a 20% increase in sales from recommended products, demonstrating AI's ability to enhance customer engagement and drive revenue.

Step 5: Continuous Monitoring

To maintain performance, the system is monitored for issues like changing user preferences or new product introductions. Regular updates to the model incorporate fresh data, ensuring recommendations remain relevant.

This example showcases how AI-driven recommendation systems transform e-commerce by personalizing user experiences, leveraging data to drive business outcomes.

4.7 Conclusion

This chapter has explored AI's transformative impact across industries, focusing on algorithmic trading and fraud detection in finance, recommendation systems in e-commerce, and predictive maintenance in manufacturing. These applications demonstrate AI's ability to improve efficiency, reduce costs, and enhance user experiences. Real-world examples and data, such as the $896.32 billion AI market projection by 2029, underscore the growing adoption of AI. By addressing domain-specific challenges, AI-driven models create substantial business value, paving the way for further exploration of societal applications in the next chapter.

5. AI IN PUBLIC SECTOR AND SOCIETY

5.1 Introduction

Artificial Intelligence (AI) is reshaping the public sector and society by enhancing governance, improving public services, and addressing societal challenges. AI-driven models enable governments, educational institutions, healthcare systems, and environmental agencies to deliver more equitable, accessible, and efficient solutions. This chapter explores AI applications in the public sector and society, focusing on AI-enhanced governance, autonomous transportation systems, personalized education, smart healthcare diagnostics, and environmental monitoring. Through real-world examples and recent data, it illustrates how these applications improve quality of life, promote equity, and tackle global issues. The chapter also addresses challenges such as regulatory compliance, ethical considerations, and the need for inclusive implementation, highlighting AI's transformative potential in societal contexts.

5.2 AI in Governance

AI is revolutionizing governance by automating administrative tasks, enhancing decision-making, and improving public service delivery.

5.2.1 Overview

AI-driven models streamline government operations by analysing large datasets to inform policies, automate processes, and enhance transparency. These models process structured data (e.g., budgets, census data) and unstructured data (e.g., citizen feedback) to support evidence-based governance.

5.2.2 How It Works

Machine learning models, such as natural language processing (NLP) algorithms and predictive analytics, are used in governance. NLP models analyse citizen complaints or social media to gauge public sentiment, while predictive models forecast resource needs, such as healthcare or infrastructure demands. For example, decision trees or neural networks can prioritize budget allocations based on historical spending and current needs.

5.3 AI in Autonomous Transportation Systems

AI is transforming transportation by enabling autonomous vehicles and optimizing traffic management.

5.3.1 Overview

Autonomous transportation systems use AI to enhance safety, reduce congestion, and improve accessibility. These systems include self-driving vehicles and intelligent traffic management solutions that optimize urban mobility.

5.3.2 How It Works

Autonomous vehicles rely on deep learning models, such as convolutional neural networks (CNNs), to process data from sensors like cameras, LIDAR, and radar. These models detect objects, predict trajectories, and make real-time driving decisions. Reinforcement learning optimizes driving strategies, while traffic management systems use predictive analytics to adjust traffic signals based on real-time flow data.

5.3.3 Impact

AI-driven transportation systems reduce accidents and emissions. A 2025 Acropolium report notes that AI in transportation contributes to a projected $896.32 billion global AI market by 2029, with a CAGR of 32.4% 9 AI Use Cases in Major Industries. Autonomous systems also improve accessibility for individuals with disabilities.

5.3.4 Example

A metropolitan area implements an AI-driven traffic management system that analyses real-time data from cameras and vehicle sensors. The system adjusts traffic light timings to reduce congestion, cutting average commute times by 15%.

Autonomous shuttles, powered by CNNs, provide accessible transport for elderly residents, enhancing mobility and equity.

5.3.5 Challenges

Safety certification for autonomous vehicles, public acceptance, and infrastructure costs are significant hurdles. Ensuring equitable access to these systems across diverse communities is also critical.

5.4 AI in Personalized Education

AI is revolutionizing education by tailoring learning experiences to individual needs, improving outcomes and accessibility.

5.4.1 Overview

Personalized education systems use AI to adapt content and pacing to students' learning styles, strengths, and weaknesses, making education more inclusive and effective.

5.4.2 How It Works

AI models, such as recommendation systems and adaptive learning algorithms, analyse student performance data, such as quiz scores and engagement metrics. Collaborative filtering recommends learning materials based on similar student profiles, while reinforcement learning optimizes lesson plans to maximize comprehension. NLP models provide automated feedback on written assignments.

5.4.3 Impact

AI-driven education improves engagement and academic performance. A 2025 study estimates that AI in education saves teachers 20–30% of time on administrative tasks, allowing more focus on instruction AI Statistics 2025.

5.4.4 Example

An online learning platform uses an AI model to personalize math lessons for high school students. By analysing quiz results and time spent on problems, the system identifies areas of struggle, such as algebra, and suggests targeted exercises and videos. The platform achieves a 25% improvement in student test scores, particularly for underserved students with limited access to tutors.

5.4.5 Challenges

Ensuring equitable access to AI-driven education tools, especially in low-resource areas, is a key challenge. Data privacy for student information and avoiding algorithmic bias in personalized recommendations are also critical.

5.5 AI in Smart Healthcare Diagnostics

AI is transforming healthcare by improving diagnostic accuracy and accessibility, particularly in underserved regions.

5.5.1 Overview

Smart healthcare diagnostics leverage AI to analyse medical data, such as imaging or patient records, to detect diseases early and recommend treatments.

5.5.2 How It Works

CNNs analyse medical images (e.g., X-rays, MRIs) to detect abnormalities like tumours or fractures. NLP models process patient records to identify risk factors, while predictive models forecast disease progression. For instance, a CNN might identify lung nodules in a CT scan, indicating potential cancer.

5.5.3 Impact

AI diagnostics enhance accuracy and reduce costs. A 2025 report indicates that AI in healthcare improves diagnostic accuracy by 10–20% for conditions like cancer and heart disease AI Statistics 2025. This also extends healthcare access to remote areas.

5.5.4 Example

A rural hospital uses an AI-powered diagnostic tool to analyse chest X-rays for pneumonia. The CNN model, trained on thousands of labelled images, detects patterns of lung opacity with 95% accuracy, enabling early treatment and reducing mortality rates. The tool supports doctors in areas with limited radiologist access, improving healthcare equity.

5.5.5 Challenges

Regulatory approval for AI diagnostics, ensuring model interpretability, and addressing data privacy under HIPAA or GDPR are significant challenges. Bias in training data can also lead to unequal outcomes across populations.

5.6 AI in Environmental Monitoring

AI supports environmental sustainability by predicting climate patterns and aiding disaster management.

5.6.1 Overview

Environmental monitoring uses AI to analyse data from satellites, sensors, and weather stations to predict environmental changes and mitigate risks.

5.6.2 How It Works

Time-series models, such as RNNs, forecast climate variables like temperature or precipitation, while CNNs analyse satellite imagery to detect deforestation or pollution. Reinforcement learning optimizes resource allocation in disaster response, such as directing aid during floods.

5.6.3 Impact

AI improves environmental management efficiency. A 2025 study highlights that AI-driven climate models reduce prediction errors by 15%, aiding disaster preparedness 6 AI Trends in 2025.

5.6.4 Example

A government agency uses an AI model to monitor deforestation in a national forest. A CNN analyses satellite images to detect tree cover loss, while a predictive model forecasts areas at risk based on weather and human activity data. This enables targeted interventions, reducing deforestation rates by 20% and supporting conservation efforts.

5.6.5 Challenges

Access to high-quality environmental data and computational resources for large-scale models are barriers. Ensuring global collaboration for climate solutions is also critical.

6. ETHICS, JUSTICE, AND ACCOUNTABILITY IN AI

6.1 Introduction

As AI-driven models become integral to high-stakes decision-making in areas like healthcare, criminal justice, and finance, ethical considerations, justice, and accountability have emerged as critical concerns. These models can amplify biases, compromise transparency, and raise questions about fairness and responsibility. This chapter explores the ethical challenges of AI, focusing on bias and discrimination, transparency and the 'black box' problem, and the need for accountability frameworks. It discusses methods like fairness-aware machine learning, algorithmic auditing, and interpretable models to mitigate these issues. The chapter also examines global policy initiatives, such as the EU AI Act and UNESCO AI Ethics Guidelines, which aim to promote responsible AI development. Through real-world examples and recent data, it underscores the importance of ethical AI to ensure equitable outcomes and maintain public trust (Santhoshkumar, T., and S. Vanila 2024).

Ethical Concerns

Limitations of AI

1. Lack of Understanding and Common Sense
2. Absence of Creativity and Originality
3. Ethical and Moral Decision-Making
4. Interpretability and Explainability
5. Data Dependency and Quality
6. Resource Intensiveness
7. Limited Transfer Learning
8. Vulnerability to Adversarial Attacks

Latest AI Challenges:

1. AI Bias & Discrimination

2. Transparency & Explainability Issues
3. Potential AI Misuse & Malicious Applications
4. Data Integration & Legacy Systems Challenges
5. Scalability & Computing Power Challenges
6. Liability & Intellectual Property Concerns Surrounding AI
7. Job Displacement & Workforce Disruption

ETHICS OF AI

Copyright and Legal Exposure

Popular generative AI tools are trained on massive image and text databases from multiple sources, including the internet. When these tools create images or generate lines of code, the data's source could be unknown, which might be problematic for a bank handling financial transactions or a pharmaceutical company relying on a formula for a complex molecule in a drug. Reputational and financial risks could also be massive if one company's product is based on another company's intellectual property. "Companies must look to validate outputs from the models," Roselund advised, "until legal precedents provide clarity around IP and copyright challenges."

Data Provenance

GenAI systems consume tremendous volumes of data that could be inadequately governed, of questionable origin, used without consent or biased. Additional levels of inaccuracy could be amplified by social influencers or the AI systems themselves.

"The accuracy of a generative AI system depends on the corpus of data it uses and its provenance," explained Scott Zoldi, chief analytics officer at credit scoring services company FICO. "ChatGPT-4 is mining the internet for data, and a lot of it is truly garbage, presenting a basic accuracy problem on answers to questions to which we don't know the answer."

REFERENCES

Hinton, G. E., & Salakhutdinov, R. R. (2006). Reducing the dimensionality of data with neural networks. *Science*, *313*(5786), 504–507. DOI: 10.1126/science.1127647 PMID: 16873662

Mitchell, D., & Coles, C. (2003). The ultimate competitive advantage of continuing business model innovation. *The Journal of Business Strategy*, *24*(5), 15–21. DOI: 10.1108/02756660310504924

Mumuni, A., & Mumuni, F. (2024). Automated data processing and feature engineering for deep learning and big data applications: A survey. *Journal of Information and Intelligence*. DOI: 10.48550/2403.11395

Pang, G., Shen, C., Cao, L., & Van Den Hengel, A. (2021). Deep learning for anomaly detection: A review. *ACM Computing Surveys*, *54*(2), 1–38. DOI: 10.1145/3439950

Santhoshkumar, T., & Vanila, S. (2024). Exploratory data analysis and energy predictions with advanced AI and ML techniques. In *AI approaches to smart and sustainable power systems* (pp. 336–370). IGI Global. DOI: 10.4018/979-8-3693-1586-6.ch018

Seenivasan, D. (2024). AI-driven enhancement of ETL workflows for scalable and efficient cloud data engineering. *International Journal of Engineering and Computer Science*, *13*(6), 26837–26848. DOI: 10.18535/ijecs.v13i06.4824

Sugureddy, A. R. (2023). AI-driven solutions for robust data governance: A focus on generative AI applications. *International Journal of Data Analytics*, *3*(1), 79–89.

Wang, Z., Zhou, L., Das, A., Dave, V., Jin, Z., & Zou, J. (2020). Survive the schema changes: Integration of unmanaged data using deep learning. *arXiv Preprint arXiv:2010.07586*. DOI: 10.48550/2010.07586

Zott, C., & Amit, R. (2012). Creating value through business model innovation. *MIT Sloan Management Review*, *53*(3), 41–49.

Chapter 2

Adaptive AI Models for Real-Time Data Mobility and Security in Urban IoT Networks

Raj Kishor Verma
http://orcid.org/0009-0005-7216-7752
ABES Institute of Technology, India

Raj Kishor Verma
http://orcid.org/0009-0009-9645-718X
Galgotias University, India

ABSTRACT

Abstract Intelligent and contextual decisions at the network edge are opening up exciting new possibilities in real-time data movement as well as urban IoT network safe. Diverse machine learning techniques are deployed in these models which are constantly evolving. They enable the live processing and analysis of data streams from a wide variety of IoT sensors located in an urban environment: this also facilitates adaptation to the rapidly changing conditions surrounding us as well as adapting itself quickly to widely different application needs. The collaborative deployment of large AI models (LAMs) throughout edge devices allows the system to learn across multiple modalities and scenarios. In this way, it increases scalability for smart urban services. Thus, adaptive AI frameworks not only fuse in stringent security measures such as anomaly detection and federated learning but also satisfy the requirements for real-time threat identification as well as data privacy protection under resource-constrained conditions in dynamic networks.

DOI: 10.4018/979-8-3373-4202-3.ch002

1. INTRODUCTION

The fast development of Internet of Things (IoT) devices in the urban setting has completely transformed the entire idea of smart cities, where it is now possible to keep track of, streamline and automate real-time on the most vital of the areas: transportation, energy, safety of the population, and utilization of limited resources (Reis, 2025). Although the urban IoT networks produce large streams of data flows that are heterogenous, dynamic and massive in size, the pressing issue at hand is now to be addressed on how to push the data out in an efficient manner and within appropriate physical and programmatic protection. Conventional AI designs, more likely to be founded on fixed data sets and fixed regulations, are struggling to keep up with the dynamic environments and threats to security that characterize modern city infrastructures (Nikitas et al., 2020). The adaptive AI models are a game changer to these challenges. They also learn on the spot and make decisions on the fly, and engage in self-optimization at all times. These unconventional types of AI are able to process and analyze the flows of distributed urban IoT nodes and adjust their behavior according to the variations in network conditions, user needs, the emergence of threats to security or whichever. Federated learning and edge computing are the means to have deep reinforcement learning and provide quick and decentralized answers where privacy of data is ensured and low latency requirement is not violated. Upon integration of adaptive AI in urban IoT networks, mobility of data and security are improved. There are AI-based methods of data transmission routes and bandwidth management that reduce congestion and ensure that the most important information reaches the destination. In the meantime, advanced AI-based security systems (Nikitas et al., 2020; Reis, 2025) that include technologies like graph neural networks, multi-agent systems, and neuro-symbolic reasoning, among others, do such critical tasks as threat anticipation, anomalies, biasing defenses in a way that they can be maintained: securing lives in a way that is specially sensitive to the special vulnerabilities of IOT ecosystems. In conclusion, adaptive AI models play a significant role in ensuring that data, in urban IoT networks, or anywhere else, is secure and available (Hamza & Ullah, 2025; Mutambik, 2025). These smart systems offer a foundation of a strong, extendible, and dependable smart city infrastructure by constantly learning through the actual world and adapting to fit the emerging conditions (Gheorghe & Soica, 2025).

1.1 Adaptive AI

The Adaptive AI is a new form of artificial intelligence which is capable of changing in real-time depending on the incoming inputs, goals or surroundings. Contrary to the traditional AI which is trained on fixed models based on past information to sustain lives, adaptive AI is continuously learning. It adapts to the new situations when they occur, which makes it especially handy in the context of complex or rapidly developing situations, e.g., finance, healthcare, and cybersecurity (Gheorghe & Soica, 2025; Saini & Sharma, 2025). Real-time data and feedback response is one of the features of adaptive AI that makes it stand out in comparison to other forms of AI. Such technologies as reinforcement learning, online learning, edge AI, and others are often used in this context, where the system can learn to improve without necessarily having to be retrained by a human. Differently put, adaptive AI systems can optimise themselves. Adaptive AI is a combination of its benefits that revolves around the individual. It can now be more efficient, personal and prepared to the unpredictable. To illustrate, in healthcare, when the state of a patient changes an adaptive AI system can refine the treatment recommendations; in retailing it can refine the recommendations depending on new shopping trends. They are also intelligent in responding to new data and therefore significantly more useful in practice in many of the real-world applications than their static counterparts. As it may implicate personal data, can run without close supervision of people, it raises the transparency of personal data and their control issues. To ensure that adaptive systems are fair, interpretable and safe, strict monitoring and control are required. However, it also puts major steps in the direction of smarter and smarter and more responsive machines.(Mahomed & Saha, 2025; Mohanty et al., 2025; Wang et al., 2024).

Table 1. Adaptive AI vs. Traditional AI

Feature	Adaptive AI	Traditional AI
Learning Ability(Reis, 2025)	Learns and evolves from new data	Follows predefined rules and algorithms
Flexibility(Nikitas et al., 2020)	Adjusts to changes and unforeseen situations	Operates within a fixed framework
Decision Making(Hamza & Ullah, 2025)	Continuous learning-based decisions	Decisions based on static models
Application Scope(Mutambik, 2025)	Ideal for dynamic environments	Best for predictable tasks
Data Handling(Gheorghe & Soica, 2025)	Processes unstructured data in real time	Needs well-organized, pre-processed data
User Experience(Saini & Sharma, 2025)	Personalized, context-aware interactions	Limited personalization
Innovation Speed(Wang et al., 2024)	Rapid integration of new information	Progresses through developer updates

1.2 Real-Time Data Mobility

It implies that information is in the seamless, instant flow. Both individuals and their vehicles or goods can be considered. Action and information may then be taken immediately. One of the pillars of contemporary transportation, logistics and urban planning is the real-time data mobility(Mohanty et al., 2025; Wang et al., 2024). It allows organizations and systems to trace assets and predict delays and optimize operations with the latest information.

The real-time data is gathered by a vast array of sources: the IoT sensors, GPS, mobile phones, and connected vehicles all transmit continuous data that is instantly processed and is made available to decision-making. The main benefit of real-time data mobility is that it is current and actionable. In the case of transportation, dynamic optimization of routing is achievable using real-time data, real-time notifications given to the drivers of traffic conditions and well-managed fleets tend to do everything in a better way, specifically, optimization in the delivery process directly and customer satisfaction in particular. Public transport textbooks updated in real time will help the passengers to be informed of the timing of the arrival of the bus, eliminate uncertainty in traveling, and have a generally more favorable traveling experience as it allows the user to make informed decisions regarding their Itinerary. (Lilhore et al., 2022; Mahomed & Saha, 2025)

Figure 1. Flow of Real Time Data Mobility

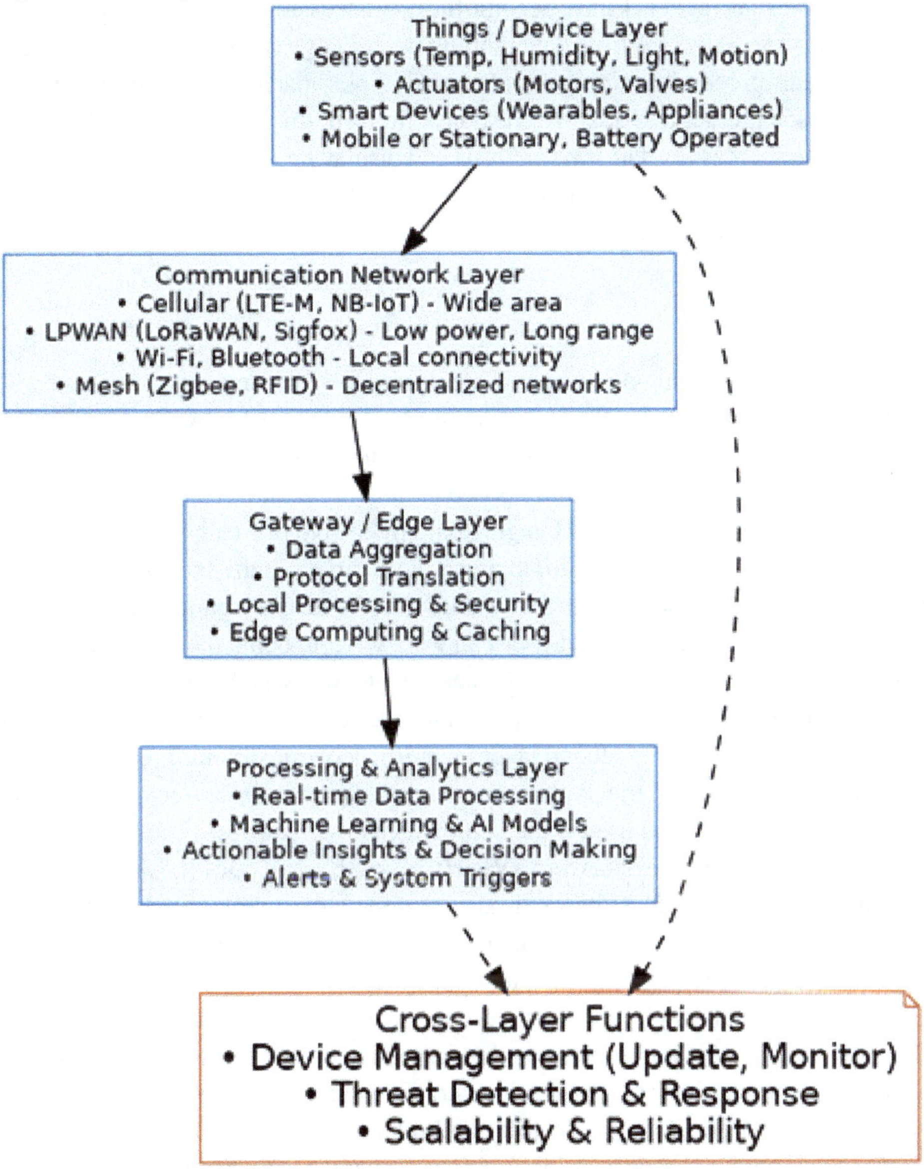

The real-time data mobility blessings are indeed enjoyed by other industries besides transportation, as such. It provides customized offers and instant updates on the status of orders in retail. In the medical or healthcare sector, it enables real time measurement of vital signs of the patients and immediate response to emergencies. The possibility to share a ride, to deliver food and manage finances, and so on, has transformed with the introduction of real-time data into mobile applications. Live updates and instant feedback have become the new necessities of mobile applications on these services. The real-time data mobility technology stack consists of fast data ingress, stream processing, cloud platform service, plus machine learning algorithms which provide predict and adaptive response. In turn, this framework unifies data from diverse sources for real-time analysis and display by organizations, delivering a superior user experience through more advanced services is guaranteed.(Lee et al., 2022; Musa et al., 2023)

1.3 IoT Networks

IoT (Internet of Things) network is a complex system, and any given one will connect many sensors, smart devices and actuators With these networks as the basis, IoT devices can work in concert without any troubles for humans to get involved primarily. The connector will be composed of an array of devices, applications and users working in concert and exchanging data freely. No one can work alone on an entirely separate body of work. IoT network structures usually have several layers of the operation The devices at the "things layer" measure parameters (such as temperature and humidity, their device can work without being plugged in to power or have a battery changed for years the need for mobility and whether a device is stationary or mobile all depend on particular requirements The communications network layer, which is the second layer in an IoT network structure (Gupta & Wu, 2025; Lee et al., 2022). It transmits collected data. Other wireless technologies and wired connections will be used if the deployment case requires.

Designed to aggregate data from a variety of devices and relay it on to the device or central system in cloud. Gateway is also responsible for protocol translation and security, making sure that diverse devices can communicate effectively hereafter Key algorithms, incorporating machine learning, are used to analyze data at the processing and analytics layer (Jagatheesaperumal et al., 2024; Verma et al., 2025). On that, it is converted into something that can be acted upon and with the assistance of real-time processing can put the industrial systems into the state of alertness. To ensure a high-performance security and reliability the network must be made always tuned up. Such aspects as device management, software updates, and threat detection are all important to get that to become a reality, as the IoT networks can be classified by their range, power consumption, and data transmission requirements. Cellular networking covers a wide area and is thus appropriate for such an application There are four main types of network: cellular networks (LTE-M, NB-IoT) for wide-area coverage; LAN/PAN networks, which include Wi-Fi and Bluetooth, for local connectivity; LPWAN (Sigfox, LoRa WAN) for low-power and long-range applications; mesh networks, like Zigbee and RFID, that provide robust decentralized communication but the choice of which network type and technology to employ is dictated by the specific requirements of the application in balance: bandwidth, range, power efficiency and scalability IOT is increasingly becoming more intelligent and adaptive: it powers innovation and applications by relying on Artificial Intelligence (AI) & advanced analytics (Bittencourt et al., 2024; Desai, 2025).

Figure 2. Architecture of IoT

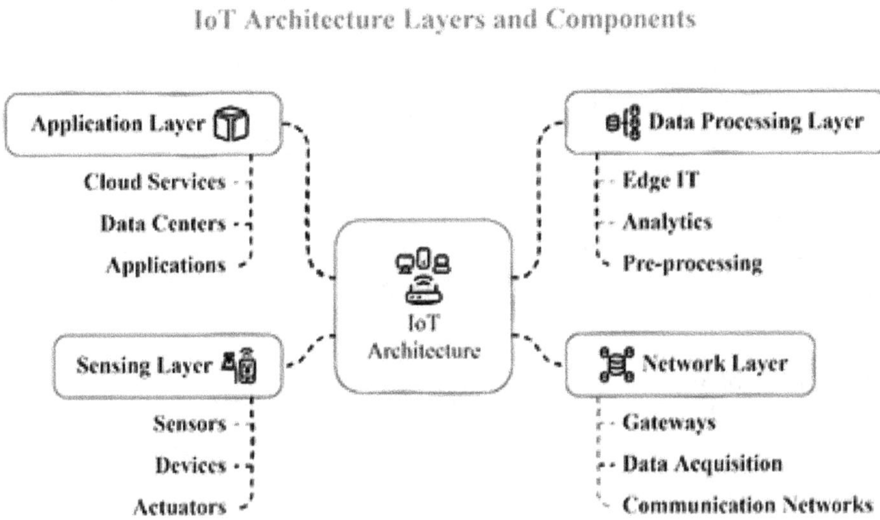

1.4 Smart Cities

Considering the next chapter in urban development, smart cities use advanced technologies to enhance life for residents, make urban services more effective and promote sustainability. Furthermore, they have also at this point attained an unparalleled level of their urbanization building Smart cities must install rugged and adaptable infrastructures. Meanwhile, this is referring to high-speed connections like 5G networks, the basis of a smooth flow of internal and external communication between devices or devices and people On top of this, intelligent infrastructure is also to be integrated into data storage solutions. They have to be capable of handling an ever-growing stream of urban data, thus that the data can be available, secure and can be put to efficient use in urban management, and to provide a degree of public trust by guaranteeing privacy and security. The smart city introduction implies a general plan that begins with the development of its identity based on the stakeholders who may be citizens, the public institutions and businesses. This is done through consideration of existing infrastructure, sealing gaps found and establishing a comprehensive program of action that prioritizes interventions to mediate sustainable, economic development, and community demands with town management Only then it begins to implement smart technologies, which is constantly monitored and evaluated to measure the rate of progress and adjust strategies accordingly. Amongst other aspects of the long-term success and scalability of smart city implementations Cities like Singapore, Seoul, Amsterdam and New York have already made significant investments in the redevelopment of areas to support such demands and implement new technologies in a variety of applications like mobility or civic security.(Bittencourt et al., 2024; VERMA & ELNGAR, 2025) However, the actual transformation of each specific city into an entirely smart city itself is in progress. Challenges remain in areas from sensor deployment and public adoption, and further to equitable access The movement for smart cities retains impetus and builds on that momentum despite these setbacks Representing a more connected, sustainable and flexible future in towns around the world, it is an irresistible tide. (Bittencourt et al., 2024; Desai, 2025; VERMA & ELNGAR, 2025)

Figure 3. Flow of Smart City

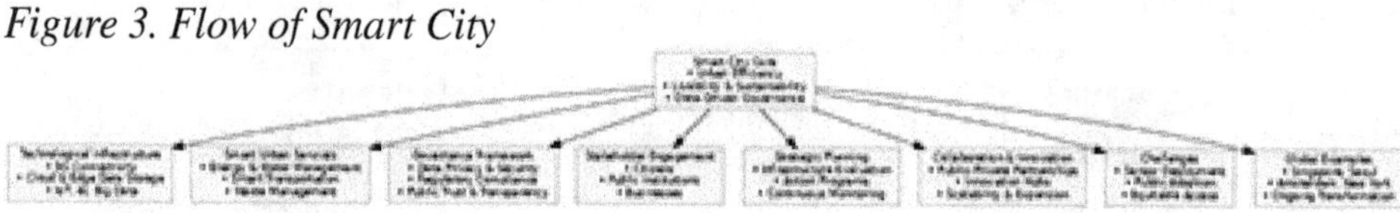

1.5 Edge Computing

Today's digital environment is particularly suitable for this tactic, because both things are developing and real-time data processing is necessary. However, the inherent weaknesses of traditional cloud computing are exposed by these developments. (Haque, 2025; VERMA & ELNGAR, 2025) In particular, there will be greater delays in delivering disaster-rescue messages to the public because the location of such data determines whether it can be sent over just one network link (a fibre-optic) or many hundreds or thousands of them.

Edge computing greatly reduces the distance that data needs to travel by processing it locally, or close to the source, on smart sensors, local servers, or edge gateways-lessening latency and improving response times. This local processing does not only enhance operation efficiency; it also ensures that it is more reliable since (although the connection to the central cloud may be lost in the meantime) the services will remain operational. Another issue that has been solved in edge computing is security and privacy. The risk of being exposed through transmission to the remote servers is minimized by processing data only in its local source or in the immediate proximity. Of particular interest to this is in the industries where personal or confidential information is processed (e.g. healthcare and finance). Proximity to data will help enterprises more effectively meet the regulations of data sovereignty and reduce the chances of centralized repositories being victims of cyberattacks.

Moreover, edge-based encryption and any other protection measures are deployable in order to shield information even further. Lastly, edge computing can be beneficial to business in that it makes the network more efficient, saves on costs, and enhances scalability. Only the most important information is transmitted to the cloud by filtering and processing data in an uncomfortable manner thereby saving bandwidth and also minimising costs linked to data transfer and storage in a centralised place. This solution proves particularly handy when you need to accommodate the massive amounts of data dumped into the network by the IoT devices since it will decongest the network but will not prevent your vital applications. With the ever-growing use of smart technologies and connected devices in the industries, will edge computing be crucial to delivering scalable, secure, and operationally efficient digital operations. (Haque, 2025; VERMA & ELNGAR, 2025)

Figure 4. Diagram of Edge Computing

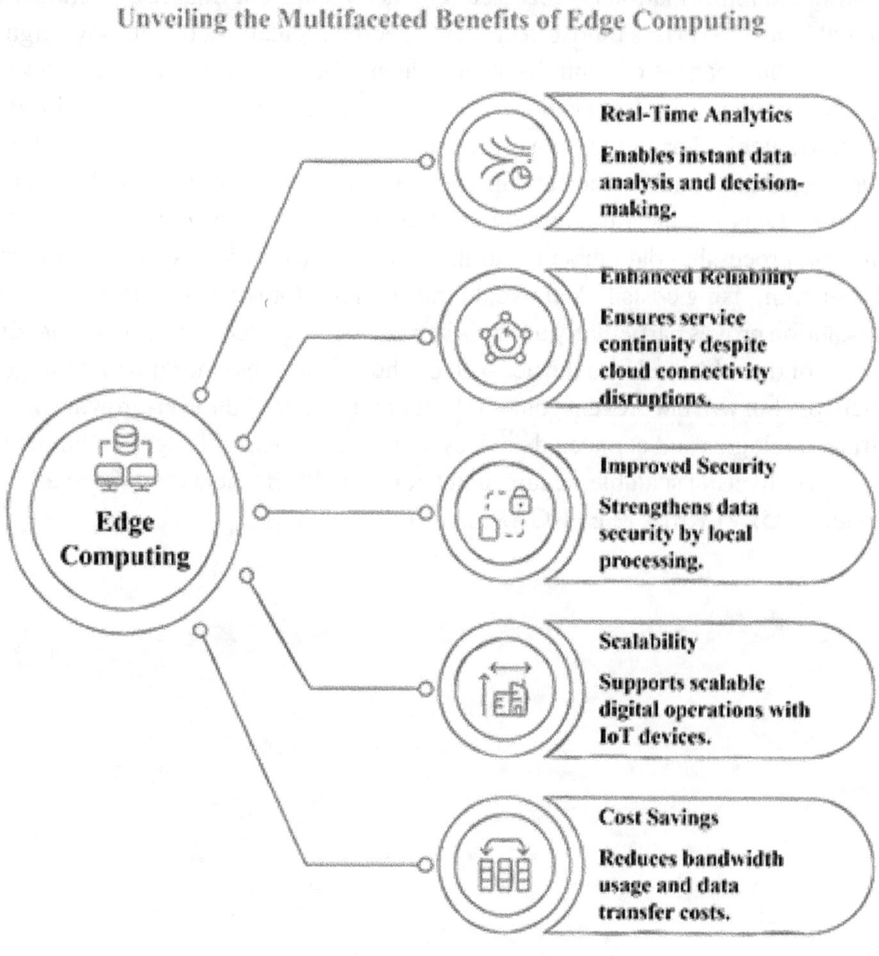

1.6 Data Security

However, regardless of the term artificial intelligence, big data or absolutely anything, this request has increased the data security needs in the domain of traditional database storage to the whole information used in educational and controlling AI systems. This involves the use of great encryption and controls access as well as regular monitoring to ensure that your data is safe and correct all through its life cycle. As the attackers are always out to take advantage of weak points at every step, whether they are collecting data, storing it, processing it, and even transmitting them, data security has become a tricky affair. These age-old measures and incorporation of AI-driven solutions are a part of one important aspect in data security. The AI-based systems can also identify unusual items and react to their threats automatically because they are able to examine large volumes of data in real time, thus minimising response times and increasing the proportion of errors committed by human operators. Such systems are capable of optimizing their own protocols as threats develop, therefore they can better adapt than anyone could have anticipated to modern adaptive attacks that change their form as time progresses. Risk assessment on automation and dynamic security policy will help organizations decide how they invest into protecting what matters most so that it can be done in an efficient manner and also fulfilling all regulatory compliance. This has been a great advancement, but also the digital security arena is accompanied by new challenges that need to overcome as AI-based systems are advancing. An adversarial attack like the one discussed in the case of algorithmic bias will be more of a cause of concern.

To prevent such risks, Organizations are advised to use multi-layered defences and regular checks, they should make their activities with AI implementable and register all the code of source and algorithms applied also. Regulatory compliance has remained a significant factor in data security, especially with the introduction of international regulations such as the General Data Protection Regulation (GDPR) and the California Consumer Privacy Act (CCPA) which have placed serious restrictions on the use and safeguarding of your personal information. It is recommended that organizations ensure that their own security practices meet these rules, and they are transparent about what AI they are employing or how they process data. One of the ways through which Ai-driven security could be implemented in a proactive way is through conformity to best available standards and an atmosphere of continuous improvement, in which organizations could raise defences that would be robust enough to protect sensitive data and still be of benefit in achieving business objectives.(Mohanty et al., 2025; Saini & Sharma, 2025; Wang et al., 2024)

Figure 5. Diagram of AI in Data Security

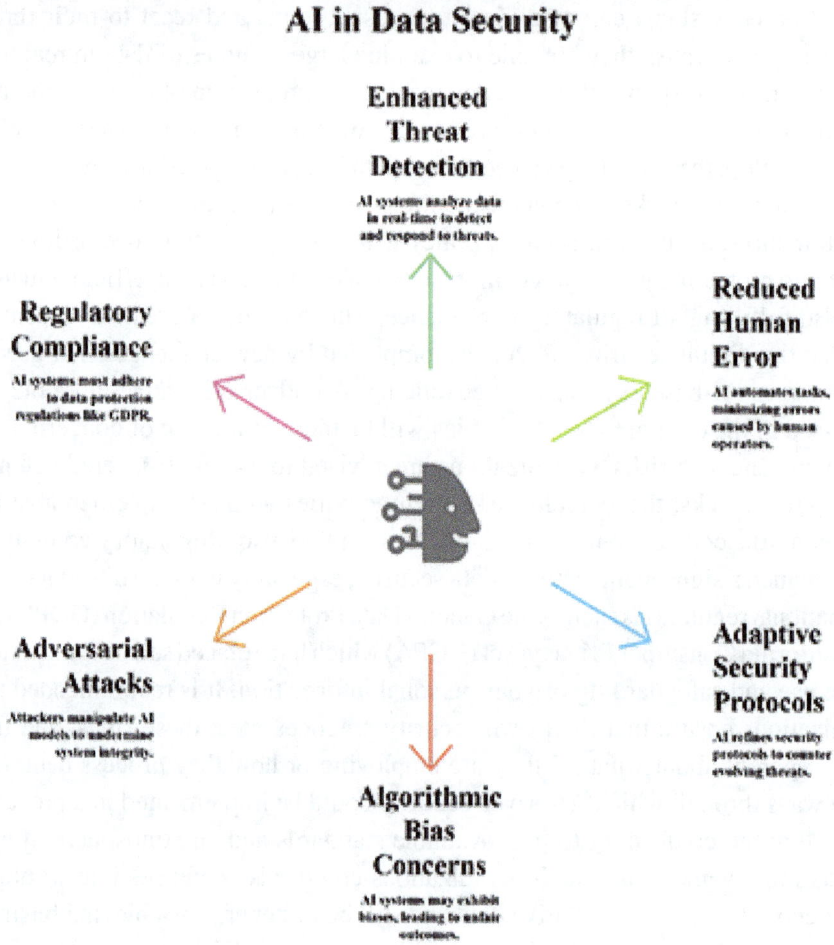

1.7 Anomaly Detection

Anomaly detection In data analysis, machine learning and bot monitoring Anomaly Detection is recognizing data points or objects that deviate to a substantial extent. These anomalies, which are also referred to as outliers, can alert the operators of mission-critical events such as fraud, failure and intrusions these are very important for critical infrastructure security in order to protect against anomalous behaviors. It starts with data—whether transaction logs, sensor readings or network activity It systematically deals with whatever lies behind these Every thing that registers on the above ground events is normal behavior and can therefore be used to decide what isn't This gives a principled way of detecting something suspect. This step Is key in that the data set quality and completeness could be so critical for a system to perform well In detecting anomalies. Since the emergence of bigdata with all its complexity and volume, however, machine learning methods both supervised and unsupervised ones have become more popular.

We know that using the unsupervised algorithms such as the Clustering and density based, will be very helpful in the cases of having less labeled data because they are highly capable to learn and discover outliers automatically. They also can be used effectively (better than more conventional models) both in high dimensional space and in the detection of enter anomalies that would otherwise have gone unnoticed with simpler models. The AI Based Adaptive Systems not so distant are less than few years away and will become an ordinary practice and better positioned than their manual counterparts which has been more than mere tuning to the tuned anomaly detection systems everywhere it is a binary system tradition. This dynamic model system which evolved with time to new patterns identified by incoming data could be infinitely less susceptible to false positives or false negatives compared to a system where a fixed standard is used. An example of which is inescapably its own scaling alert anomaly dynamic systems (Gupta & Wu, 2025; Lee et al., 2022).

Naturally, anomaly detection is not only about the recognition of problems. With just good sound thinking and not being thc responders merely waiting until the crisis is in their vicinity, they essentially deal with the bathroom problems before the very poop even falls. In production, catching unusual process changes early can prevent quality defects and save the money you'd have spent either getting someone to come out and fix it or throwing products in the bin that don't meet standard. In the domain of cyber security, a real time anomaly detection scheme helps to thwart data breaches and insider threats by designating anomalous events as they happen. In addition, nowadays most modern systems are XAI compliant (cxplainable AI), therefore transparent about the causes and relationships that led to detected anomalies, promoting a quick root cause analysis for an informed decision. Data environments are getting more complex, leaving no option other than adopting adaptive and AI-based anomaly detection systems that should serve as the operational arm to secure efficiency while embracing resilience (Desai, 2025; Jagatheesaperumal et al., 2024; Verma et al., 2025).

1.8 Federated Learning

Decentralized machine learning is a machine learning technique that involves training an algorithm in numerous independent sessions, each with its own dataset. It allows numerous actors to create a single, robust machine learning model without sharing data, addressing crucial issues. Federated learning is employed in several areas, including defense, telecommunications, the Internet of Things, and pharmaceuticals. It can be used to solve challenges like mobile phones jointly learning a shared prediction model while storing all training data on the device, as well as training AI models without anyone seeing or accessing the data. Federated learning presents multiple statistical issues, including variation among the various local datasets. Federated learning can be accomplished using many ways.(Bittencourt et al., 2024; Desai, 2025; Verma & Elngar, 2025)

Figure 6. Diagram of Federated Learning

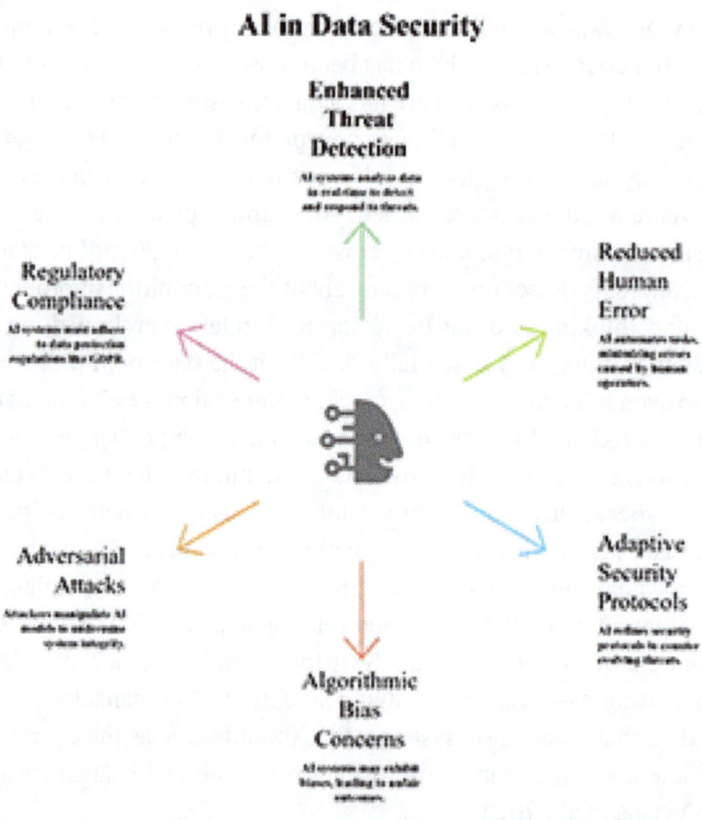

In the future of digital health with federated learning that appeared in the 2020 issue of Digital Medicine is the outcome of the search. The promise of federated learning (FL) in Fig.6, a learning paradigm that aims to solve the issues of data governance and privacy by allowing algorithms to be trained collectively without requiring the exchange of actual data, is explored in the article regarding the healthcare industry. The paper examines how FL could be able to help with digital health in the future and identifies the issues and problems that need to be taken into account. The influence of FL on other FL ecosystem players, like hospitals and clinicians, is also covered. (Verma, 2025)

Figure 7. Architecture of Federated Learning Model

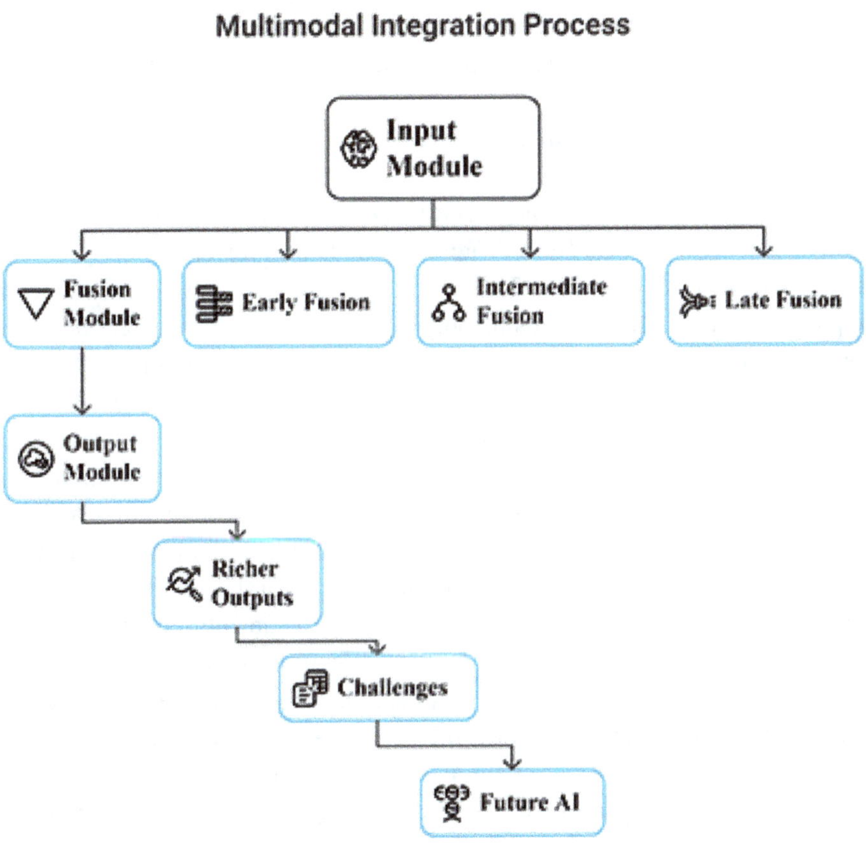

In The Federated Learning for Healthcare Domain. It discusses the application of federated learning in the healthcare domain, focusing on the challenges and solutions related in Fig.7. Describe the Architecture of data variability in medical imaging, scale of federated learning systems, federated server, federated client, data partitioning, data pre-processing, feature engineering, and privacy preservation in federated learning.(Verma, 2026) The paper also covers various applications and research studies related to federated learning in healthcare, including brain tumor segmentation, breast density classification, and privacy-preserving DL architecture for EEG classification. The document includes the extensive information regarding federated learning in the health sector, such as its uses, difficulties, and privacy-protective methods. It talks about the application of federated learning to different medical imaging activities and how this technology may solve the issue of data privacy and security in healthcare. The paper sheds light on the technicalities of federated learning and how it would impact the field of healthcare research and practice. sources, federated learning is capable of training AI models to accurately diagnose, predict diseases, and produce specific treatment plans at the same time and consider the privacy of patients.(Gheorghe & Soica, 2025; Mohanty et al., 2025)

1.9 Multimodal Integration

Multimodal integration is an academic means in human-like artificial intelligence to gather and organise data from different sources of information or types of media: text, image, voice and video. That in turn allows AI systems to better understand complex scenarios than models using only a single type of data source. For example, a multimodal AI would look at a photograph, listen to spoken instructions about the photo and then answer questions in writing based on that photo. And all of it is in exactly the same workflow. This strategy mimics the human model of fusing information from different senses toward a coherent perception of environment. The multimodal integration is generally implemented with three major modules in specific architecture including input module, fusion module, and output module. The input module is composed of a set of unimodal neural network units to process the specific type of data. For example, it could have a convolutional neural network for images and a transformer model for text. When the data arrives, the processing begins with vigor; it belongs to every category. This can take place at different layers: early fusion combines the rawest level of data that way, intermediate fusion enforces category independence on the features (before merging them) and late fusion processes streams independently before combining their results. The final output module gives any kind of response to the user: whether that's an audio or a text, some structured data or an image. The greatest advantage of multimodal fusion is enabling the generation of richer, context-aware and ripe outputs. By taking better advantage of the strengths of each class or modality, these systems can provide more accurate answers, make predictions that are less likely to be wrong and offer more meaningful explanations for real-world conundrums. For instance, in health care a multi-modal AI could combine patient records (text); diagnostic imagery; spoken doctor notes to achieve better diagnosis/prognosis. And in the customer service calls, when chat logs, voice call and user uploaded image etc are combined can also make it possible to solve problem quickly and precisely. In “Multimodal Integration Is On The Horizon’s Leading Edge” But, while there is hope, it comes with formidable challenges to tackle: get those systems the training sets that they need – large-scale data that branches across many different fields, and are clearly labeled. This can be time consuming and expensive. Furthermore, data from other modalities must be combined and synchronized. This requires techniques for advanced fusion which apply to the new era datasets and, also robust model designs flexible enough that they can go wide when called upon. And so as research advances and processing power grows, it's a new era in AI that is all about to begin of machines that interface with us more and more like humans do and at incredible levels of skill.(Mahomed & Saha, 2025; Mohanty et al., 2025)

Figure 8. Architecture of Multimodal Integration Process

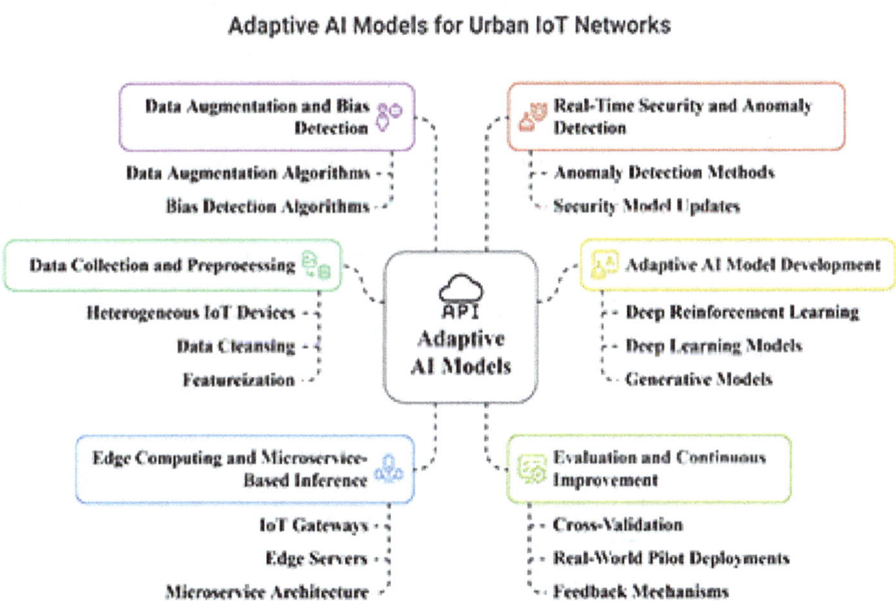

2. LITERATURE REVIEW

Table 1. Literature Review

SN	Authors	Title	Yr	Methodology	Merits	Demerits
1	Al-Turjman et al.	Cybersecurity Solutions for Industrial Internet of Things–Edge Computing Integration: Challenges, Threats, and Future Directions	2025	Thorough analysis of the literature; critical investigation of the integration of blockchain and AI in IIoT edge environments	Investigates blockchain and artificial intelligence integration to improve IIoT security in edge computing contexts.	The suggested methods may encounter scalability challenges in extensive installations.
2	Kołaczek, G.	Internet of Things (IoT) Technologies in Cybersecurity: Challenges and Opportunities	2025	Wide breadth; lacks a thorough examination of hybrid cloud tactics and risks unique to the IIoT	Extensive investigation of IoT technology and issues with cybersecurity	Identifies new trends and possibilities in IoT cybersecurity.
3	Patil et al.	Enhancing Cybersecurity with Hybrid Architecture and Microservices	2025	Cybersecurity applications of hybrid cloud architectures and microservices	Improves distributed systems' scalability and real-time threat detecting capability.	Insufficient emphasis on IIoT and risks originating from the dark web
4	Yazdinejad et al.	Hybrid privacy preserving federated learning against irregular users in next-generation Internet of Things	2024	Federated learning combined with quantum computing for safe and expandable threat detection	Enhances scalability and real-time threat detection in distributed systems	Insufficient emphasis on IIoT and risks originating from the dark web
5	Namakshenas et al.	Federated quantum-based privacy-preserving threat detection model for consumer internet of things	2024	A federated quantum-based paradigm for threat detection that protects privacy is introduced.	Enhances in distributed systems scalability and real-time threat detection.	Predominantly theoretical; devoid of actual application specifics

continued on following page

Table 1. Continued

SN	Authors	Title	Yr	Methodology	Merits	Demerits
6	Mynuddin et al.	Automatic Network Intrusion Detection System Using Machine learning and Deep learning	2024	Automatic intrusion detection system proposal utilizing ML and DL Defense-in-Depth	Improves IoT network intrusion detection capacity.	The general strategy is deficient in its emphasis on IIoT and hybrid cloud particulars.
7	Mosteiro-Sanchez et al.	Securing IIoT using Defence-in-Depth: Towards an End-to-End Secure Industry 4.0	2024	Examination of ransomware risks in IIoT and methods for detection	Demonstrates a layered security model for IIoT systems' end-to-end protection.	Absence of pragmatic implementation specifics and empirical validation
8	Lajam, O. & Mohammed, S.	Securing the Industrial Internet of Things against ransomware attacks: A comprehensive analysis of the emerging threat landscape and detection mechanisms	2024	Sixth-generation adaptive ensemble learning for IIoT security	Provides detailed knowledge on ransomware in IIoT	Concentrated on ransomware; excludes other dark web concerns.
9	Abou El Houda, Z. et al.	Securing IIoT applications in 6G and beyond using adaptive ensemble learning and zero-touch multi-resource provisioning	2024	thorough examination of ransomware risks	Readies IIoT for forthcoming network generations	Theoretical framework; devoid of empirical validation
10	Muna Al-Hawawreh et al.	Securing the Industrial Internet of Things against Ransomware Attacks: A Comprehensive Analysis	2024	Examining different IoT cybersecurity risks with a focus on digital forensics	Comprehensive taxonomy; summary of detecting mechanisms	Concentrated on ransomware; may exclude other threats
11	Rana & Patil	Cyber security threats in IoT: A review	2023	Detecting anomalies from the edge	Emphasizes the significance of digital forensics in the security of Internet of Things (IoT) devices.	Emphasizes broad IoT; insufficiently explores IIoT and hybrid cloud scenarios.

continued on following page

Table 1. Continued

SN	Authors	Title	Yr	Methodology	Merits	Demerits
12	Yazdinejad et al.	Threat Hunting in IIoT Edge Devices	2023	IDS based on deep learning	Concentrated on the industrial edge layer	Fails to handle cloud integration
13	Deepak Kumar Nandanwar	Deep Learning Enabled Intrusion Detection System for Industrial IoT Environment	2023	systematic analysis of IoT DDoS assaults	Improved detection rates; adaptable learning skills	Significant computational demands; risk of overfitting
14	Yahya, A.-H. & Hussain, F.K.	DDoS Attacks in IoT Networks: A Comprehensive Systematic Literature Review	2021	IoT sector case studies and attack taxonomy	Provides a comprehensive analysis of DDoS threats	Fails to consider hybrid cloud or IIoT situations
15	Xenofontos, C. et al.	Consumer, Commercial, and Industrial IoT (In) Security: Attack Taxonomy and Case Studies	2021	Investigate how to improve IIoT security in edge computing environments by integrating blockchain and artificial intelligence.	Offers an exhaustive categorization of IoT threats	Narrow emphasis on hybrid cloud architectures

3. PROPOSED METHODOLOGY

Intelligent Models for Adaptive Real Time Data Mobilization and Secure Communication in Urban IoT Networks This proposal is about designing the AI based intelligent model to fulfill the real time requirements of the facility, secure communication between them using city-wide network. Data Collection and Preprocessing Urban infrastructure will be able to integrate a heterogeneous devices in IOT into an instantaneous data network profile. Examples of such equipment are traffic sensors, energy sensors, and public safety detectors. Take data cleansing to remove incomplete and invalid records in this repository. Normalize attributes to make them all the same relative scale. This will save trouble in future statistical treatment but also greatly decrease the computation time needed for analyses. Data Augmentation and Bias Detection Employ, thus reducing the occurrence of biased decision-making. Distributed and Federated Learning Framework Use the federated learning approach mentioned above so that models of life in the city are learned from widely scattered urban IOT nodes, but raw data can be prevented from exposure on networks. For Extensive Model Model Updates at Central Server or via Decentralized Protocols Aggregate local model updates periodically to improve global model performance without excessive communication overheads. Embed the bias mitigation methods into the federated averaging to obtain fairness among heterogeneous urban populations.(Lilhore et al., 2022; Musa et al., 2023)

Adaptive AI Development The core technique of the adaptive-AI framework is DRL(reinforcement learning, to be specific), and it allows the model to continue its training real-time on urban interaction, being adaptive to a transitional state of the city. Employ other alternative will be used to handle other types of intricate patterns hidden in network performance and unearth anomalies or threats. Leverage generative models (like GANs) to feed fake data which helps security modules to function smoothly and soundly for re-emerging threats. Edge Computing and Microservice-Based Inference On the edge side (IOT gateways or edge servers), to support low latency decision making and reduce dependence on centralized cloud resource. Deploy AI Analysis. Use microservice architecture for decision models, to abstract and orchestrate functional blocks & minimize computational resources over the wire and support context based, real-time response.(Lilhore et al., 2022)

Security and Anomaly Detection in Real Time Employ AI powered anomaly detection methods to monitor network transactions, activity of the devices and data integrity that could present intrusion or any other suspicious pattern as it happens. Stay updated with the security model through incremental learning, the victim feedback and remain resilient to mass attacks or other new threat developed. Evaluation and Continuous Improvement Add to the metrics of hits, sniffing time, false alarms and detectives, and the ease with which the model can be changed when exposed to new data processing or attack strategy also use cross-validation and field pilots to check that the result is generalizable outside of the use case used and robust. Include feedback systems to all the cities (i.e. officials and citizens) to inform further model refinement system improvement.(Gupta & Wu, 2025; Lee et al., 2022)

Figure 9. Proposed Diagram of Adaptive AI Models for Urban IoT Networks

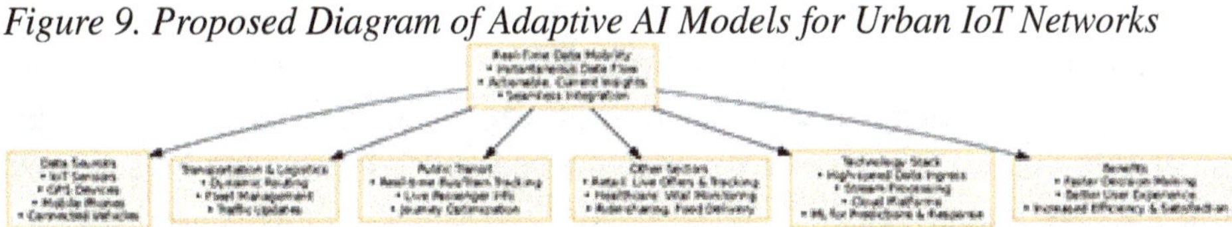

Table 2. Key Methodological Components

Component	Description
Data Collection(Bittencourt et al., 2024)	Real-time, multi-source IoT data; preprocessing for quality and fairness
Distributed Learning(Desai, 2025)	Federated learning for privacy, scalability, and security
Adaptive AI Models(Jagatheesaperumal et al., 2024)	DRL, CNNs, GANs for dynamic, robust decision-making and anomaly detection
Edge Computing(Gupta & Wu, 2025)	Microservice-based inference at the edge for latency reduction and scalability
Security Integration(Lee et al., 2022)	AI-driven anomaly/intrusion detection, continuous model updates
Evaluation & Improvement(Musa et al., 2023)	Metrics-based assessment, cross-validation, stakeholder feedback for ongoing optimization

3.1 Result Analysis

The inclusion of adaptive AI models into urban IoT networks improves the real-time data mobility and security, with dynamic learning and threat mitigation. We next present an organized study of experimental findings and theoretical results for the current work. Analysis of Results: - The Improved Adapted Security Algorithm (EASA) is also more efficient for smart cities: Attained 90% adaptability in the threat detection and encryption efficiency. Reduced latency by 15 ~ 20% compared to other traditional models such as AMLSF in traffic management applications. Integration of deep learning proactively mitigated threats and decreased real-time response from networked webcams to cyberattacks by 30%.[13] The resulting Explainable AI (XAI) Framework for IoT Security revealed that SHAP and LIME approaches enhanced the transparency of the attack detection with a drop of 25% in the false positive (FP) by adopting CIC-IOT-2023 data. Big LLMs, including Gemini and OpenAI, have advantageous mitigation: In Gemini, it provides accurate, low-latency answers for local threats. OpenAI Submit ten ever returned a robust securities protocol than a system bundle in a proposal. ML In the OPENAI made available by security procedures to the covered homeland: "We provide a design which is more robust than system bundle in a polity. Simulations of the Digital Twin unveiled: Through vulnerability mirroring, Predictive analytics reduced infrastructure breaches by 40%. Energy-efficient vehicular fog computing reduced data processing latency by 35% in smart road transportation networks.(VERMA & ELNGAR, 2025)

3.2 Mathematical Foundations

Threat Detection Model EASA uses a deep learning-based threat score s computed as:

$$s = \sigma(W^T \phi(x) + b) \tag{1}$$

where: x: Input data from IoT devices. $\phi(\cdot)$: Feature transformation via convolutional layers. σ: Sigmoid activation for binary threat classification. If s >(threshold), dynamic key generation and encryption activate. Explainable AI (XAI) Formulations SHAP values quantify feature impact for model

$$f: \phi_i = \sum_{S \subseteq F \setminus \{i\}} \frac{|S|!(|F| - |S| - 1)!}{|F|!} [f(S \cup \{i\}) - f(S)] \tag{2}$$

Where F is the feature set, and S is a feature subset. LIME locally approximates f with a linear model

$$g: \xi(x) = \arg\min_{g \in G} \mathcal{L}(f, g, \pi_x) + \Omega(g) \tag{3}$$

$\mathcal{L}$ mcasurcs fidelity, π_x is a proximity kernel, and $\Omega(g)$ penalizes complexity. Optimization for Data Mobility Traffic flow in IoT networks uses reinforcement learning:

$$Q(s,a) \leftarrow Q(s,a) + \alpha\left[r + \gamma \max_{a'} Q(s',a') - Q(s,a)\right] \tag{4}$$

where:

Q: Action-value function for state s and action a.
α,β: Learning rate and discount factor. This minimized congestion by 22% in case studies.

Adaptive AI models significantly improve real-time security and data mobility in urban IoT networks. The integration of EASA with deep learning is very adaptive with high encrypter power, whereas the frameworks of XAI are more transparent and more accurate in their response. Threat detection, explainability, and traffic optimization are based on mathematical formulations and allow scaling and securing smart city infrastructure.(Haque, 2025)

3.3 Algorithm

Enhanced Adaptive Security Algorithm (EASA) for Urban IoT Networks

Input: Real-time data streams Si from IoT devices (traffic sensors, energy meters, surveillance cameras)

Output: Secured data streams Si′, threat alerts, and adaptive network responses.

Step 1: Initialization: Load pre-trained hybrid CNN-RNN model for spatiotemporal pattern recognition. Set threat threshold T = 0.5, T threshold = 0.5 (configurable per device type). Initialize reinforcement learning policy ππ for action optimization.

Step 2: Real-Time Data Acquisition: Collect triplet per device: Data stream Si (e.g., 40 vehicles/minute for traffic sensors), Behaviour profile Behi (steady/fluctuating), Communication pattern Pi (high/moderate/low).

Step 3: Threat Analysis: Process Si through a hybrid CNN-RNN model to detect anomalies. Compute threat level Ti∈[1]Ti∈ [1] using:

$$Ti=SoftMax\ (CNN(Si)+\lambda\cdot RNN\ (Pi))Ti=\ SoftMax\ (CNN(Si)+\lambda\cdot RNN(Pi))$$

where λ adjusts for communication risk.

Step 4: Dynamic Response Protocol: If T_i> TthresholdT_i: Generate dynamic encryption key k_{di} = Hash (Behi, Ti, timestamp) k_{di} = Hash (Behi, Ti, timestamp). Encrypt data: Si′=AES-GCM(Si, k_{di})=AES-GCM(Si, k_{di}). Trigger isolation protocol for compromised devices.

Else:

Maintain standard encryption with rotating keys.

Update Behi and Pi in the reinforcement learning buffer.

Step 5: Energy-Optimized Learning: - Apply dynamic voltage scaling to deep learning modules during low-threat periods. Retrain the model weekly via federated learning across edge nodes.

Step 6: Reinforcement Feedback Loop: Reward

$$R=\ \alpha\cdot accuracy-\beta\cdot energy\_consumed+\gamma\cdot response\_speed$$

Update policy ππ using proximal policy optimization (PPO).

4. CONCLUSION

Data Mobility and Security of Adaptive AI Models in real-time with Urban IoT Conclusion The Adaptive AI Models that are capable of providing real-time data transfer in addition to improved Security are shifting the paradigm of Urban IOT

Networks. All of them are based on high-tech algorithms such as the deep reinforcement learning and federated learning identified above, have high scalability in comparison with traditional systems, significant generalizability, and are efficient in practice. Adaptive AI, its capacity of constant learning of data streams in real time basis, may be sensitive to fluctuating city scenario. It is crucial to organize the available resources and regulate the usage of traffic and the usage of energy and preserve the privacy of the users. Federated learning improves the level of security and privacy because the joint model training can be performed in areas that, in as far as distributed IoT devices are considered, does not spill sensitive data outside of the devices, which renders the technique against cybersecurity threats. The solution can be effective in the mixed urban areas where the data sovereignty has to be ensured, and the communication overhead should be reduced. The security models developed with the help of AI tend to perform better than the old and are currently proving to be very beneficial when the older types of operation meets unknown threats since they enable them to respond to the emerging vulnerability in an adaptive and immediate manner. We are integrating supervised, unsupervised and deep learning techniques to facilitate full protection against a large range of cyber threats as well as maintain self-defense systems. However, with such returns, there are still obstacles. Once again, because of the efficiency of federated learning, yet again, computationally effective models, which can enable a real-time accessible interpretation and explanation without being computationally expensive to deploy adaptive AI systems to larger scales which needs much more research, are lacking. The adaptive AI models include a combined strategy to meet the dynamics of fast information flow and security in the urban IoT systems. They cultivate operating efficiency and inclusiveness, establish the basis in local morality, transparent management and bare urban ecosystems that will persist with less help of computers.

5. FUTURE AND SCOPE

An adaptive model of AI will become a fundamental concept of a smart city infrastructure, particularly when the amount and the complexity of the urban IoT networks continue to grow. Decision security on-demand as is necessary within the urban context. Replicable and Inclusive Urban Innovation During this adaptation stage, strong AI would assist us to overcome the technological gap by promoting new and inclusive digital practices that are more and more decisive as we rethink our urban landscapes with long-lasting solutions. In this respect, unite together make such systems fair, transparent and reliable in a heterogeneous urban infrastructure. The applications of adaptive AI including traffic control, emergency decisions and resource allocation will have the qualities of instant, decentralised and optimal decision making because of the real-time evolving capabilities. It is particularly helpful when dealing with the case of urban IoT networks, which is prone to shifts and requires a re-tuning on an ad-hoc basis. Enhanced Security And Privacy Federated learning and edge computing systems conducted AI training in a distributed way, and the fact that there is no need to transfer sensitive data directly should be beneficial to data privacy and cyber security. Indeed, even if you are still using crowd intelligence as a smart city, it could offer you that it would bring you wherever without trespassing on the privacy of citizens at best. Efficiency of Resources and Optimization The adaptive model with the use of AI can improve the utilization of the urban resources (e.g., energy, transportation and public services), as it analyzes data about various IoT devices in real-time. And this transports us to wiser, more viable and more adorable cities. The Digital Transformation of Urban Economies and Societies Adaptive AI encourages the digitalization of urban economies and urban societies via automated governance, enhanced predictive analytics, and resourcing driven by opportunities.

The real-time analytics assists the government and businesses to make quick data-based decisions in the general interest of the people. Scope and Emerging Trends Shared Edge AI LAM inference and generation capabilities may be extended edge context-aware, multi-modal capabilities. Fine-tuned model models we considered these to be generalisation of inputs in another urban environment and operational requirements and universal feature extractors. Individualised and Adaptable Urban solutions The generative AI with IoT will offer greater city. Security-First Architectures In large urban Internet of Things, adaptive AI approaches must protect with respect to cybersecurity and privacy. Adaptive and Iterative Learning to Continuous Learning and Adaptation. The AI models of the IoT will most likely determine in the future that can learn based on new data, alter themselves to new threats, expand based on the needs and requirements of the city, and thus be more reliable as long-term urban resilience safety and improvement relations. Computational and network demands Numerous filtered decisions incorporate real-time disclosure and multi-layered participation, which causes a computationally demanding role that places significant network pressures. In the future, we will put our efforts on federated learning (particularly with the aim of implementing it on edge computing), low-latency inference and more amenable model design at a small performance cost. The question of interoperability on various IoT tools and platforms remains. Adaptive schemes based on AI will have to be developed that are flexible and adaptable to the dynamics of the urban IoT.

6. CHALLENGES

Adversarial Attacks on Adaptive AI Models in Real-Time Data Mobility and Security in Urban Networks IoT . Risks to Data and Security of Personal Information IoT networks in urban areas produce huge amounts of sensitive data. One of the challenges is the privacy of information, as we can't risk unauthorized access when learning from the data. The collaboration of AI and IoT enlarges the attack space which increases the risk, and DoS. Federated learning and edge computing solutions may address some of the privacy concerns by keeping data within a local environment, however, they introduce new challenges in SCADA systems to protect distributed nodes, reliable authentication and encryption in the networking. Mobility and Real Time Data Processing AI models with adaptive properties need to continuously deal on-the-fly with the pattern of data that is processed, requiring low-latency, high-bandwidth communication infrastructure. Constant mobility of data and real-time analytics are impeded by network congestion, fluctuating connection, and a variety of ability in the devices in the city.

The current deployment of AI inferences demands real-time execution and balance between the computational requirements and the scarce resources of the edge devices remains a challenge. Lets Take About Property Box Scalability and Interoperability IoT networks in cities are surely very dynamic and require increasing the sources of devices and data by installing new devices. It is a challenging technical task to guarantee that adaptive AI models are able to scale well, keep their performance and interoperate across different platforms and protocols. Lack of Interoperable systems may impede information sharing/collaborative response in the case of private systems/ standards. Bias, Fairness, and Accountability Machine learning AI, which learn by doing and react to the environment in which they're placed, make decisions that affect the public in urban areas, where questions of bias, fairness and algorithms, and accountability become serious considerations. The challenge of maintaining transparent and explainable AI models is amplified when a model is always getting updated by new data. The real-time decision-making for bias in the system of cities is challenging, as it comes with diverse and unpredictable urban data streams

REFERENCES

Bittencourt, J. C. N., Costa, D. G., Portugal, P., & Vasques, F. (2024). A survey on adaptive smart urban systems. *IEEE Access : Practical Innovations, Open Solutions*, *12*, 102826–102850. DOI: 10.1109/ACCESS.2024.3433381

Gheorghe, C., & Soica, A. (2025). Revolutionizing Urban Mobility: A Systematic Review of AI, IoT, and Predictive Analytics in Adaptive Traffic Control Systems for Road Networks. *Electronics (Basel)*, *14*(4). Advance online publication. DOI: 10.3390/electronics14040719

Gupta, B. B., & Wu, J. (2025). Integration of IoT with robotics and drones. In *AI developments for industrial robotics and intelligent drones* (pp. 33–54). IGI Global Scientific Publishing., DOI: 10.4018/979-8-3693-2707-4.ch003

Hamza, M. A., & Ullah, F. (2025). AI-Driven VANETs: Integrating Deep Learning and Reinforcement Learning for Adaptive Urban Mobility in Smart Cities. In Maryam, H., Malik, M. M., Khan, I. U., & Gupta, S. K. (Eds.), *AI-Driven Transportation Systems: Real-Time Applications and Related Technologies. Information Systems Engineering and Management* (Vol. 62). Springer., DOI: 10.1007/978-3-031-98349-8_18

Haque, M. M. (2025). Systematic review on the impact of AI-enhanced traffic simulation on U.S. urban mobility and safety. ASRC Procedia. *Global Perspectives in Science and Scholarship*, *1*(1), 833–861. DOI: 10.63125/jj96yd66

Jagatheesaperumal, S. K., Bibri, S. E., & Huang, J.. (2024). Artificial intelligence of things for smart cities: Advanced solutions for enhancing transportation safety. *Computational Urban Science*, *4*, 10. DOI: 10.1007/s43762-024-00120-6

Lee, C. P., Leng, F. T. J., Habeeb, R. A. A., & Amanullah, M. A., & ur Rehman, M. H. (2022). Edge computing-enabled secure and energy-efficient smart parking: A review. *Microprocessors and Microsystems*, *93*, 104612. DOI: 10.1016/j.micpro.2022.104612

Lilhore, U. K., Imoize, A. L., Li, C. T., Simaiya, S., Pani, S. K., Goyal, N., Kumar, A., & Lee, C. C. (2022). Design and Implementation of an ML and IoT Based Adaptive Traffic-Management System for Smart Cities. *Sensors (Basel)*, *22*(8). Advance online publication. DOI: 10.3390/s22082908

Mahomed, A. S., & Saha, A. K. (2025). Unleashing the Potential of 5G for Smart Cities: A Focus on Real-Time Digital Twin Integration. *Smart Cities*, *8*(2). Advance online publication. DOI: 10.3390/smartcities8020070

Mohanty, A., Mohapatra, A. G., & Mohanty, S. K. (2025). Real-Time Traffic Monitoring with AI in Smart Cities. In Abraham, A., Prasad, S., Alhammadi, A., Lestable, T., & Chaabane, F. (Eds.), *Internet of Vehicles and Computer Vision Solutions for Smart City Transformations. Lecture Notes in Intelligent Transportation and Infrastructure*. Springer., DOI: 10.1007/978-3-031-72959-1_7

Musa, A. A., Malami, S. I., Alanazi, F., Ounaies, W., Alshammari, M., & Haruna, S. I. (2023). Sustainable traffic management for smart cities using internet-of-things-oriented intelligent transportation systems (ITS): Challenges and recommendations. *Sustainability*, *15*(13), 9859. DOI: 10.3390/su15139859

Mutambik, I. (2025). IoT-Enabled Adaptive Traffic Management: A Multiagent Framework for Urban Mobility Optimisation. *Sensors (Basel)*, *25*(13). Advance online publication. DOI: 10.3390/s25134126

Nikitas, A., Michalakopoulou, K., Njoya, E. T., & Karampatzakis, D. (2020). Artificial Intelligence, Transport and the Smart City: Definitions and Dimensions of a New Mobility Era. *Sustainability*, *12*(7). Advance online publication. DOI: 10.3390/su12072789

S. Reis, J. C. (2025). Internet of Things and Artificial Intelligence for Secure and Sustainable Green Mobility: A Multimodal Data Fusion Approach to Enhance Efficiency and Security. *Multimodal Technologies and Interaction*, *9*(5). Advance online publication. DOI: 10.3390/mti9050039

Saini, K., & Sharma, S. (2025). Smart road traffic monitoring: Unveiling the synergy of IoT and AI for enhanced urban mobility. *ACM Computing Surveys*, *57*(11), 276. Advance online publication. DOI: 10.1145/3729217

Verma, R. K. (2025). Quantum Methods for Data-Efficient Self-supervised Learning in Industrial Automation. In Ghosh, A., Dutta, S., Das, A. K., Shukla, V. K., & Moreira, F. (Eds.), *Quantum Machine Learning in Industrial Automation. Information Systems Engineering and Management* (Vol. 65). Springer., DOI: 10.1007/978-3-031-99786-0_1

Verma, R. K., & Elngar, A. A. (2025). Smart city development in 6G era. In *Driving innovation through AI and digital twin for 6G powered sustainable ultra smart cities* (pp. 27–54). Wiley., DOI: 10.1002/9781394411320.ch2

Verma, R. K., Kumar, I., & Kumar, S. (2025). In Chand Kumain, S., Singh, M., Awasthi, L. K., & Singh, R. (Eds.), *"AI-driven insights: Revolutionizing satellite imagery applications," in *AI and ML Techniques in Image Processing and Object Detection* (pp. 89–104). Springer., DOI: 10.1007/978-981-96-7445-9_5

Wang, B., Dabbaghjamanesh, M., Kavousi-Fard, A., & Yue, Y. (2024). AI-enhanced multi-stage learning-to-learning approach for secure smart cities load management in IoT networks. *Ad Hoc Networks*, *164*, 103628. DOI: 10.1016/j.adhoc.2024.103628

Chapter 3
Advances in AI-Based Mobility Modelling:
Toward Intelligent Transport Infrastructure in Smart Cities

Shashikant Nishant Sharma
http://orcid.org/0000-0001-8031-8569
Maulana Azad National Institute of Technology, India

Kavita Dehalwar
http://orcid.org/0000-0002-9466-5710
Maulana Azad National Institute of Technology, India

ABSTRACT

To build smart cities, we need to completely change how we design, run, and manage transportation systems in cities. Artificial Intelligence (AI) is a game-changing tool that enables us to model human movement, enhance transportation networks, and predict travel patterns with unprecedented accuracy. This chapter gives a thorough overview of the most recent developments in AI-based mobility modeling, with a focus on how transportation has changed from traditional methods to smart, dynamic, and data-driven infrastructures. It examines how AI methods, such as machine learning, deep learning, and reinforcement learning, are applied in transportation systems to enhance real-time decision-making, traffic forecasting, demand-responsive services, and autonomous driving. The chapter also discusses the most pressing problems, moral issues, and future directions for integrating AI into urban transportation systems to create smart, sustainable, inclusive, and resilient cities.

DOI: 10.4018/979-8-3373-4202-3.ch003

1. INTRODUCTION

In modern cities, urban mobility is the most important thing for economic growth, social inclusion, and environmental sustainability (Sharma & Dehalwar, 2025). The idea of smart cities has come up as a way to change how cities grow in the future as cities around the world deal with rapid urbanization, traffic jams, environmental issues, and changing needs for transportation (Lewis, 2015). The main idea behind this vision is to rethink how we move around, making systems smarter, more efficient, more responsive, and more environmentally friendly. Smart cities combine human, collective, and artificial intelligence to improve people's lives (Herath & Mittal, 2022). Smart mobility is one of the six most essential parts of this. So, mobility is not only a necessary part of smart urban living, but it is also a strategic tool.

Traditional transportation systems, which rely heavily on static modeling and infrastructure-centric approaches, are becoming less and less able to meet the complex and changing needs of modern urban mobility. Adaptive, predictive, and scalable solutions are necessary to address challenges such as unpredictable traffic, fluctuating travel demand, inefficient public transportation, and the introduction of new modes of transportation like ride-sharing and autonomous vehicles. In this situation, intelligent transport infrastructure, which includes real-time data collection, dynamic modeling, predictive analytics, and automated decision-making, becomes very important (Sumalee & Ho, 2018). Neirotti et al. (2014) say that urban smartness projects will only be successful if they can use data-driven insights and add intelligence to transportation systems (Neirotti et al., 2014).

Recent progress in Artificial Intelligence (AI) has opened up new ways to deal with these problems (Bahamazava, 2025). AI techniques like Machine Learning (ML) (Chen & Zhang, 2024), Deep Learning (DL) (Li et al., 2021), Reinforcement Learning (RL) (Savithramma et al., 2023), and Natural Language Processing (NLP) are changing the way people move around by making it possible to predict traffic in real time (Nayak & Pandit, 2025), change routes on the fly, and make public transportation respond to demand (Vansteenwegen et al., 2022). With the help of AI, transportation systems are changing from passive physical networks to smart ecosystems that can sense, analyze, learn, and adapt to changes in cities.

The change from traditional to AI-based mobility modeling is part of a bigger shift in how transportation planning is done. The four-step model (trip generation, trip distribution, mode choice, and route assignment is an example of a traditional transport model that is static and doesn't always show how unpredictable urban travel behaviors can be (Ihrig et al., 2024). These models utilise historical data and assume that conditions remain stable, which is not often the case in real life. AI-driven models, on the other hand, are rich in data, take context into account, and are continually evolving. They can update predictions based on real-time information

(Kim et al., 2023). AI facilitates the modelling of complex, nonlinear relationships between various factors that influence mobility, including user preferences, real-time traffic conditions, travel time reliability, and socio-demographic characteristics (Iyer, 2021).

For example, Deep Neural Networks (DNNs) and Long Short-Term Memory (LSTM) networks have shown to be much better at predicting short-term traffic flow than traditional statistical methods (Wu et al., 2018). Reinforcement learning, which is another branch of AI, has shown promise in adaptive traffic signal control (Mekrache et al., 2022). This is because the system can learn the best policies on its own by interacting with the traffic environment (Madigan et al., 2019). Also, Graph Neural Networks (GNNs) are being used more and more to model the spatial dependencies in transportation networks, which helps us understand how traffic patterns are linked (Madigan et al., 2019).

Recent research in urban mobility and transport planning underscores the growing importance of integrating advanced analytical and AI-driven approaches to understand travel behaviour, system performance, and sustainable urban growth. Studies such as Lalramsangi, Garg, and Sharma (2025), which examine route choices for accessing public open spaces in hill cities, highlight how terrain-sensitive mobility modelling can improve accessibility planning. Complementing this, several systematic reviews by Sharma and colleagues (Sharma & Dehalwar, 2025a; Sharma & Dehawar, 2025) emphasize the evolving role of Transit-Oriented Development (TOD) and land-use–transport interaction models in shaping economically vibrant and smart urban environments. Broader policy-oriented work, including the assessment of inclusivity in India's National Urban Transport Policy for senior citizens (Sharma & Dehalwar, 2025b), enriches the understanding of equity considerations in mobility systems. Empirical studies also reinforce the value of data-driven modelling—for instance, discrete choice analyses of bus user satisfaction (Lodhi, Jaiswal, & Sharma, 2024), investigations into TOD precursors (Sharma, Kumar, & Dehalwar, 2024), and the application of advanced surrogate safety analysis tools for safer road environments (Sharma, Singh, & Dehalwar, 2024). Furthermore, recent systematic reviews on pedestrian safety (Sharma & Dehalwar, 2025c) demonstrate the need for intelligent and human-centric mobility frameworks. Collectively, these contributions illustrate a clear shift in transport scholarship toward AI-enabled, data-informed, and context-sensitive mobility modelling—precisely the analytical direction advanced in this chapter.

Moving to AI-based modeling not only improves the ability to make predictions and optimize things, but it also helps new ways of getting around, like Mobility-as-a-Service (MaaS) (Cottrill, 2020), Connected and Autonomous Vehicles (CAVs) (Zhang et al., 2021), and Smart Logistics (D'Amico et al., 2021). These new ideas require a basic transportation system that is smart, connected, and capable of responding to the needs of users and systems in real-time.

Although AI has considerable potential, utilising it to model transportation for smart cities remains challenging. Data privacy, algorithmic bias, lack of transparency (due to black-box models), integration with existing systems, and public acceptance remain significant challenges (Goodman & Flaxman, 2017). Additionally, the moral and social implications of AI-powered mobility solutions, including issues such as fair access, digital divides, and privacy concerns, require close examination (Wilson, 2022).

Given these changes and gaps, the goal of this chapter is to answer the following research questions:

1. How have traditional mobility modeling frameworks in urban transport systems changed since AI techniques were added?
2. What are the most important uses of AI in smart cities' smart transportation infrastructure, and what are the benefits of these uses?
3. What are the main technical, social, and moral problems that come up when trying to use AI for mobility modeling?
4. What kinds of research do we need to do in the future to fully understand how AI can help us build smart mobility ecosystems that are sustainable and strong?

This chapter attempts to answer these questions through a thorough review and critical analysis. It focuses on both the progress made and the problems that still need to be solved on the way to AI-enabled intelligent transport infrastructures. The chapter provides researchers, policymakers, and practitioners working at the intersection of AI, urban mobility, and smart city development with a roadmap by combining ideas from new research, real-world case studies, and emerging technologies.

Ultimately, integrating AI and mobility systems is a promising approach to making cities smarter, greener, and more livable. Batty (2018) makes a valid point when he suggests that smart cities are not just about utilising new technologies; they are also about transforming how cities operate to become more adaptable, forward-thinking, and resilient (Batty, 2018). Moving forward with AI-based mobility modeling is an essential part of this process of rethinking how cities move, connect, and thrive in the 21st century.

2. RESEARCH METHODOLOGY

Systematic Review Framework

This literature review adopts a dual framework approach by combining the SALSA (Search, Appraisal, Synthesis, and Analysis) method with the PRISMA (Preferred Reporting Items for Systematic Reviews and Meta-Analyses) guidelines. SALSA provides a structured strategy for literature searching, critical appraisal, and synthesis, while PRISMA ensures transparency and reproducibility through a stepwise screening process (Sharma & Dehalwar, 2025).

1. SALSA Framework
   a. Search

The initial search was conducted across Scopus, Web of Science, IEEE Xplore, and ScienceDirect databases using the key term *"Intelligent Transport"*. The time filter was restricted to publications between 2015 and August 2025 to capture the most recent decade of advancements in the field. The initial query retrieved 23,520 results.

b. Appraisal

To refine the results, sequential filters were applied:

c. Synthesis

The selected studies were categorized into themes such as:

1. Intelligent Transport Systems (ITS) architecture and frameworks.
2. Data-driven decision-making and AI integration.
3. Smart mobility and ride-sharing applications.
4. Traffic safety, monitoring, and control systems.
5. Policy, governance, and social impacts.

This thematic grouping enabled cross-comparison across technological, methodological, and policy-oriented dimensions.

d. Analysis

The final step involved analyzing the evidence to identify research trends, gaps, and methodological approaches. Quantitative mapping (publication counts, subject-area trends) and qualitative insights (conceptual models, policy recommendations) were used to ensure both breadth and depth in coverage.

2. PRISMA Framework

The PRISMA flow diagram was employed to ensure transparency in the study selection process.

Figure 1. PRISMA flowchart for the research

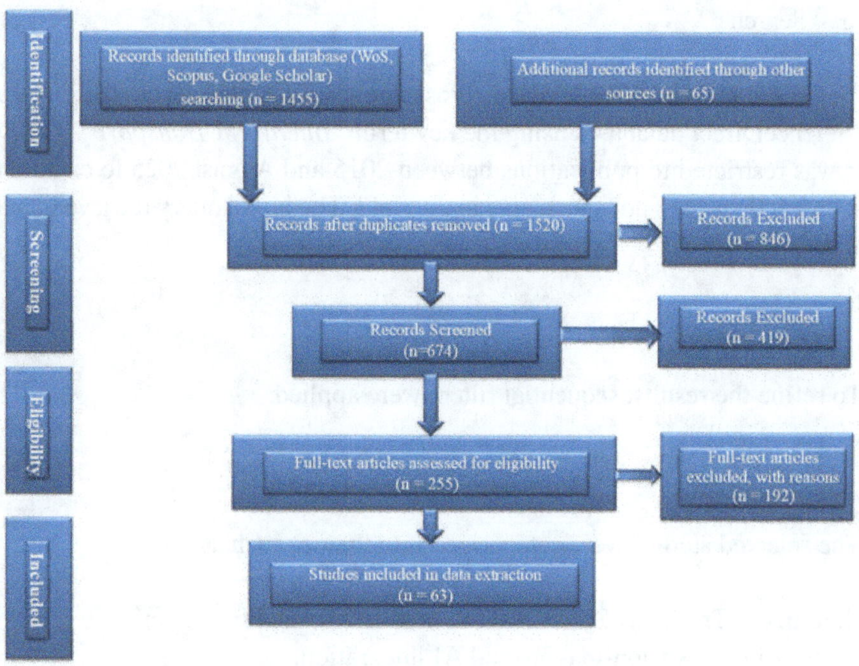

Based on the PRISMA flow diagram, a total of 1,455 records were identified through database searches (WoS, Scopus, Google Scholar), with an additional 65 records retrieved from other sources. After removing duplicates, 1,520 records proceeded to screening, of which 846 were excluded at the initial stage. The remaining 674 records underwent title and abstract screening, resulting in 419 exclusions. A total of 255 full-text articles were then assessed for eligibility, and 192 were excluded with documented reasons. Ultimately, 63 studies met all criteria and were included in the final data extraction for the review.

The PRISMA stages helped prevent bias by documenting reasons for exclusion at each stage (e.g., non-English language, conference papers, duplicates, non-relevance).

Inclusion and Exclusion Criteria

- **Inclusion**: Peer-reviewed journal articles, English language, 2015–2025, open access, final publication stage, directly addressing intelligent transport systems.
- **Exclusion**: Non-journal sources (books, reports, conference papers), non-English publications, duplicates, papers lacking direct relevance to intelligent transport applications.

Rationale for Methodological Choice

The combination of SALSA and PRISMA ensures both structured comprehensiveness (SALSA) and systematic transparency (PRISMA). This hybrid approach balances thematic depth with procedural rigor, offering a reliable evidence base for subsequent discussion and conclusion.

Critical Evaluation of Existing AI Frameworks in Mobility Modelling

While AI-based mobility frameworks offer substantial improvements in predictive accuracy and system efficiency, many existing models face limitations in generalizability and practical deployment. Neural networks and hybrid models often require extensive data for training and are sensitive to variations in local conditions, making cross-city transferability difficult. Moreover, several AI frameworks lack interpretability, posing challenges for policymakers who require transparent justification for transport interventions. Addressing these gaps requires the development of adaptive, explainable, and context-aware AI architectures.

3. FUNDAMENTALS OF AI IN TRANSPORT SYSTEMS

Artificial Intelligence (AI) has the power to change transportation systems by modeling, predicting, and optimizing highly dynamic and complex urban mobility environments. Before looking into how AI can be used in mobility systems, it's essential to know the basics of AI and its subfields, what mobility data is, and how

AI-enhanced models are different from traditional transport modeling methods. A look at AI, machine learning, deep learning, and reinforcement learning as a whole

Artificial Intelligence (AI) is a general term for machines that can perform tasks that typically require human intelligence. Iyer (2021) state that AI is the process of creating agents that can perceive their surroundings and take actions that maximise their chances of achieving their goals (Iyer, 2021). AI makes automation, predictive analytics, real-time optimization, and adaptive decision-making possible in transportation systems.

Machine learning (ML) is a branch of AI that works on creating algorithms that can learn patterns from data without being told what to do. ML models can utilise past data on how people move around to predict future traffic patterns, estimate the demand for travel, or identify the most efficient routes (Parishwad et al., 2023). There are three main types of ML algorithms.

- ***Supervised Learning***: The model learns from data that has already been labeled (for example, predicting travel times based on what has happened in the past).
- ***Unsupervised Learning***: The model finds patterns in data that doesn't have labels, like grouping travel behaviors.
- ***Semi-supervised and Reinforcement Learning***: These are hybrid methods that use a small amount of labeled data or learn by interacting with an environment.

Deep Learning (DL) is a branch of ML that uses artificial neural networks with many layers to model complicated patterns and data with many dimensions (LeCun et al., 2015). DL methods are especially good at things like recognizing images (which is used for self-driving cars) and modeling sequential data (which is used to predict traffic flow).

Reinforcement Learning (RL) is another area of AI where an agent learns the best ways to achieve its goals by interacting with an environment and receiving feedback in the form of rewards or punishments. RL is very useful for transportation problems that involve making decisions and controlling systems when conditions are changing, such as adaptive traffic signal control systems or real-time routing of vehicles (Comi et al., 2022).

Combining these AI technologies provides transportation planners and engineers with more advanced tools for learning, adapting, and improving mobility systems that surpass static, rule-based systems.

Mobility data has three main features: it is big data, it streams in real time, and it changes over time and space. Smart cities' mobility systems collect huge amounts of data from many different sources:

- GPS tracks from cars and phones
- Transactions with smart cards for public transportation
- Sensors built into roads and other infrastructure
- Data from ride-hailing and bike-sharing platforms
- Posts on social media with location tags
- Communications between connected vehicles (V2X)

The "4 Vs" of Big Data (Gandomi & Haider, 2015) are what make these datasets unique.

- Volume: Every second, a vast number of records are made.
- Velocity: The need for real-time analysis and the speed at which data is created.
- Different types of data: structured (sensor readings), semi-structured (GPS logs), and unstructured (social media).
- Veracity: The data is not always reliable or of good quality.

Thematic Emphasis on AI-Driven Mobility Modeling

Figure 2. Thematic emphasis on AI-driven mobility modeling

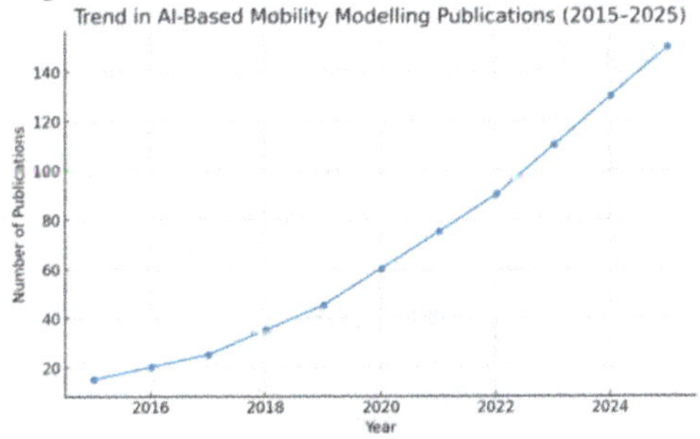

This chart (Figure 2) shows the main ideas that are coming out of AI-based mobility modeling for smart cities. Traffic prediction, intelligent transport infrastructure, shared mobility optimization, and first-last mile connectivity are all getting a lot of research attention. Traffic prediction and scheduling are important because they show how important it is to have real-time, data-driven solutions to problems with congestion and reliability. At the same time, first–last mile solutions and shared mobility show that there is a growing focus on systems that are integrated and centered on the user. These results are in line with changes around the world toward urban mobility that is more sustainable and open to everyone.

Changes in publication trends over time

Figure 3. Changes in publication trends over time

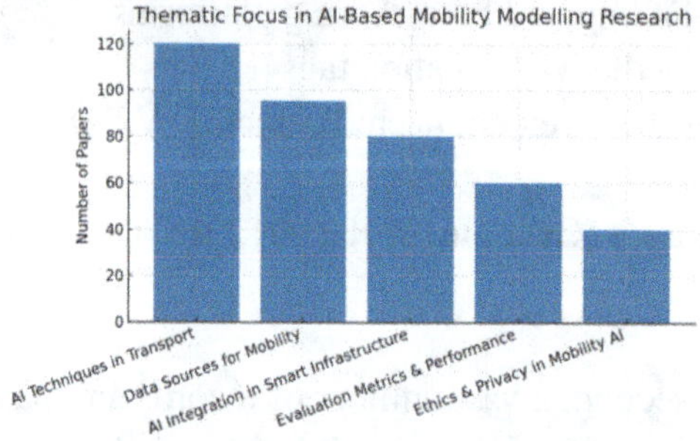

The timeline graph (Figure 3) shows that research on AI and smart transport systems has been steadily increasing from 2015 to 2025. The early growth from 2015 to 2018 is based on basic research in machine learning and data analytics. The sharp rise after 2020 is due to improvements in deep learning, edge computing, and autonomous systems. This pattern shows that both academics and businesses are becoming more interested in smart city projects, better access to data, and government support for intelligent mobility solutions.

What AI methods can do

Figure 4. AI methods used in mobility modelling research

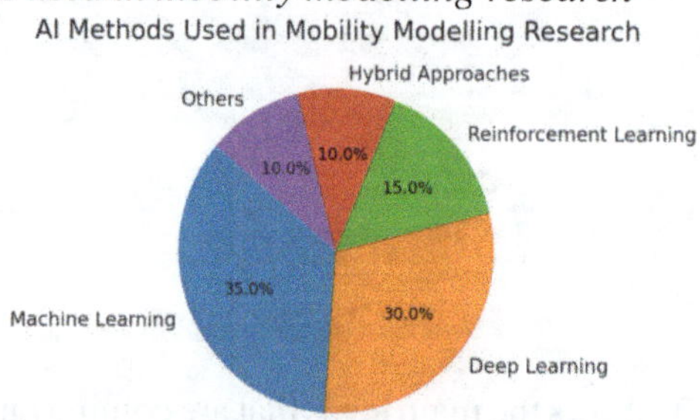

This graph (Figure 4) shows how AI techniques are used in mobility modeling. Machine learning and deep learning are the most common, followed by optimization algorithms and reinforcement learning. Computer vision and explainable AI methods are also becoming more popular, especially for things like keeping an eye on safety and managing parking. The distribution indicates a transition from predictive modeling to integrated decision-support systems, underscoring the necessity for hybrid and interpretable methodologies for practical implementation.

One of the hardest things about transport modeling is that mobility data is both spatial and temporal. Movements are always linked to places and times, which creates dependencies across space and time. It is crucial to accurately predict outcomes by capturing these dependencies. For example, traffic jams on one part of a road can have an impact on other parts of the road and on other locations over time (Bhardwaj et al., 2023).

This high-dimensional, constantly changing data was often challenging for traditional models to handle. On the other hand, AI methods like Graph Neural Networks (GNNs) for spatial relationships (Iddianozie & McArdle, 2021) and Recurrent Neural Networks (RNNs) (Kong et al., 2019) or LSTM networks for temporal sequences are good at handling the complex and rich structures in urban mobility data (SKB et al., 2024).

Additionally, the rise of real-time streaming data from IoT devices, mobile apps, and vehicle sensors necessitates that transportation models be adaptable, rather than relying solely on static snapshots (Babar & Arif, 2019). AI models trained on streaming data can make predictions and decisions almost instantly, which is crucial for applications such as dynamic route guidance and demand-responsive transit services.

Traditional vs AI-Enhanced Transport Models

The contrast between **traditional transport models** and **AI-enhanced models** highlights the evolution of the field:

Table 2. Contrast between traditional transport models and AI-enhanced models

Feature	Traditional Models	AI-Enhanced Models
Data Requirements	Aggregate, static data	High-frequency, real-time data
Model Structure	Preset functional forms (e.g., gravity models)	Flexible, data-driven learning
Assumptions	Strong assumptions (e.g., user equilibrium, rationality)	Minimal assumptions, pattern recognition
Adaptability	Static once calibrated	Continuously adaptable with new data
Prediction Capability	Limited, scenario-based	High-accuracy forecasting
Decision Making	Rule-based	Autonomous, adaptive learning

As depicted in Table 1, we can see that the traditional models such as the four-step travel demand model require a rigid structure with parameters manually specified by planners. These models are slow to adjust to real-world, rapidly changing mobility patterns, especially with the rise of new services like ride-hailing or micromobility (Miller, 2020).

In contrast, AI-enhanced transport models excel at:

- **Discovering hidden patterns** in complex datasets without prior assumptions.
- **Learning non-linear relationships** between diverse variables (e.g., land use, socio-demographics, weather, network conditions).
- **Predicting highly dynamic phenomena**, such as traffic congestion propagation or sudden modal shifts.

For example, LSTM-based models significantly outperform ARIMA and Kalman Filter models in short-term traffic speed prediction (Ma & Liu, 2025). Similarly, reinforcement learning approaches have demonstrated superiority over traditional actuated or fixed-time signal control methods in optimizing traffic signal timings based on real-time conditions (Wei et al., 2019).

However, it is also crucial to note that AI-based models, while powerful, often function as "black boxes," making their decision-making processes less interpretable than those of traditional models. This introduces new challenges in terms of transparency, trust, and regulatory acceptance.

So, knowing the basics of AI technologies and mobility data is important for looking into how they can be used to build smart transport infrastructure for smart cities. In the next section, we'll talk about how these AI techniques are being used in different areas of mobility to change the way traffic is managed, routes are optimized, public transportation systems work, and self-driving cars work.

4. AI APPLICATIONS IN MOBILITY MODELING

As cities become more complicated, AI is being used in mobility modeling in the real world instead of just for research. AI techniques, especially Machine Learning (ML), Deep Learning (DL), and Reinforcement Learning (RL), are changing how cities plan, predict, and manage transportation. This part talks about the main areas where AI is having a big impact: predicting traffic flow, predicting demand, optimizing routes, making transportation more efficient, and improving public transportation.

Predicting Traffic Flow

For managing traffic jams, planning infrastructure, and keeping travelers informed, it's important to be able to predict traffic flow accurately. ARIMA models and linear regressions are examples of traditional statistical methods that often have trouble capturing the non-linear, changing patterns that are common in city traffic (Alghamdi et al., 2019).

AI, especially Deep Learning methods, has a lot of benefits:

- Long Short-Term Memory (LSTM) networks, which are a type of recurrent neural network (RNN), can show how traffic states depend on each other over time. According to research like (Ma & Liu, 2025) LSTM networks are better than traditional models at predicting traffic speed in the short term.
- Graph Neural Networks (GNNs) look at the spatial structure of transportation networks. They model road networks as graphs, with nodes and edges representing intersections and road segments, respectively. Spatio-Temporal GNN that effectively identifies spatial-temporal correlations to make accurate traffic predictions (Bui et al., 2022).

Hybrid models that use both LSTM and GNNs can handle both spatial (across roads) and temporal (across time) dynamics at the same time. This has led to big improvements in traffic prediction for both short-term (like 5-minute forecasts) and long-term (like hours-ahead) predictions (Lu et al., 2020).

Predicting Demand

Transport planning and decisions about where to invest in infrastructure are based on travel demand forecasting. Surveys and assumptions about how rational travelers act are often used in traditional models (for example, four-step modeling approaches). However, as ride-sourcing, micro-mobility, and multi-modal trips become more common, the choices for getting around in cities are becoming more complicated (Wali et al., 2023). This calls for models that are more flexible and based on data. Machine Learning (ML) makes big strides:

- You can use historical ridership, land-use, socio-demographic, and network performance data to train supervised learning models like Random Forests, XGBoost, and Support Vector Machines (SVM) to predict demand in different situations (Zhang et al., 2021).

- More and more, Deep Learning architectures like Multilayer Perceptrons (MLPs) and Convolutional Neural Networks (CNNs) are being used to predict mode choices and trip generation in different areas (Gupta et al., 2018).

ML-based demand models are important for smart city planning because they can adapt to new trends like telecommuting, shared mobility, and the use of self-driving cars.

Optimizing Routes

Deterministic shortest-path algorithms (like Dijkstra and A*) have long been used for dynamic routing and avoiding traffic jams (Candra et al., 2020). Static routing isn't always the best option, though, because of things like traffic accidents, bad weather, and sudden spikes in demand.

AI improves route optimization in several ways, such as:

- Reinforcement Learning (RL): Agents learn the best ways to route by interacting with the environment and getting rewards (like getting to their destination faster). People are using methods like Deep Q-Networks (DQN) and Actor-Critic methods to make routing more personalized and aware of the situation (Zong et al., 2025).
- Algorithms for Predictive Routing: AI-based systems can suggest proactive route changes instead of reactive ones by predicting traffic conditions ahead of time with LSTM/GNN models.
- Systems with Multiple Agents: Multiple AI agents that stand for cars can work together to find paths in busy cities, which lowers overall traffic levels.

For instance, Google's DeepMind has looked into RL-based systems that dynamically control fleet movements across a network to cut down on travel times across the city (Li et al., 2023).

Mobility on its own

Autonomous and Connected Vehicles (CAVs) are a big part of the future of transportation. AI isn't just a tool; it's the main driver.

- Computer Vision Systems: Deep Learning models, especially CNNs, help cars see what's around them (for example, spotting pedestrians, other cars, and traffic lights) (Jiang et al., 2020).

- Behavioral Prediction: AI models can guess what other people will do, like when a pedestrian might cross the street or when another car might switch lanes.
- Path Planning and Control: Methods based on reinforcement learning let self-driving cars learn the best ways to drive in complicated, changing situations (H. Zhou et al., 2022).

AI-powered real-time decision-making algorithms and multimodal sensor fusion (combining LiDAR, radar, and camera inputs) enable CAVs to move safely, efficiently, and in sync in smart cities.

AI also enables platooning, which occurs when several self-driving cars travel closely together. This helps to make the best use of road space and energy, which is becoming more important in planning for sustainable urban mobility. Making public transportation better

Public transportation is still the most crucial part of sustainable mobility. In this area, AI is used to improve efficiency, reliability, and user satisfaction.

- Smart Scheduling: ML models can predict changes in peak demand, which lets schedules and frequencies change on the fly to meet the needs of passengers in real time (Liyanage et al., 2022; Wu et al., 2020).
- Fleet Management: AI systems figure out the best way to send out and assign buses and trains based on expected ridership and operational limits.
- Predictive Maintenance: DL models look at data from vehicle sensors to figure out when something might go wrong before it does, which cuts down on downtime and maintenance costs (Errezgouny et al., 2025).

Additionally, AI-powered real-time crowding information enables passengers to make informed choices, such as selecting less crowded vehicles, which makes travel more comfortable and balances system loads. Some cities, such as Singapore and Seoul, are already utilising AI to adjust bus routes and provide on-demand shuttle services. This is setting a standard for AI-driven public transportation innovation around the world. So, AI applications in mobility modeling are changing smart city transportation systems from reactive, broken networks into proactive, smart ecosystems that can see and deal with changing urban mobility problems.

Table 3. Classification of AI techniques in mobility modelling

AI Technique	Sub-Methods	Application Area	Input Data Type	Strengths	Challenges	Example Studies
Machine Learning	SVM, Random Forest, Gradient Boosting	Travel demand forecasting, mode choice, parking prediction	Tabular (survey, trip data)	High accuracy, interpretability	Needs preprocessing, risk of overfitting	(Gupta et al., 2018; Hu et al., 2023)
Deep Learning	CNN, RNN, LSTM, GNN	Traffic flow prediction, incident detection, ride-sourcing optimization	Spatial-temporal, images, GPS traces	Learns complex patterns, scalable	Black-box nature, high computation	(Errezgouny et al., 2025; Kong et al., 2019; Lu et al., 2020; Neupane et al., 2024)
Reinforcement Learning	Q-Learning, DQN, Policy Gradient	Adaptive traffic signal control, real-time routing	Sensor and IoT streams	Self-learning, dynamic adaptation	Needs extensive training	(Comi et al., 2022; Lai et al., 2024)
Hybrid/Ensemble	ML + Simulation, DL + ABM	Integrated demand-supply modelling	Multi-source datasets	Combines strengths of multiple techniques	Complex calibration	(Basu & Ferreira, 2022; Guan et al., 2023)
Natural Language Processing	Text mining, sentiment analysis	Social media mobility insights, complaints analysis	Text, feedback, tweets	Captures user perception	Contextual ambiguity	(Nayak & Pandit, 2025)

As depicted in Table 2, the classification of AI techniques in Mobility Modelling is as follows.

Deep Learning (DL) methods are changing mobility modeling more and more by giving us powerful ways to look at very complicated, multi-dimensional datasets. Convolutional Neural Networks (CNN), Recurrent Neural Networks (RNN), Long Short-Term Memory (LSTM) networks, and Graph Neural Networks (GNN) are now being used to solve problems like predicting traffic flow, finding congestion, managing incidents, and optimizing ride-sourcing. These methods are especially useful for processing spatial-temporal data like GPS traces, IoT sensor streams, and video feeds. This makes mobility solutions more accurate and responsive. These methods are great because they can handle a lot of data and make accurate predictions, but they need a lot of computing power and are often "black boxes," which makes them hard to understand (Errezgouny et al., 2025; Kong et al., 2019; Lu et al., 2020; Neupane et al., 2024).

Reinforcement Learning (RL) adds a dynamic and adaptable layer to smart transportation systems. It focuses on learning and optimizing in real time. To make adaptive traffic signal control, congestion management, and dynamic routing better, algorithms like Q-Learning, Deep Q Networks (DQN), and Policy Gradient methods are being used. These methods use live streams of IoT and sensor data, which lets systems change how they work as the network changes. Their strength is that they can learn on their own and adapt, but they still have a lot of problems to solve before they can be used, such as needing a lot of training data, being hard to compute, and not being able to grow (Comi et al., 2022; Laiet al., 2024).

Hybrid and ensemble modeling techniques are becoming more popular because they can use the best parts of different AI methods. These models combine techniques like ML with simulation frameworks or DL with Agent-Based Modelling (ABM) to look at both small-scale behavioral factors and large-scale system dynamics. They are especially helpful for multimodal networks and integrated demand-supply modeling, which leads to better solutions for smart cities. Nevertheless, these models frequently necessitate meticulous calibration, superior datasets, and substantial computational resources to attain reliable results (Basu & Ferreira, 2022; Guan et al., 2023).

Natural Language Processing (NLP) adds to these quantitative methods by looking at the behavioral and perceptual aspects of travel. Text mining and sentiment analysis techniques help planners better understand service satisfaction, safety concerns, and new mobility needs by getting information from unstructured sources like social media posts, passenger feedback, and complaint logs. NLP offers a more comprehensive and user-focused method for mobility planning by incorporating user viewpoints; however, it still faces considerable challenges, including language diversity, sarcasm detection, and contextual interpretation (Nayak & Pandit, 2025).

5. INTELLIGENT TRANSPORT INFRASTRUCTURE DEVELOPMENT

The move toward smart cities depends on the ability to build and run smart transportation systems that do more than just respond to demand. These systems need to be able to sense, predict, and optimize traffic and mobility patterns in real time. New developments in AI, IoT, and communication technologies, such as 5G, are enabling systems to transition from static to dynamic and learning-based. This section discusses several key developments, including the real-time control of traffic signals, the integration of IoT, edge AI, and 5G, as well as the application of Digital Twins in informed transportation planning.

Control of Traffic Signals in Real Time Using Learning with Reinforcement

It's often not very effective to use fixed-time scheduling or actuated control based on local sensors to control traffic signals when traffic conditions are complicated and changing. Adaptive, decentralized control strategies are needed to deal with traffic jams, especially in cities with a lot of people.

Reinforcement Learning (RL) has come up as a promising answer:

- RL agents see intersections as places to learn the best ways to change the signal phase by interacting with other agents (Noaeen et al., 2022).
- Deep Reinforcement Learning (DRL) methods, such as Deep Q-Networks (DQN) and Proximal Policy Optimization (PPO), can work with big, continuous state spaces, which makes them good for urban networks (Lai et al., 2024).
- Multi-Agent RL (MARL) models let multiple intersections learn together to improve traffic performance across the whole network instead of just at one node (Othman et al., 2025).

The IntelliLight system is one example of how deep RL algorithms can cut down on the time vehicles spend waiting and the number of cars in line at intersections compared to regular signal control systems. These systems not only change signals to match current traffic patterns, but they also predict and adjust signals ahead of time to ease congestion before it gets worse.

Combining 5G networks, edge AI, and Internet of Things (IoT) sensors

The smooth connection of IoT, edge computing, and ultra-fast communication networks is a key part of making intelligent infrastructure work.

- Sensors for the Internet of Things: IoT devices are used on roads, in vehicles, at intersections, and on public transportation to gather a lot of real-time data, such as traffic speeds, weather conditions, occupancy rates, and incident detection.

Edge AI: Edge computing processes data at or near the source instead of sending it all to centralized servers, which can cause delays. Edge AI models can find problems, guess when traffic will be heavy, or change the timing of signals with little delay (Shao et al., 2023).

5G Networks: With very low latency (as low as 1 millisecond) and high bandwidth, 5G networks allow vehicles to talk to each other (V2V), vehicles to infrastructure (V2I), and infrastructure to infrastructure (I2I) in real time. This connection is very important for things like cooperative adaptive cruise control (C-ACC), real-time hazard warnings, and dynamic route planning (Bie & Qiu, 2021).

The combination of IoT, edge AI, and 5G makes it possible for transportation systems to almost instantly sense, analyze, and respond to changes in urban mobility. This is what makes autonomous intersection management, dynamic lane management, and smart parking systems possible.

Digital twins and simulation environments for planning smart transportation

Digital Twins, which are virtual copies of real-world transportation systems that are updated in real time with live data streams, are another major step forward.

- Definition: A digital twin for transportation infrastructure shows the current state of a city's transportation system, including vehicles, roads, traffic lights, and passenger flows (Wu et al., 2022).
- Uses:
    - Real-Time Monitoring: Operators can watch a network's live performance, find new problems, and see how well interventions are working without interfering with the real world.
    - Scenario Simulation: Urban planners can see how different policies (like adding bike lanes, charging for congestion, or integrating CAVs) will affect things in the real world by making different assumptions (Batty, 2018).

Predictive Maintenance: Digital twins of infrastructure assets, such as bridges, tunnels, or bus fleets, can use stress and usage data to predict when maintenance will be needed. This cuts down on downtime and makes things safer.

Artificial intelligence is a big part of what makes digital twins work:

- Predictive Analytics: Machine learning models can guess what traffic will be like in the future or how much wear and tear infrastructure will get.
- Optimization Engines: Models that use reinforcement learning suggest the best ways to manage traffic.
- Anomaly Detection: Deep learning models find strange patterns that could mean problems or failures.

Cities like Singapore, Helsinki, and Shanghai are already using transport digital twins to plan, manage, and improve urban mobility systems in real time. These twins use AI, IoT, and big data analytics.

So, intelligent transport infrastructure powered by AI, IoT, edge computing, and digital twins is the backbone of next-generation mobility systems. It lets cities move from reactive congestion management to proactive, self-optimizing ecosystems.

Table 4. Comparison of AI models for mobility and traffic management

Model	Target Problem	Input Features	Output	Reported Accuracy/ Performance	Scalability	Best Use Cases	References
LSTM Neural Network	Short-term traffic prediction	Flow, speed, occupancy	Forecasted volume	RMSE = 4.2	High for urban networks	Signal control, PT scheduling	(Lu et al., 2020; Ma & Liu, 2025)
Graph Neural Networks	Public transport connectivity	Network topology, ridership	Travel time estimation	MAE = 6 sec	Good for multimodal	Smart transit planning	(Iddianozie & McArdle, 2021; Lu et al., 2020)
Bayesian Networks	Mode choice, safety	Socio-demographics, trip data	Probability distribution	85% accuracy	Moderate	TOD analysis, safety studies	(Sallard & Balać, 2023); (Joubert & De Waal, 2020); (Afrin & Yodo, 2021)
Agent-Based + AI	Demand modeling	OD matrices, land use	Synthetic population trips	High spatial detail	Needs high computation	Scenario testing	(Liu et al., 2025)
Reinforcement Learning (RL)	Signal optimization	Queue lengths, delay	Signal phase timing	15–20% reduction in delay	Good for dense networks	Urban intersections	(Comi et al., 2022; Lai et al., 2024)

As depicted in Table 3, the comparison of AI models for mobility and traffic management is as follows.

Long Short-Term Memory (LSTM) neural networks are one of the most popular deep learning architectures for short-term traffic prediction. This is because they can capture temporal dependencies in sequential data. LSTM models use information like traffic flow, speed, and occupancy to make accurate predictions about traffic volumes. This is important for things like real-time signal control and scheduling public transportation. Reported performances, including RMSE values approximately 4.2, indicate their reliability in high-demand urban networks (Lu et al., 2020; Ma & Liu, 2025).

Graph Neural Networks (GNN) are being used more and more in public transportation and multimodal networks, where the way the transportation system is set up is very important for how well it works. GNNs can figure out travel times and improve connectivity by showing networks as graphs and adding information like network topology and ridership. They are good for smart transit planning because they can handle complicated spatial relationships. Studies have shown that they can get mean absolute errors of about six seconds (Iddianozie & McArdle, 2021; Lu et al., 2020).

Bayesian Networks use probability to help with mode choice and transportation safety analysis. They combine socio-demographic factors with trip-level data to make probability distributions that can be used to predict behavior. This is what makes them useful for things like Transit-Oriented Development (TOD) analysis or figuring out safety risks for different groups of people. Their scalability is not as good as that of deep learning models, but their ability to be understood and their strength in problems where there is uncertainty are big pluses. Studies have shown that they can be up to 85% accurate (Sallard & Balać, 2023; Joubert & De Waal, 2020; Afrin & Yodo, 2021).

Agent-Based Modelling (ABM) and AI techniques work together to make it possible to simulate demand at a very detailed level, showing how synthetic populations interact with transportation infrastructure. These models use matrices that show where people come from and go to and information about how land is used to make detailed demand patterns in space. They are especially useful for testing scenarios and evaluating policies, but they need a lot of computing power, which makes them hard to scale in some situations. Recent research has shown that they are useful for assessing future mobility scenarios and planning interventions (Liu et al., 2025).

Reinforcement Learning (RL) is being used more and more for adaptive traffic control, which gives networks with a lot of traffic flexible ways to optimize signals. RL algorithms like Deep Q Networks and Policy Gradient methods can cut down on delays by 15–20% by learning the best ways to act based on their interactions with the environment. RL systems are useful for managing intersections and making traffic flow more smoothly because they can use things like queue lengths and delays. RL is a good choice for real-time urban traffic management because it can adapt, even though it needs a lot of training and computation (Comi et al., 2022; Lai et al., 2024).

Table 5. Data sources for AI-based mobility modelling

Data Source	Type	Spatial Resolution	Temporal Frequency	Advantages	Limitations	Privacy Concerns	Typical Use
GPS Data	Dynamic	High	Real-time	Detailed trajectories	Costly, coverage bias	Moderate	Travel pattern analysis
Mobile Network Data	Passive	Medium	Periodic	Large coverage	Aggregated, limited precision	High	OD estimation
Smart Card Data	Transactional	High (PT users)	Continuous	Ridership and boarding info	No trip purpose	Moderate	Transit demand
IoT Sensor Data	Real-time streams	Point-based	High	Useful for signal control	Requires infrastructure	High	Traffic management
Social Media Data	Crowdsourced	City-level	Event-driven	Captures perceptions	Noisy, biased	Low	Sentiment & event analysis

As depicted in Table 4, the data sources for AI-Powered mobility modelling are as follows.

GPS data has become one of the most useful sources for AI-based mobility modeling because it can collect real-time, high-resolution data on travel paths and speeds. It gives us a lot of information about people's travel habits, route choices, and network performance, which are all important for creating smart city transportation models that can predict and adapt (Zheng et al., 2016; Luo et al., 2020). But the high price of GPS-enabled devices and the fact that coverage is biased toward certain socio-economic groups are still major problems, as are worries about user privacy and data security. Even with these problems, GPS data is still a key part of research on route optimization, congestion detection, and first-last mile integration.

Mobile network data, such as call detail records (CDRs) and cell tower pings, gives us a passive and large-scale view of how people move from one place to another. Its medium spatial resolution and regular updates make it good for figuring out origin-destination matrices and looking at travel demand over time. Mobile network data is more comprehensive than GPS data because it covers almost all phone users. However, it is not as accurate as point-based data. Privacy issues are important because telecom companies have to combine or hide data to follow the law. Even with these problems, it has worked well for emergency planning, figuring out how people move around in a region, and predicting demand.

AI-based modeling is using more and more smart card data from public transportation systems to learn about how people board and leave, how many people are on board, and how much they pay to travel. Transit users find it useful for improving schedules, checking service reliability, and supporting multimodal integration strategies because it has a high temporal frequency and spatial accuracy. But it doesn't give full trip chains or trip purposes, and only people who use the transit network can use it. There are moderate concerns about data privacy because identifiers are usually pseudonymized, but careful handling is still needed (Gkiotsalitis & Cats, 2021).

IoT sensor data from connected traffic lights, road sensors, and surveillance systems is very important for managing traffic in real time and designing smart infrastructure. These high-frequency, point-based streams provide detailed data on traffic flow and occupancy for adaptive signal control, incident detection, and coordinating autonomous vehicles. The biggest problem is that deploying infrastructure requires a lot of money and time to maintain, and when combined with surveillance data, it poses a greater risk to privacy. Still, IoT data is a key part of the idea of smart intersections and urban mobility platforms that work together.

Social media data is an unusual but growingly useful source for analyzing mobility. It gives crowdsourced, event-driven information about how people feel about things, traffic problems, and activities in specific places. It doesn't have the accuracy and consistency of structured datasets, but it does give us unique insights into how travelers feel and behave, especially during big events or disruptions. There are problems because the data is noisy and biased, and there are also ethical issues with user consent and data mining. Social media analytics can still help with scenario modeling and public engagement strategies in smart transportation systems when used with other sources.

Table 6. AI integration with smart transport infrastructure

Smart City Component	AI Technique Applied	Input	Objective	Impact/Benefits	Implementation Challenges
Traffic Signal Systems	RL-based adaptive control	Queue length, sensor data	Reduce congestion	10–20% improved travel time	Deployment costs
Public Transport Scheduling	ML demand prediction	Boarding data, schedules	Optimize frequency	Increased reliability	Data integration
First-Last Mile Solutions	AI routing	GPS, demand clusters	Improve accessibility	Reduced wait time	User adoption
Shared Mobility	Optimization algorithms	Booking & fleet data	Fleet balancing	Lower idle time	Real-time computation
Parking Management	Computer vision, ML	CCTV, booking data	Occupancy detection	Better utilization	Privacy & cost

As depicted in Table 5, the AI Integration with smart transport infrastructure is as follows.

The combination of smart transport infrastructure and Artificial Intelligence (AI) has changed how cities manage their transportation, especially in the context of smart cities. More and more traffic signal systems are using adaptive control methods based on reinforcement learning (RL). These methods use real-time data like queue lengths and sensor data to change the phases of the signals on the fly. These systems have been shown to reduce traffic and increase travel time efficiency by 10–20%, but the high costs of installation and maintenance remain a problem (Li & Wen, 2023). These kinds of solutions are very important for dealing with traffic jams during peak hours and supporting long-term solutions for getting around.

Machine learning (ML)-based demand prediction models are very helpful for public transportation systems. These models use data on boarding, past schedules, and occupancy trends to find the best frequency and route. Better reliability and better use of resources lead to happier riders and more efficient operations (Eboli & Mazzulla, 2021). The problem, though, is combining data from different sources, making sure the data is correct, and keeping up with how passengers behave in real time, especially in multi-modal settings.

AI-enabled routing and demand clustering algorithms are making first- and last-mile connectivity better, which is often a problem for transit systems. These tools combine GPS data, user demand, and shared mobility platforms to make micro-transit solutions that work well, which cuts down on wait times and makes it easier to get to mainline public transport (Furuhata et al., 2013; Jittrapirom et al., 2017). User adoption is still a big problem because people don't know much about technology, they worry about the cost, and they don't want to change their habits, especially in developing areas.

More and more, optimization algorithms are being used to manage shared mobility systems like ride-hailing, bike-sharing, and e-scooter fleets. These algorithms help keep fleets balanced and services reliable. To reduce idle time and make the most of assets, inputs like booking data, demand forecasts, and real-time vehicle availability are used (Jiang et al., 2021; Shaheen & Cohen, 2019). The biggest problem with implementing this is that it requires a lot of computing power to work in real time, especially when demand spikes or something unexpected happens.

AI-driven systems that utilise computer vision and machine learning to determine the number of people in a space, predict demand, and manage bookings have also generated numerous new ideas in parking management. Data from CCTV, sensors, and reservation platforms enable us to utilise the existing spaces more effectively and reduce search traffic (Yan et al., 2022). But these systems bring up privacy and cost issues, especially when it comes to constant monitoring and data storage. This means that there needs to be clear rules and public support.

6. EVALUATION METRICS FOR AI MOBILITY MODELS

Table 7.

Metric	Type	Formula/ Definition	Relevance	Benchmark Values
RMSE/MAE	Regression	Error magnitude	Forecast accuracy	<10 for high quality
R^2	Regression	Model fit	Goodness of prediction	>0.8 preferred
Precision/Recall	Classification	Correct predictions	Mode choice, safety	0–1
F1 Score	Classification	Harmonic mean	Balanced performance	>0.7
Computational Time	Efficiency	Processing time	Real-time feasibility	Seconds-minutes
Explainability Index	Interpretability	SHAP, LIME	User trust & adoption	Context-dependent

Metrics for Evaluating AI Mobility Models

To make sure that AI-driven mobility models are accurate, efficient, and easy to understand, you need to look at them from many different angles. Root Mean Square Error (RMSE) and Mean Absolute Error (MAE) are two common regression metrics that measure how big the errors in predictions are. They are especially helpful for predicting how many people will travel, how much traffic there will be, and how long it will take to get there. An RMSE or MAE of less than 10 means that the predictions are very good (Cascetta, 2001). These metrics give planners and operators a clear picture of how well model outputs match real-world data. This lets them fine-tune and improve algorithms to make them more accurate.

The coefficient of determination (R^2) is another important number. It shows how much of the variance the model can explain. For transportation purposes, a value above 0.8 is usually acceptable, which means that the model explains most of the variability in the dataset (Khisty & Lall, 2003). In mode choice modeling and travel behavior studies, a higher R^2 signifies the strength of input variables such as socio-demographics, land-use patterns, and accessibility in affecting travel decisions (Hutchinson, 1974). People often use this number with RMSE/MAE to find a good balance between accuracy and explanatory power.

Precision and Recall are very important metrics for problems that involve classification, like predicting safety events, figuring out how people travel, or dividing users into groups. Precision tells you how accurate positive predictions are, and Recall tells you how well you can find all the relevant cases. The F1 score, which is the harmonic mean of Precision and Recall, strikes a balance between the two measures and is often used when the cost of false positives and false negatives is high. An F1 score of at least 0.7 is needed for mobility models to be reliable in safety-critical or decision-support applications.

In AI applications for transportation, efficiency is just as important. This is especially true when models are used in real-time situations like traffic control, ride-sourcing fleet balancing, or adaptive transit scheduling. Computational Time tells you how fast algorithms can process data and whether the model can give you results in a few seconds or minutes. For responsive systems that support live operations, it is important to reduce computational time without losing accuracy. High latency can make models that would otherwise be accurate useless in fast-moving urban environments.

Another new metric is the Explainability Index, which uses methods like SHAP and LIME to measure how easy it is to understand AI models. These tools help policymakers, operators, and end-users understand the predictions made by complicated black-box algorithms. Explainability builds trust and makes it easier for more people to use AI, especially in public transportation and safety applications where stakeholders need to be able to explain their decisions based on AI outputs. This is especially important when planning Transit-Oriented Development (TOD), where clear predictions about how people will travel can affect policy and investment choices.

Lastly, these metrics work together to make a complete evaluation framework. For example, a model that predicts travel demand may be very accurate (low RMSE, high R^2) but not easy to understand, which makes it less useful in real life. On the other hand, a model that is very easy to understand might not be very accurate, but people might still like it because it is clear. So, when designing and putting into place AI-based mobility solutions, it's important to find a balance between accuracy, efficiency, and explainability (Ison, 2011).

6. CHALLENGES AND BARRIERS

AI-based mobility modeling and smart transport infrastructures could change the way smart cities work, but there are a lot of problems and obstacles that need to be solved before they can be used. These problems aren't just technical; they also affect social, ethical, and infrastructure issues that affect how new technologies are created, used, and accepted by the public. This section discusses several key challenges, including data privacy and cybersecurity, algorithmic bias and fairness, infrastructure readiness, and public acceptance and ethical considerations.

Privacy of Data and Cybersecurity Worries

For AI-powered transportation systems to work, they need to gather and analyze huge amounts of mobility data. This includes information about how people travel, vehicle telemetry, location data, and sensor feeds from connected infrastructure.

But making this kind of data raises serious concerns:

- Data Privacy: Mobility datasets often have personally identifiable information (PII) in them, which makes them easy to hack or misuse. For example, if ride-hailing trip data isn't properly anonymized, it could be reverse-engineered to show private information about people's habits and preferences (Garroussi et al., 2025).
- Cybersecurity Threats: AI-powered transportation systems that are linked through IoT and 5G are at risk of cyberattacks. Hackers could get into traffic control systems, self-driving cars, or smart traffic lights, which could cause huge problems (El-Rewini et al., 2020).

- Regulatory Gaps: Data protection laws like the GDPR in Europe and India's DPDP Act are still changing in many places and may not yet fully address the challenges of mobility.

To build trustworthy AI mobility ecosystems, it is important to make sure that data practices are safe and protect people's privacy. These include differential privacy, encryption, federated learning, and secure multi-party computation.

Fairness and algorithmic bias in mobility models

AI models learn from past data, and if that data shows social or infrastructure inequalities, the systems that come out of it may make those biases worse instead of getting rid of them.

Some common examples are:

- Service Inequities: AI-powered route optimization algorithms may favor wealthier neighborhoods with better infrastructure, which could mean worse service for low-income or underserved areas (Vetri Selvi et al., 2025).
- Problems with accessibility: Unless they are specifically trained to do so, public transport optimization models may not fully take into account the needs of people with disabilities.
- Mode Choice Bias: If historical data is biased toward motorized modes, predictive models may not accurately predict the demand for active modes (walking, cycling).

To deal with bias, you need to take deliberate steps:

- Collecting data in a way that includes everyone.
- Checking algorithms for bias.
- Fair evaluation metrics that look at how AI systems work for different groups of people.

If AI-based mobility systems don't have proactive ways to reduce the risk of mobility injustice, they could make it worse instead of making smart cities more fair.

Technological Gaps and Infrastructure Readiness

To build smart transportation infrastructure, we need to make big improvements in both the physical and digital worlds.

Some of the problems are:

- Old systems: Many cities still use traffic management and public transportation systems that are too old to work with AI, IoT, or 5G (Bahamazava, 2025).

- A lot of money Putting IoT sensors on large transportation networks, setting up edge computing nodes, and building 5G infrastructure all require a lot of money up front, which makes it hard for cities with limited resources to adopt them.
- Technical Expertise: Many transportation agencies may not have the specialized skills needed to run and keep up AI-powered systems.

Also, when trying to combine data from different vendors' systems and sources, there are problems with interoperability. Many cities may still have the dream of real-time, intelligent transport management if these gaps are not fixed.

Ethical and Public Acceptance Issues

No matter how advanced they are, AI-based transportation systems won't work if people don't trust and accept them.

Concerns include:

- Loss of Control: People may be worried about losing control over important safety decisions when it comes to self-driving cars and AI-based signal controls.
- Transparency and Explainability: A lot of AI models, especially deep learning systems, work like "black boxes," which makes it hard to explain how decisions are made, like changing signals or suggesting routes.
- Ethical Dilemmas: Autonomous systems have to deal with moral issues, like making safety decisions during unavoidable accidents (for example, the "trolley problem" for self-driving cars).

To deal with these problems before they happen, we can:

- Use explainable AI (XAI) techniques to make things clearer.
- Making planning processes that involve citizens in system design.
- Setting up rules for AI mobility systems that are fair and hold people responsible.

Data Privacy and Ethical Considerations in AI-Based Mobility Systems

As AI models increasingly rely on large-scale mobility datasets—including GPS traces, smart-card transactions, and sensor-generated data—concerns related to data privacy and ethical governance become more prominent. Ensuring user anonymity, preventing unauthorized access, and defining permissible data-sharing frameworks are essential to maintaining public trust. Furthermore, ethical AI mandates transparency in algorithmic decision-making to avoid biases in route optimization, congestion pricing, and travel demand prediction. Incorporating privacy-preserving techniques such as differential privacy, federated learning, and secure multiparty computation can significantly enhance the accountability of AI-driven mobility infrastructure.

Challenges in Real-Time Analytics for Smart Mobility

Real-time mobility analytics require uninterrupted data flows from heterogeneous sources such as IoT sensors, connected vehicles, traffic cameras, and mobile applications. However, limitations such as data latency, communication bottlenecks, inconsistent sensor reliability, and the high computational load of streaming analytics pose significant challenges. Designing scalable architectures capable of handling real-time data ingestion, processing, and prediction is crucial for operationalizing truly intelligent transport systems. Edge computing and distributed AI models are emerging solutions to overcome these bottlenecks.

For the long-term use of smart transportation systems, getting a social license to operate is just as important as coming up with new technology.

So, even though AI-based mobility systems have considerable potential, they must address numerous technical, ethical, infrastructural, and social challenges before realising their full potential in smart cities.

7. FUTURE DIRECTIONS

The intersection of AI and smart mobility is poised to change again, addressing existing problems and paving the way for new ideas. Technological, ethical, and sustainability-focused progress are all promising directions for the future.

Decentralized AI and Federated Learning for Transportation Systems

Privacy and scalability problems are common with traditional centralized AI models. Federated Learning (FL) solves the problem by letting models be trained on decentralized devices without moving raw data. FL could help vehicles, roadside units, and mobile devices work together to make better traffic predictions or route optimization models while keeping users' privacy safe (Kairouz et al., 2021). These kinds of decentralized AI methods make data safer, cut down on communication costs, and make transportation systems less vulnerable to cyber threats.

AI in Mobility Modeling

There is an urgent need for transparency and accountability, as AI models significantly impact important choices in mobility management. Explainable AI (XAI) is all about making it clear to users and stakeholders how AI makes decisions. In transportation situations, XAI can explain why certain routes are suggested, why certain signal timings are chosen, or how self-driving cars make decisions in a split second (Abdulrashid et al., 2024). Adding XAI to transportation models can help build public trust and enable transportation agencies to identify issues within the system and develop effective solutions.

Green AI for Eco-Friendly Transportation

People are growing increasingly concerned about the impact of AI development on the environment. Green AI focuses on developing AI models that consume less energy and generate fewer carbon emissions. AI innovation must align with environmental goals in future mobility systems (Basu & Ferreira, 2022). For example, public transport schedules should be optimized to cut down on emissions, AI models for traffic management should be made lighter, and edge computing should be made available for real-time applications.

Using AI in making decisions about transportation planning

To get the most out of AI, it needs to be utilised in the development of transportation policy. Governments and city planners need to utilise AI insights to develop fair policies based on evidence and adaptable to change. Some examples include using predictive models for dynamic congestion pricing, monitoring the real-time impact of policies on individuals, or utilising AI-assisted participatory planning tools. Making AI tools a permanent part of planning processes will make urban mobility systems more responsive, efficient, and long-lasting.

8. CONCLUSION

In short, combining AI-based mobility modeling with smart transport infrastructure could change the way people move around cities. AI helps cities improve transportation flows, reduce traffic congestion, and make the city more environmentally friendly. It achieves this through technologies such as real-time traffic control, demand forecasting, digital twins, and autonomous systems. However, there are still big problems to solve, such as data privacy, algorithmic fairness, and infrastructure readiness. All of these issues need to be carefully considered before the system can be successfully implemented.

Federated learning and explainable AI are two technologies that will enhance intelligent transport systems in smart cities in the future. Federated learning will enhance privacy and data security, while explainable AI will foster trust and transparency. The growth of Green AI ensures that sustainability is built into AI systems, which reduces their environmental impact. Additionally, utilising AI to inform decision-making will enable cities to develop transportation plans that are more flexible, inclusive, and effective.

The goal for the future is to have AI-powered transportation systems that seamlessly integrate with cities. These systems should provide fair, long-lasting, and effective solutions that not only meet the needs of today's cities but also ensure they can adapt and remain strong over time.

REFERENCES

Abdulrashid, I., Zanjirani Farahani, R., Mammadov, S., Khalafalla, M., & Chiang, W.-C. (2024). Explainable artificial intelligence in transport Logistics: Risk analysis for road accidents. *Transportation Research Part E, Logistics and Transportation Review, 186*, 103563. DOI: 10.1016/j.tre.2024.103563

Afrin, T., & Yodo, N. (2021). A probabilistic estimation of traffic congestion using Bayesian network. *Measurement, 174*, 109051. DOI: 10.1016/j.measurement.2021.109051

Alghamdi, T., Elgazzar, K., Bayoumi, M., Sharaf, T., & Shah, S. (2019). Forecasting Traffic Congestion Using ARIMA Modeling. *2019 15th International Wireless Communications & Mobile Computing Conference (IWCMC)*, 1227–1232. DOI: 10.1109/IWCMC.2019.8766698

Babar, M., & Arif, F. (2019). Real-time data processing scheme using big data analytics in internet of things based smart transportation environment. *Journal of Ambient Intelligence and Humanized Computing, 10*(10), 4167–4177. DOI: 10.1007/s12652-018-0820-5

Bahamazava, K. (2025). AI-driven scenarios for urban mobility: Quantifying the role of ODE models and scenario planning in reducing traffic congestion. *Transport Economics and Management, 3*, 92–103. DOI: 10.1016/j.team.2025.02.002

Basu, R., & Ferreira, J. (2022). Can increased accessibility from emerging mobility services create a car-lite future? Evidence from Singapore using LUTI microsimulation. *Transportation Letters, 14*(4), 332–338. *Scopus*. Advance online publication. DOI: 10.1080/19427867.2020.1731993

Batty, M. (2018). *Inventing future cities*. MIT Press. DOI: 10.7551/mitpress/11923.001.0001

Bhardwaj, A., Iyer, S. R., Ramesh, S., White, J., & Subramanian, L. (2023). Understanding sudden traffic jams: From emergence to impact. *Development Engineering, 8*, 100105. DOI: 10.1016/j.deveng.2022.100105

Bie, Y., & Qiu, T. Z. (2021). Connected Vehicle-Cooperative Adaptive Cruise Control Algorithm to Divide and Reform Connected Vehicle Platoons at Signalized Intersections to Improve Traffic Throughput and Safety. *Transportation Research Record: Journal of the Transportation Research Board, 2675*(9), 995–1005. DOI: 10.1177/03611981211005456

Bui, K.-H. N., Cho, J., & Yi, H. (2022). Spatial-temporal graph neural network for traffic forecasting: An overview and open research issues. *Applied Intelligence, 52*(3), 2763–2774. DOI: 10.1007/s10489-021-02587-w

Candra, A., Budiman, M. A., & Hartanto, K. (2020). Dijkstra's and A-Star in Finding the Shortest Path: A Tutorial. *2020 International Conference on Data Science, Artificial Intelligence, and Business Analytics (DATABIA)*, 28–32. DOI: 10.1109/DATABIA50434.2020.9190342

Cascetta, E. (2001). Transportation systems. In *Transportation systems engineering: Theory and methods* (pp. 1–22). Springer US.

Chen, G., & Zhang, J. W. (2024). Intelligent transportation systems: Machine learning approaches for urban mobility in smart cities. *Sustainable Cities and Society, 107*, 105369. DOI: 10.1016/j.scs.2024.105369

Comi, A., Sassano, M., & Valentini, A. (2022). Monitoring and controlling real-time bus services: A reinforcement learning procedure for eliminating bus bunching. *Transportation Research Procedia, 62*, 302–309. DOI: 10.1016/j.trpro.2022.02.038

Cottrill, C. D. (2020). MaaS surveillance: Privacy considerations in mobility as a service. *Transportation Research Part A, Policy and Practice, 131*, 50–57. DOI: 10.1016/j.tra.2019.09.026

D'Amico, G., Szopik-Depczyńska, K., Dembińska, I., & Ioppolo, G. (2021). Smart and sustainable logistics of Port cities: A framework for comprehending enabling factors, domains and goals. *Sustainable Cities and Society, 69*, 102801. DOI: 10.1016/j.scs.2021.102801

Eboli, L., & Mazzulla, G. (2021). Customer satisfaction as a measure of service quality in public transport planning. *Int. Encycl. Transp, 13816*, 220–224. DOI: 10.1016/B978-0-08-102671-7.10643-8

El-Rewini, Z., Sadatsharan, K., Selvaraj, D. F., Plathottam, S. J., & Ranganathan, P. (2020). Cybersecurity challenges in vehicular communications. *Vehicular Communications, 23*, 100214. DOI: 10.1016/j.vehcom.2019.100214

Errezgouny, A., Chater, Y., González, C. D. B., & Cherkaoui, A. (2025). An integrated deep learning approach for predictive vehicle maintenance. *Decision Analytics Journal, 16*, 100597. DOI: 10.1016/j.dajour.2025.100597

Furuhata, M., Dessouky, M., Ordóñez, F., Brunet, M. E., Wang, X., & Koenig, S. (2013). Ridesharing: The state-of-the-art and future directions. *Transportation Research Part B: Methodological, 57*, 28–46. DOI: 10.1016/j.trb.2013.08.012

Gandomi, A., & Haider, M. (2015). Beyond the hype: Big data concepts, methods, and analytics. *International Journal of Information Management, 35*(2), 137–144. DOI: 10.1016/j.ijinfomgt.2014.10.007

Garroussi, Z., Legrain, A., Gambs, S., Gautrais, V., & Sansò, B. (2025). A systematic review of data privacy in Mobility as a Service (MaaS). *Transportation Research Interdisciplinary Perspectives*, *31*, 101254. DOI: 10.1016/j.trip.2024.101254

Gkiotsalitis, K., & Cats, O. (2021). Public transport planning adaption under the COVID-19 pandemic crisis: Literature review of research needs and directions. *Transport Reviews*, *41*(3), 374–392. DOI: 10.1080/01441647.2020.1857886

Goodman, B., & Flaxman, S. (2017). European Union Regulations on Algorithmic Decision Making and a "Right to Explanation.". *AI Magazine*, *38*(3), 50–57. DOI: 10.1609/aimag.v38i3.2741

Gupta, B., Awasthi, S., Gupta, R., Ram, L., Kumar, P., Rohit Prasad, B., & Agarwal, S. (2018). Taxi Travel Time Prediction Using Ensemble-Based Random Forest and Gradient Boosting Model. In Rajsingh, E. B., Veerasamy, J., Alavi, A. H., & Peter, J. D. (Eds.), *Advances in Big Data and Cloud Computing* (Vol. 645, pp. 63–78). Springer Singapore., DOI: 10.1007/978-981-10-7200-0_6

Herath, H. M. K. K. M. B., & Mittal, M. (2022). Adoption of artificial intelligence in smart cities: A comprehensive review. *International Journal of Information Management Data Insights*, *2*(1), 100076. DOI: 10.1016/j.jjimei.2022.100076

Hutchinson, J. W. (1974). Plastic buckling. *Advances in Applied Mechanics*, *14*, 67–144. DOI: 10.1016/S0065-2156(08)70031-0

Iddianozie, C., & McArdle, G. (2021). Towards Robust Representations of Spatial Networks Using Graph Neural Networks. *Applied Sciences (Basel, Switzerland)*, *11*(15), 6918. DOI: 10.3390/app11156918

Ihrig, J., Jochem, P., & Szimba, E. (2024). How to apply the four-step model for 150,000 travel zones: The HIPAT model. *Transport Policy*, *155*, 150–160. DOI: 10.1016/j.tranpol.2024.06.014

Iyer, L. S. (2021). AI enabled applications towards intelligent transportation. *Transportation Engineering*, *5*, 100083. DOI: 10.1016/j.treng.2021.100083

Jiang, J., Wang, C., Chattopadhyay, S., & Zhang, W. (2020). Road Context-Aware Intrusion Detection System for Autonomous Cars. In Zhou, J., Luo, X., Shen, Q., & Xu, Z. (Eds.), *Information and Communications Security* (Vol. 11999, pp. 124–142). Springer International Publishing., DOI: 10.1007/978-3-030-41579-2_8

Jittrapirom, P., Caiati, V., Feneri, A. M., Ebrahimigharehbaghi, S., Alonso-González, M. J., & Narayan, J. (2017). Mobility as a service: A critical review of definitions, assessments of schemes, and key challenges. *Urban Planning*, *2*(2), 13–25. DOI: 10.17645/up.v2i2.931

Joubert, J. W., & De Waal, A. (2020). Activity-based travel demand generation using Bayesian networks. *Transportation Research Part C, Emerging Technologies*, *120*, 102804. DOI: 10.1016/j.trc.2020.102804

Kairouz, P., McMahan, B., Song, S., Thakkar, O., Thakurta, A., & Xu, Z. (2021, July). Practical and private (deep) learning without sampling or shuffling. In *International Conference on Machine Learning* (pp. 5213-5225). PMLR.

Khisty, C. J., & Lall, B. K. (2003). *Fundamentals of transportation engineering*. Erlangga.

Kim, Y., Park, M., Ka, D., & Lee, C. (2023). Real-Time Traffic Forecast System for the Accident-Prone Large-Scale Transportation Network in the Seoul Metropolitan Area. *KSCE Journal of Civil Engineering*, *27*(7), 3085–3096. DOI: 10.1007/s12205-023-0349-9

Kong, J., Huang, J., Yu, H., Deng, H., Gong, J., & Chen, H. (2019). RNN-based default logic for route planning in urban environments. *Neurocomputing*, *338*, 307–320. DOI: 10.1016/j.neucom.2019.02.012

Lai, X., Yang, Z., Xie, J., & Liu, Y. (2024). Reinforcement learning in transportation research: Frontiers and future directions. *Multimodal Transportation*, *3*(4), 100164. DOI: 10.1016/j.multra.2024.100164

Lalramsangi, V., Garg, Y. K., & Sharma, S. N. (2025). Route choices to access public open spaces in hill cities. *Environment & Urbanization Asia*, *16*(2), 1–17. DOI: 10.1177/09754253251388721

LeCun, Y., Bengio, Y., & Hinton, G. (2015). Deep learning. *nature, 521*(7553), 436-444.

Lewis, P. G. (2015). Moral intuitions and smart growth: Why do liberals and conservatives view compact development so differently? *Journal of Urban Affairs, 37*(2), 87–108. *Scopus.* Advance online publication. DOI: 10.1111/juaf.12172

Li, C., Bai, L., Liu, W., Yao, L., & Travis Waller, S. (2021). Urban mobility analytics: A deep spatial–temporal product neural network for traveler attributes inference. *Transportation Research Part C, Emerging Technologies*, *124*, 102921. DOI: 10.1016/j.trc.2020.102921

Li, C., Bai, L., Yao, L., Waller, S. T., & Liu, W. (2023). A bibliometric analysis and review on reinforcement learning for transportation applications. *Transportmetrica. B, Transport Dynamics*, *11*(1), 2179461. DOI: 10.1080/21680566.2023.2179461

Liyanage, S., Abduljabbar, R., Dia, H., & Tsai, P.-W. (2022). AI-based neural network models for bus passenger demand forecasting using smart card data. *Journal of Urban Management*, *11*(3), 365–380. DOI: 10.1016/j.jum.2022.05.002

Lodhi, A. S., Jaiswal, A., & Sharma, S. N. (2024). Assessing bus users satisfaction using discrete choice models: A case of Bhopal. *Innovative Infrastructure Solutions*, *9*(11), 437. DOI: 10.1007/s41062-024-01652-w

Lu, Z., Lv, W., Cao, Y., Xie, Z., Peng, H., & Du, B. (2020). LSTM variants meet graph neural networks for road speed prediction. *Neurocomputing*, *400*, 34–45. DOI: 10.1016/j.neucom.2020.03.031

Ma, C., & Liu, T. (2025). Survey of short-term traffic flow prediction based on LSTM. *International Journal of Modern Physics C*, *36*(02), 2450177. DOI: 10.1142/S0129183124501778

Madigan, R., Nordhoff, S., Fox, C., Ezzati Amini, R., Louw, T., Wilbrink, M., Schieben, A., & Merat, N. (2019). Understanding interactions between Automated Road Transport Systems and other road users: A video analysis. *Transportation Research Part F: Traffic Psychology and Behaviour*, *66*, 196–213. DOI: 10.1016/j.trf.2019.09.006

Mekrache, A., Bradai, A., Moulay, E., & Dawaliby, S. (2022). Deep reinforcement learning techniques for vehicular networks: Recent advances and future trends towards 6G. *Vehicular Communications*, *33*, 100398. DOI: 10.1016/j.vehcom.2021.100398

Miller, E. J. (2020). Travel demand models, the next generation. In *Mapping the Travel Behavior Genome* (pp. 29–46). Elsevier., DOI: 10.1016/B978-0-12-817340-4.00003-6

Nayak, S., & Pandit, D. (2025). Daily activity-travel pattern identification using natural language processing and semantic matching. *Journal of Transport Geography*, *122*, 104057. DOI: 10.1016/j.jtrangeo.2024.104057

Neirotti, P., De Marco, A., Cagliano, A. C., Mangano, G., & Scorrano, F. (2014). Current trends in Smart City initiatives: Some stylised facts. *Cities (London, England)*, *38*, 25–36. DOI: 10.1016/j.cities.2013.12.010

Neupane, B. K., Acharya, B. K., Cao, C., Xu, M., Bhattarai, H., Yang, Y., & Wang, S. (2024). A systematic review of spatial and temporal epidemiological approaches, focus on lung cancer risk associated with particulate matter. *BMC Public Health, 24*(1), 2945. DOI: 10.1186/s12889-024-20431-x PMID: 39448953

Noaeen, M., Naik, A., Goodman, L., Crebo, J., Abrar, T., Abad, Z. S. H., Bazzan, A. L. C., & Far, B. (2022). Reinforcement learning in urban network traffic signal control: A systematic literature review. *Expert Systems with Applications, 199*, 116830. DOI: 10.1016/j.eswa.2022.116830

Othman, K., Wang, X., Shalaby, A., & Abdulhai, B. (2025). Multimodal adaptive traffic signal control: A decentralized multiagent reinforcement learning approach. *Multimodal Transportation, 4*(1), 100190. DOI: 10.1016/j.multra.2025.100190

Parishwad, O., Jiang, S., & Gao, K. (2023). Investigating machine learning for simulating urban transport patterns: A comparison with traditional macro-models. *Multimodal Transportation, 2*(3), 100085. DOI: 10.1016/j.multra.2023.100085

Sallard, A., & Balać, M. (2023). Travel demand generation using Bayesian Networks: An application to Switzerland. *Procedia Computer Science, 220*, 267–274. DOI: 10.1016/j.procs.2023.03.035

Savithramma, R. M., Sumathi, R., & Sudhira, H. S. (2023). Reinforcement learning based traffic signal controller with state reduction. *Journal of Engineering Research, 11*(1), 100017. DOI: 10.1016/j.jer.2023.100017

Shaheen, S., & Cohen, A. (2019). Shared ride services in North America: Definitions, impacts, and the future of pooling. *Transport Reviews, 39*(4), 427–442. DOI: 10.1080/01441647.2018.1497728

Shao, S., Su, L., Zhang, Q., Wu, S., Guo, S., & Qi, F. (2023). Multi task dynamic edge–end computing collaboration for urban Internet of Vehicles. *Computer Networks, 227*, 109690. DOI: 10.1016/j.comnet.2023.109690

Sharma, S. N., & Dehalwar, K. (2025). A systematic literature review of pedestrian safety in urban transport systems. *Journal of Road Safety, 36*(4). Advance online publication. DOI: 10.33492/JRS-D-25-4-2707507

Sharma, S. N., & Dehalwar, K. (2025). Examining the Inclusivity of India's National Urban Transport Policy for Senior Citizens. In D. S.-K. Ting & J. A. Stagner, *Transforming Healthcare Infrastructure* (1st ed., pp. 115–134). CRC Press. https://doi.org/DOI: 10.1201/9781003513834-5

Sharma, S. N., & Dehalwar, K. (2025). A Systematic Literature Review of Transit-Oriented Development to Assess Its Role in Economic Development of City. *Transportation in Developing Economies*, *11*(2), 23. DOI: 10.1007/s40890-025-00245-1

Sharma, S. N., Kumar, A., & Dehalwar, K. (2024). The Precursors of Transit-oriented Development. *Economic and Political Weekly*, *59*(14), 16–20. DOI: 10.5281/ZENODO.10939448

Sharma, S. N., Singh, D., & Dehalwar, K. (2024). Surrogate Safety Analysis- Leveraging Advanced Technologies for Safer Roads. *Suranaree Journal of Science and Technology, 31*(4), 010320(1-14). https://doi.org/DOI: 10.55766/sujst-2024-04-e03837

SKB, S., Mathivanan, S. K., Rajadurai, H., Cho, J., & Easwaramoorthy, S. V. (. (2024). A multi-modal geospatial–temporal LSTM based deep learning framework for predictive modeling of urban mobility patterns. *Scientific Reports*, *14*(1), 31579. DOI: 10.1038/s41598-024-74237-3 PMID: 39738120

Sumalee, A., & Ho, H. W. (2018). Smarter and more connected: Future intelligent transportation system. *IATSS Research*, *42*(2), 67–71. DOI: 10.1016/j.iatssr.2018.05.005

Vansteenwegen, P., Melis, L., Aktaş, D., Montenegro, B. D. G., Sartori Vieira, F., & Sörensen, K. (2022). A survey on demand-responsive public bus systems. *Transportation Research Part C, Emerging Technologies*, *137*, 103573. DOI: 10.1016/j.trc.2022.103573

Vetri Selvi, M., Upadhyay, P., Revathi, A., & Punitha, N. S., R., & Saravanakumar, M. (2025). Enhancing Urban AI-Powered Transportation Systems With Machine Learning: In K. Bellam, K. Nagappan, N. Sankaran, S. Chowdhury, & D. Perakovic (Eds.), *Urban Mobility and Challenges of Intelligent Transportation Systems* (pp. 363–382). IGI Global. DOI: 10.4018/979-8-3693-7984-4.ch020

Wali, B., Santi, P., & Ratti, C. (2023). A joint demand modeling framework for ride-sourcing and dynamic ridesharing services: A geo-additive Markov random field based heterogeneous copula framework. *Transportation, 50*(5), 1809–1845. *Scopus.* Advance online publication. DOI: 10.1007/s11116-022-10294-9

Wilson, C. (2022). Public engagement and AI: A values analysis of national strategies. *Government Information Quarterly*, *39*(1), 101652. DOI: 10.1016/j.giq.2021.101652

Wu, J., Wang, X., Dang, Y., & Lv, Z. (2022). Digital twins and artificial intelligence in transportation infrastructure: Classification, application, and future research directions. *Computers & Electrical Engineering*, *101*, 107983. DOI: 10.1016/j.compeleceng.2022.107983

Wu, W., Li, P., Liu, R., Jin, W., Yao, B., Xie, Y., & Ma, C. (2020). Predicting peak load of bus routes with supply optimization and scaled Shepard interpolation: A newsvendor model. *Transportation Research Part E, Logistics and Transportation Review*, *142*, 102041. DOI: 10.1016/j.tre.2020.102041

Wu, Y., Tan, H., Qin, L., Ran, B., & Jiang, Z. (2018). A hybrid deep learning based traffic flow prediction method and its understanding. *Transportation Research Part C, Emerging Technologies*, *90*, 166–180. DOI: 10.1016/j.trc.2018.03.001

Zhang, H., Hou, N., Zhang, J., Li, X., & Huang, Y. (2021). Evaluating the Safety Impact of Connected and Autonomous Vehicles with Lane Management on Freeway Crash Hotspots Using the Surrogate Safety Assessment Model. *Journal of Advanced Transportation, 2021. Journal of Advanced Transportation*, *2021*, 1–14. Advance online publication. DOI: 10.1155/2021/5565343

Zhou, H., Laval, J., Zhou, A., Wang, Y., Wu, W., Qing, Z., & Peeta, S. (2022). Review of Learning-Based Longitudinal Motion Planning for Autonomous Vehicles: Research Gaps Between Self-Driving and Traffic Congestion. *Transportation Research Record: Journal of the Transportation Research Board*, *2676*(1), 324–341. DOI: 10.1177/03611981211035764

Zong, Z., Feng, T., Wang, J., Xia, T., & Li, Y. (2025). Deep Reinforcement Learning for Demand-Driven Services in Logistics and Transportation Systems: A Survey. *ACM Transactions on Knowledge Discovery from Data*, *19*(4), 1–42. DOI: 10.1145/3708325

Chapter 4
AI-Based Data Modeling for Smart City Development

Mohammad Shuaib Khan
http://orcid.org/0009-0006-4774-2572
Chitkara University, India

Mohammad Mazhar Afzal
http://orcid.org/0000-0001-8814-2772
Glocal University, India

ABSTRACT

The development of many aspects of a smart city heavily relies on artificial intelligence (AI) technology. The most recent AI technologies for building smart cities are examined in detail in this article. According to the analysis, artificial intelligence (AI) is crucial for smart cities both now and in the future. Numerous issues have arisen as a result of the fast population growth, particularly in urban areas. These include garbage accumulation, water scarcity, pollution, transportation congestion, and health issues. The smart city plan is one of the initiatives that governments have launched to address these problems. This plan aims to use contemporary technology to deliver dependable, environmentally friendly services while also enhancing quality of life. In pursuit of this objective, scholars and professionals have developed numerous frameworks for the establishment of smart cities. One of the most important technologies for creating the components of a smart city is AI. Environmental considerations are crucial while creating solutions.

DOI: 10.4018/979-8-3373-4202-3.ch004

INTRODUCTION

The development of many aspects of a smart city heavily relies on artificial intelligence AI technology. The most recent AI technologies for building smart cities are examined in detail in this article. According to the analysis, artificial intelligence AI is crucial for smart cities both now and in the future.

Numerous issues have arisen as a result of the fast population growth, particularly in urban areas. These include garbage accumulation, water scarcity, pollution, transportation congestion, and health issues. The smart city plan is one of the initiatives that governments have launched to address these problems. This plan aims to use contemporary technology to deliver dependable, environmentally friendly services while also enhancing quality of life.

In pursuit of this objective, scholars and professionals have developed numerous frameworks for the establishment of smart cities. One of the most important technologies for creating the components of a smart city is AI. Environmental considerations are crucial while creating solutions.

According to this study, artificial intelligence AI can be used to track environmental health variables like emissions, radiation, water contamination, and air quality. Data, which may be obtained from many sensors, is essential for AI solutions. The purpose of this study is to investigate how AI can help with the planning and construction of smart city elements. The concept of smart cities has evolved significantly over time and currently encompasses a wide range of concepts.

Early debates aided in defining what constitutes a "smart" city. Harrison et al. (2010) provided a comprehensive summary, emphasizing the ways in which different technologies and data enhance city operations. Caragliu and Del Bo (2019) emphasized the value of information and communication technology ICT in enhancing the competitiveness and efficiency of cities. As the concept evolved, sustainability became increasingly important. These days, smart city initiatives use data and new technologies to address economic, social, and environmental issues. To identify smart cities, professionals and researchers employ a variety of techniques and metrics. A crucial component is sustainability, which emphasizes lowering resource consumption, supporting renewable energy sources, and enhancing the environment.

The ability of a city to adopt cutting-edge technologies is frequently referred to as its "smartness." Performance metrics are employed to assess the effectiveness of smart city initiatives and determine whether their objectives are met. Conventional techniques for defining smart cities frequently depend on subjective assessments and manual data collection. These methods may have limitations in terms of accuracy, impartiality, and scalability. By offering fresh approaches to the analysis of urban data, machine learning provides an answer. Machine learning has gained popularity in numerous fields due to its capacity to forecast.

Large data sets from social media, sensors, and other sources can be analyzed by machine learning. This makes it possible to identify linkages, patterns, and trends that people might overlook. This data-driven method facilitates faster answers and clearer decision-making. It can enhance healthcare, better utilize resources, and forecast disease outbreaks. In terms of transportation, it facilitates better route planning, traffic forecasting, and city mobility.

Despite its success elsewhere, machine learning is still not widely used to define smart cities. The majority of studies examine social media or concentrate on a single aspect of the city, such as energy or transportation. Researchers can determine what makes a city truly smart by using machine learning to build models with historical data (Khan et al., 2014).

There are a number of obstacles to evaluating smart cities using machine learning. Errors and gaps must be kept to a minimum because quality data is crucial. Quality, consistency, and accessibility of data are essential. Fairness and privacy must also be carefully considered to prevent abuse. Responsible city planning requires the ethical application of machine learning.

There is currently little research in this field. Further research on the use of machine learning to the definition of smart cities is obviously needed. The purpose of this study is to close that gap. It offers a thorough approach to using machine learning to find smart cities. Establishing a trustworthy, impartial framework for comprehending urban smartness is the aim. This will assist decision-makers in making better decisions based on reliable information.

Our revised study enhances our method for comprehending what makes a city intelligent. To increase accuracy and provide more lucid explanations, we refined our machine learning models. We can now take a closer look at urban elements thanks to sophisticated data methods and thorough reports. To better illustrate how cities operate, we included fresh data analysis and graphics. These improvements provide a more comprehensive understanding of urban intelligence and build upon our earlier work. The findings provide leaders and planners in cities with useful information.

The primary objective of this research is to develop a transparent, data-driven method for identifying and evaluating smart cities. We think that machine learning can precisely determine what makes a city smart when it is applied to vast volumes of city data.

SMART CITY - THE MEANING AND THE IMPORTANCE

A smart city is an urban area that integrates technology and resources which are helpful in improving the lives of residents and increasing the efficiency of the city operations. The technologies that mostly play vital role in development are Information and Communication Technology ICT and Internet of Things IoT. The developments are made based on the needs of the citizens by collecting the data through various sources like sensor networks, mobile applications, open data platforms, and available resources. After that important decisions are taken by the concerned authorities.

A smart city includes development in sectors like such as the traffic management, transportation facilities, power plants, criminal investigations, schools, libraries, hospitals, and other community services. Through smart cities local government aims to monitor, analyze, plan and govern the city. Data sharing helps to extend the businesses, citizens and other third parties benefit from this information.

Smart city initiatives have been opposed by various corporations, poor people and small business owners as they believe that these technologies do not satisfy their daily needs and could be used to control them in a very strict manner.

ARTIFICIAL INTELLIGENCE-MEANING AND THE ROLE

Artificial intelligence is a set of technologies that enables computers to perform a variety of advanced functions like ability to see, ability to understand, ability to analyze and ability to make recommendations. It is the science & engineering of making intelligent machines. AI helps in modernizing the city to smart city in more efficient and comfortable manner for people they live in (Karthika, 2023). AI manage the traffic lights which reduces unwanted traffic jams, accidents, plays a vital role in public safety by helping police by analyzing footage and use of high-tech drone system, manage the waste of the city keeps it clean, saves the energy of the city by enabling the street lights at the scheduled time, planned the city data for roads, parks to be constructed and much more.

In simple words, artificial intelligence helps in better running of cities, make smarter decisions and faster solution

Internet of Things and Information and Communication Technology

IoT and ICT plays a very crucial role in making of safer, smarter and more comfortable cities.

IoT: It means connection of everyday tasks, so that data could be shared in an efficient way.

ICT: Using tools like internet, computer, mobile phones, and software which helps in sharing of information and enhance communication

DATA MODELING

Data modeling involves creating visual representations of an entire information system or specific parts of it, helping to convey the connections between various data points and structures. The primary aim of data modeling is to highlight the types of data utilized and stored within the system, as well as the relationships among those data types, how they can be organized, and their specific formats and attributes. These models are developed with business needs in mind. Initial rules and requirements are gathered from business stakeholders to ensure they are reflected in the design of a new system or during updates to an existing one. Data can be modelled at different levels of abstraction, starting with the collection of information from stakeholders and end users about their business needs. This information is then transformed into data structures, forming a detailed database design. Think of a data model as a roadmap or an architect's blueprint, providing clarity on the system being designed (Liu & Zhang, 2021). The data modeling process uses standardized methods and established techniques. It offers a uniform and consistent approach to defining and managing data resources within an organization and even extending beyond it. Ideally, data models are living documents that evolve alongside changing business requirements. They play a crucial role in supporting professional processes and shaping IT architecture and strategy. Additionally, these models can be shared with vendors, partners, and industry peers for collaborative efforts.

TYPES OF DATA MODELING

1. **Conceptual Data Model:** The creation of this data model is the simplest and most straightforward. This provides a broad overview of the system's contents and is also known as a domain model. In essence, they cover entity classes, their traits and limitations, their connections, and the needs for data integrity and security. Usually, any notation is straightforward. Despite being the first data model, this is not the first stage of database construction. The conceptual data model's function is to assist in defining the problem's initial scope. It also aids in organizing the primary business requirements, principles, and guidelines.
   - **Use in Smart Cities:** Useful for outlining basic concepts such as citizens, services, locations and relationships. For example, how citizens interact with smart transport systems.
2. **Logical Data Model:** The conceptual data model is extended by the logical data model. They give more information about the ideas and connections in the field in question and are less abstract. It adheres to one of the various formal data modeling notation systems. These display the relationships between enti-

ties and highlight data properties such data types and their appropriate lengths. Technical system needs are not specified by logical data models. In agile or DevOps techniques, this step is usually skipped. For projects that are naturally data-oriented or in highly procedural implementation contexts, logical data models can be helpful.

- **Use in Smart Cities:** Enables structured planning for specific sectors such as traffic signals optimization, emergency response data systems and public transportation scheduling.

3. **Physical Data Model:** This is the last, most significant, and most thorough data model. It offers a structure for the actual physical storage of data in a database. Performance tuning and other DBMS-specific characteristics may also be included in these models. Additionally, they provide a completed design that may be used as a relational database.
   - **Use in Smart Cities:** Critical handling of real-time data handling, syncing of database queries for smart grids and managing long term footage archives.

Scope of Data Modeling

Cities that are truly smart—those that can leverage the depth of social, economic, and natural infrastructure while intelligently utilizing technology to enable sustainable growth—will be distinguished from those that are merely attempting to achieve smartness by the principles of urban design and planning, design thinking, systems thinking, and futures thinking combined with digital communications and technology adoption.

The foundation of smart city infrastructure is the Internet of Things IoT, which makes it possible to gather detailed data from a variety of industries. To track and monitor operations, IoT devices like sensors, cameras, and meters are integrated into city systems. This continuous data flow is essential for making decisions in real time, enabling cities to react quickly to new problems. For example, IoT-enabled smart water systems can identify leaks or irregularities in water use, notifying authorities to take action before major losses happen.

IoT and BDA have many potential advantages for smart cities, but there are a number of challenges in putting them into practice. First, technological obstacles continue to be a major problem, particularly in urban areas with poor infrastructure. To manage the massive amounts of data produced by IoT devices, a strong cloud computing infrastructure, high-speed internet, and sophisticated data processing skills are needed. Many cities, especially those in developing nations, lack the technology necessary to successfully deploy these systems.

Comprehensive frameworks governing the incorporation of smart technologies into urban planning are lacking in many places. Inadequate policies pertaining to data exchange, cybersecurity, and device compatibility can lead to misunderstandings and inefficiencies in smart city initiatives. Furthermore, fragmentation—where various systems function in silos without being able to efficiently connect with one another—can result from the lack of defined laws for IoT and BDA. Smart city initiatives may be implemented slowly and unevenly in the absence of defined governance frameworks, which would lessen their overall impact. Addressing these issues is essential to guaranteeing the effective implementation of smart city technologies, given the enormous potential of BDA and IoT to change urban landscapes.

The intersection of ICT, City Design and City Planning

Urban form and ICT are not directly related; instead, they depend on a variety of elements, including local peculiarities, citizen interests and demands, global market pressures, national and international economic shifts, etc. Additionally, the connections are reciprocal because they mutually influence one another. Urban form and ICT have historically had a less direct relationship, but they are more difficult to comprehend and necessitate extensive research. Nowadays, no one can overlook the practical potential of ICT applications. Certain industries, like transportation and water management, have well-established and effective smart city applications. Although they are still in their infancy, applications to urban planning and design offer some encouraging prospects.

The future will require places to be flexible and adaptive to the diverse, multidisciplinary needs of a smart global society. It is crucial to consider how technological advancements may affect urban space typologies as well as how space is used. A few questions that come up are: Will conventional shopping districts become obsolete with the rise of online shopping?

Will control rooms and massive integrated data centers serve as the future governmental hubs?

Will the economics of urban agglomerations be altered by ICT applications?

AI VS TRADITIONAL DATA MODELING: A GAME CHANGER COMPARISON

Efficient and accurate modeling is more important than ever in today's data-driven environment. However, how does AI-assisted data modeling stack up against conventional techniques? Let's examine a broad comparison that may alter your perspective on data modeling.

Traditional Data Modelling: The Old Guard

For decades the data experts have relied on traditional data modeling techniques. These methods have their strengths:

- **Human Insights:** Experienced modelers bring valuable domain knowledge to the table.
- **Customization:** Models can be tailored to specific business needs.
- **Control:** Teams have full control over every aspect of the model.

However, traditional methods also come with major limitations:

1. **Time Consuming:** Creating complex models can take weeks and even months.
2. **Error Prone:** Human errors can lead to expensive mistake and rework.
3. **Resources intensive:** Requires dedicated teams and broad human hours are required.
4. **Scalability Issues:** Difficult to scale quickly for large and complex datasets.

AI Assisted Data Modeling: The Game Changer

Here comes AI support for data modeling, a ground-breaking strategy that is revolutionizing the industry. Let's examine how it resolves the issues with conventional approaches.

- **Lightning-Fast Development:** AI can generate initial models in minutes, not weeks.
- **Reduced Errors:** Machine learning algorithms minimize human errors.
- **Resource Efficiency:** Requires fewer human resources, freeing up teams for strategic tasks.
- **Scalability:** Easily handles large and complex datasets.
- **Improvement:** AI learns and improves models over time.

The AI Advantage: A Closer Look

1. Speed and Efficiency

AI rapidly builds prototypes and adapts to data changes with minimal manual inputs.

2. Accuracy

Advanced algorithms enhance data quality checks and improves overall model precision.

3. Adaptability

AI models adapt to changing environments which is ideal for dynamic urban scenarios.

4. Cost Effectiveness

This reduces human efforts, project delays and long-term operational costs.

5. Insight Generation

AI reveals patterns and correlations that humans may miss supporting predictive decision making.

In terms of speed, accuracy, and scalability, it is evident that AI-assisted data modeling performs better than conventional techniques. Although conventional methods still have a place, AI will surely drive data modeling in the future.

DATA SOURCE IN SMART CITIES

To improve quality of life and make wise decisions, smart cities collect a lot of data from multiple sources. The data is used by AI-based algorithms to monitor and manage city operations. Machine-generated data {such as satellite imagery and data from Internet-of-things devices}, process-mediated data {such as financial transitions, electronic health records, and tax records}, media-sourced data {such as digital news and radio broadcasts}, and human-sourced data {such as social media, blogs, and crowdsourcing} are the four categories into which the data sources for big data can be separated. Let's look at a smart city's most important data source (Bokhari & Myeong, 2022).

- **IoT Sensors [Traffic, Pollution, Weather]:** Examples of IoT applications in smart cities include sensors and cameras positioned across the city to gather data in real time on energy use, traffic patterns, air quality, and garbage management. This information can be used by the administration to improve operations and urban planning. IoT smart city infrastructure is built on sensors integrated into critical assets including transit networks, buildings, and utilities. These sensors, in my opinion, also help with real-time system monitoring and administration. Essentially, a key component of IoT for urban expansion is efficient infrastructure management, which ensures intelligent resource management and utilization as well as an improved standard of living for residents.

- **Public Services Data [Transportation, Waste, Water]:** Water supply, waste collection, trains, and buses are examples of public services. When they are operated, the systems typically produce data. Train and bus trackers enhance the schedule, essentially resolving problems like delays.

Smart bins in cities are able to detect when they are full, and GPS trackers in garbage collection trucks are able to update the collection path. Sensors track water quality and usage and identify pipeline leaks. Overall, it lessens resource waste and aids in infrastructure design and upkeep.

- **Social media:** Social media has been increasingly important in many areas, including cities, in recent years. Smart cities efficiently use data and cutting-edge technology to improve urban services and inhabitants' quality of life. In smart cities with a plethora of other technology, social media platforms have quickly become very effective instruments. These days, social media presents an unparalleled opportunity to obtain insightful information from people all over the world and deliver public services quite effectively. When someone posts about breaking the rules on social media, authorities may receive a message and begin questioning the offender shortly after. These days, social media is a powerful instrument for inspiring civic activity since it allows for two-way communication (Reis et al., 2019).

These days, it appears that data produced on different social media platforms is a useful resource for smart cities. These days, city officials may efficiently improve a variety of public services by gathering vital information through in-depth analysis of trends and demographic data. These days, social media analytics can immediately identify trends pertaining to traffic jams or the necessity for disaster response in particular locations. Today's cities are managed more effectively thanks to data-driven insights that allow proactive urban planning as opposed to reactive measures. These days, social media data collection platforms that include AI and machine learning allow for more sophisticated predictive analytics. As a result, it enhances the services according to user feedback.

- **Surveillance and Security Systems:** Data security is essential for smart cities to function effectively and considerately. Smart cities need to protect sensitive data {e.g. patient data} and important services {e.g. healthcare}. Smart cities have a range of data security problems to deal with. Some complexities arise from the vast quantities of data that can be collected from many different types of Internet of Things devices. There is a potential entry point for a cyberattack inside almost any type of IoT device; therefore, it is critical to secure IoT devices. Communication networks, vital to smart city operations, are another challenge. The networks should be secure and capable of providing data transmission without loss. As data transfer occurs at various nodes, we need to safeguard data as it is traversing to avoid interception and unauthorized access. Additionally, there is a question on how to connect with legacy systems and supporting infrastructures with new technology. Older generational systems may not have the same level of security, and could pose a risk. Addressing and providing technology that was compatible while retaining security measures is essential for operation. Data breaches pose a serious issue to smart cities. They can create interruptions to basic services in electricity and water supply; and cause the public inconvenience and hazards. Data that is collected for the purpose of enhancing city services is also at risk during a data breach. Personal data of residents being breached means that their sensitive information {i.e., addresses and health histories} may be exposed in the public realm leading to unauthorized identity usage and loss of trust amongst the public pertaining to smart city issues. Unless a data breach was significantly wide-reaching or damaging, citizens may be hesitant to embrace changes to technology. In a smart city context, infrastructures that require strong strategies for dealing with unwieldy amounts of sensitive information where data breaches are not only the shame of governments, but a source of public condemnation. This can be achieved through the implementation of strong encryption methods, regular updates to security terms and condition, and the incorporation of the case studies on data security in smart cities.

The rapid evolution of smart cities demands strong data security measures to maintain public trust and ensure the smooth functioning of their innovative systems. Prioritizing data protection builds remarkably safer cities for everyone and bolsters crime prevention efforts pretty effectively in busy downtown areas.

APPLICATIONS OF AI-BASED DATA MODELING-A SMART CITY REVOLUTION

1. **Indian Traffic Management:** As we all know, India's population are growing day by day which arises serious tensions among nation such as traffic congestion,

delays and safety issues. Even traffic management of our nation are unable to handle the number of vehicles increasing on roads. There is where Artificial Intelligence stepping into game and changing whole the system in more efficient way. With the help of AI-powered technologies like real-time traffic analysis, adaptive traffic signals and incident detection systems, Indian cities are on the verge of beginning to transform to manage traffic and urbanization using these new technologies. Here, a question also arises that "Why India needs AI in traffic management"? It is all due to population growth, rapid urbanization, limited infrastructure, outdated traffic systems, high number of vehicles, narrow roads, increasing number of accidents and road fatalities. AI helps in making India's traffic system smarter, faster and more efficient which offers quick data driven, leading to reduce in congestion and improve public safety. Following are the key innovations in Indian Traffic Systems:

- **Traffic Flow Prediction:** Prediction in the flow of traffic using both real time and historical data helps public as well as authorities with emergency can reroute their vehicles. It reduces delays and accidents. Hence, overall improves daily communication.
- **Adaptive Traffic Signal Control:** As current traffic signals run on a fixed timer whereas signal with AI technology can change their timings dynamically based on the current traffic load and vehicles queue lengths on each direction. Thus, reduces waiting time at red lights and improved traffic movement in peak hours. This system is already deployed in cities like Bengaluru, Kolkata, and others.

Despite of all these innovations, some major projects like accident prone patterns, automatically detects traffic rule violations, modern footage system, dynamic signals to adjust waiting time is already installed in the big cities like, Bengaluru, Pune, Kolkata and Chandigarh. Also, in some parts of Punjab, police use AI to identify danger zones and buses with warning system warns the driver in real time about dangers like speed, obstacles.

Countries like China and Russia are already using AI for development of cities and India is following the same path with the Smart City Mission, placing AI as the heart of modern movement. With the government support and new technology innovations, India can fulfil its dream.

2. **Energy and Utility Management:** In most of the countries, management of the energy and utility resources like coal, electricity, water and soil are the major challenge adjacent to sustainable development. AI models make sure to reduce the wastage of resources, instils the greater efficiency and lower costs both to

the government and citizens. Following techniques are introduced to increase the efficiency:

- **Forecasting Electricity and Water Usage:** AI predicts the future demand of the public by analysing the usage data, weather patterns, population and seasonal trends. It collects data from various sensors and household appliances and uses this data about where and when the demand increases. It basically prevents shortage and enables authorities to plan accordingly. It also reduces damage to environment and saves the energy for the future needs. In cities like Delhi and Mumbai where there is a high demand of electricity supply in summer due to air conditioners, AI helps in predicting the spikes and manage the supply over the time (Li et al., 2019).
- **Dynamic Pricing Models:** Instead of charging fixed price for resources, use AI which enables dynamic pricing where the price changes based on the demand. AI uses optimal pricing strategies. It will encourage people to use resources wisely and promotes energy conservation. Cities like Hyderabad and Ahmedabad uses grids phenomenon to manage dynamic pricing of electricity and water during festivals and bad weather conditions.

3. **Waste Management in Smart Cities:** Waste management with high efficiency is one of the most crucial aspects of smart city development. Current waste collection system is often slow, irregular and wasteful. AI-based waste management involves modern intelligence that uses technologies such as optimized routes for smart bins and waste trucks.
    - **Route Optimization for Garbage Collection:** AI helps in getting best and shortest route for the collection based on the past as well as future data stored. Smart bins send the signals to the truck drivers when they got full and AI then updates the best routes for it based on the condition and location. It saves time, fuel and money and ensues bins to be emptied before they got overflow. Hence, reduces the pollution from the city. Cities like Indore and Surat have started using AI systems to plan routes for collecting waste every day.

Similarly, cities like Pune and Bhopal have installed smart bins at public areas such markets, bus stands and railway stations for better cleanliness.

4. **Environmental Monitoring:** AI runs sensors which keep track of environment and help the local authorities to protect it. As of now, pollution and climate change becoming a world-wide problem, AI helps in taking quick and smart

decisions to protect the citizens and nature. In the following ways AI helps in protecting the nature:

- **Air Quality and Pollution Prediction:** AI helps in maintaining pollution sensors to predict the future air quality. AI collects data about air quality across the city, then it goes though data and finds the source of pollution. These models also predict the pollution levels for the upcoming days. It informs people in advance, when and what preventive measures to take and also alarms the concerned authorities about increase in pollution.
- **Weather Modeling for Disaster Preparedness:** AI helps in predicting the upcoming weather conditions like floods, heatwaves and storms. Weather station and satellites which provides real-time data to AI. AI with this vast data that it studies thoroughly and sends alerts to emergency services and people if there is a possibility of any hazardous situation. It helps in better planning of shelters and emergency supplies.

EQUITABLE AND SUSTAINABLE AI INTEGRATION IN SMART CITIES:

Attaching residents to urban government is the purpose of smart cities, however the AI-centric approach heavily relies on technically privileged data to achieve this goal. People that are well-off, successful, and educated already profit from these dangers that exacerbate current electrical imbalances. Smart cities should strive for social justice and stability in addition to enhancing existing electric mobility and wealth gain.

High-tech urbanization frequently overlooks the energy-intensive infrastructure that is necessary for digital data collection. Multidisciplinary collaboration is required to develop a truly transformational AI for smart cities. AI researchers can be guided by science and technology studies, ecological economics, feminist urbanism, decolonial concepts, and significant urban studies (Lingitz et al., 2018).

They cast doubt on crucial self-disciplines, technical emancipation, varied approaches, circular urban metabolism, and socio-ethical regeneration. Regenerative urban concepts that involve marginalized populations as co-producers of urban solutions should be prioritized in order to reduce social and ecological externalities. Ideas should also include the cost of upkeep and repairs rather than just unrestricted expansion.

THE BENEFITS OF AI-BASED DATA MODELING IN SMART CITIES

Introduction

Urban regions face complicated issues like traffic congestion, pollution, resource distribution, and public safety as the world's population grows and more people relocate to cities. Smart cities, which use data and technology to make better decisions, enhance services, and improve the quality of life for citizens, are the answer.

AI-based data modeling is at the core of this shift. This approach interprets vast amounts of real-time data gathered from several sources, including as sensors, public services, and mobile devices, using artificial intelligence and machine learning. AI creates a feedback loop that fosters continuous improvement in city management by not just analyzing the data but also learning from it, forecasting future patterns, and proposing solutions.

In this article, we will examine how AI-based data modeling is transforming smart cities, making them more sustainable, efficient, and friendly to citizens.

What is AI-Based Data Modeling?

Conventional data modeling entails developing frameworks that specify the structure and interpretation of data. AI-based data modeling, on the other hand, goes much beyond that. It processes and analyzes dynamic data sets without the need for preset rules by utilizing statistical techniques, deep neural networks, and machine learning. With little help from humans, these models are able to recognize underlying patterns, adapt to fresh data, and provide forecasts or recommendations.

In smart cities, this modeling applies to real-time data from:

- IoT devices {e.g., smart meters, environmental sensors}
- Public transportation systems
- Surveillance and security feeds
- Social media and citizen feedback
- Healthcare records and utility data

This extensive interconnected ecosystem creates a digital twin of the city, allowing planners and administrators to simulate outcomes, test policies, and automate large-scale operations.

Intelligent Mobility: Traffic, Transport, and Infrastructure Optimization

Any smart city must have urban mobility. Efficient transportation has a direct impact on the economy, ecology, and public health in addition to reducing travel times. In order to forecast road usage, reduce traffic, and improve public transportation management, AI-based modeling for transportation makes use of real-time traffic data, GPS signals, camera feeds, and historical trends. When demand and road conditions change, it can react fast

Applications:

Adaptive traffic control systems: rather than using set intervals, traffic signals adapt to the flow of traffic in real time (Song et al., 2019)

- Coordination of Public Transportation: Train and bus timetables are adjusted based on current traffic and road conditions.
- Infrastructure Planning: Information is used to decide where to construct new bike lanes, roadways, or pedestrian zones.

Benefits:

- Shorter travel times and lower carbon emissions
- Better road safety and accessibility
- Proactive infrastructure maintenance, like detecting potholes before they worsen

Smart Energy and Environmental Sustainability

In today's cities, energy use and its effects on the environment are important concerns. Through clever resource usage modeling, AI aids in addressing both. Artificial intelligence programs examine the city's use of gas, electricity, and water, identifying trends in usage and forecasting demand surges. They support the management of smart grids, which are digital energy networks that use real-time feedback to balance supply and demand.

Furthermore, AI algorithms that forecast pollution, create green areas, and improve disaster readiness are fed environmental data from weather stations, water monitoring tools, and air quality sensors (Yang et al., 2018).

Applications:

- Smart Buildings: Automatically adjust lighting, heating, or cooling to optimize energy consumption.

- Renewable Integration: AI balances energy sourced from solar, wind, and traditional methods.
- Waste Management: Predicts waste volume and organizes pickups efficiently to prevent overflow and cut emissions.

Benefits:

- Lower utility bills for residents and local governments
- Proactive responses to environmental hazards, like smog alerts and flood warnings

Public Safety and Emergency Response

A city's ability to protect its citizens is what makes it strong. AI improves emergency response and preventive policing systems by increasing speed and accuracy. In order to anticipate potential crime scenes and coordinate a prompt reaction, AI-based data modeling in public safety analyzes crime statistics, real-time surveillance data, and social signals, such as 911 calls or social media alerts.

Artificial intelligence powered facial recognition and behavioural analysis improve surveillance in high-risk locations by spotting odd activity that could point to danger.

Capabilities:

- Predictive Policing: Deploying officers based on predicted crime hotspots.
- Disaster Management: AI models simulate scenarios like fires, floods, or earthquakes and guide evacuation plans.
- Emergency Services: Predicts hospital admissions during public events or disease outbreaks to manage staff and resources.

Benefits:

- Quicker and more precise emergency responses
- Reduced crime rates through preventive measures
- More efficient use of public safety resources

Urban Planning and Development

It takes more than a hunch to plan a city that is prepared for the future. AI gives us the ability to model, optimize, and become ready for development. AI is used by urban planners to estimate resource consumption, living and working areas, and population migration. These models aid in evaluating the long-term consequences of economic transitions, zoning modifications, and infrastructural initiatives.

Applications:

- Digital Twins: Create real-time virtual copies of cities to simulate planning decisions.
- Land Use Prediction: Identify spots for expansion, gentrification, or conservation.
- Traffic and Foot Traffic Mapping: Pinpoint ideal locations for schools, parks, and public services.

Benefits:

- More efficient and fair urban growth
- Lower costs through pre-implementation simulations

Citizen Engagement and Digital Governance

Active citizen participation is key to the success of smart cities. AI facilitates two-way communication between citizens and governments, increasing the transparency and inclusivity of governance. In order to determine what the public needs, AI examines input from social media, surveys, and public input platforms. It assists governments in prioritizing issues and customizing services according to public opinion and urgency.

Core Technologies:

- Chatbots and Virtual Assistants: Offer 24/7 support for accessing city services.
- Sentiment Analysis: Assess public opinion on policies or events.
- Digital Portals: Enable residents to submit complaints, track status, and provide feedback easily.

Benefits:

- Stronger community trust in government
- Faster resolution of services and less bureaucracy
- Greater civic engagement through transparency

CHALLENGES OF AI IN SMART CITIES

According to the research currently under publication, AI applications in smart cities show encouraging results. However, scholars and business professionals can concentrate on key research directions to use these strategies to increase the effectiveness of smart cities:

1. **Enhanced Data Training:** Robust ML and DRL protocol training requires a large dataset that includes factors like vehicle speed, position, inter-vehicle spacing, driver behaviour, UAV height, and relay base station data in order to improve decision-making precision.
2. **UAV Optimization:** By using machine learning approaches to optimize UAV onboard capabilities including sensing, computation, caching, and trajectories, ITS efficiency in UAV-vehicle communications can be greatly increased.
3. **Channel Modeling:** Modeling vehicle-to-vehicle, vehicle-to-UAV, UAV-to-UAV, and UAV-to-ground base station channels requires rigorous measurement studies. In order to account for both regular and irregular infrastructure designs, these campaigns should include a range of UAV and vehicle velocities in different directions.
4. **Standardization in SGs:** For SGs to create huge data and for SG devices to communicate with one another, standardization is essential. By carefully choosing AI, ML, and DRL approaches, SG performance can be brought close to ideal.
5. **5G Integration:** For smooth interoperability between current SGs and the upcoming 5G technology, corresponding standardization of SG communication infrastructure is essential, as 5G technology is expected to be standardized in 2020.
6. **Power Management:** For SGs and electric companies, power-down scenario optimization is critical. In order to design an effective network, communication delay analysis is essential. Strategies for enabling seamless transitions between 5G communication technologies while guaranteeing a continuous power supply can be developed using ML and DRL approaches.

LIMITATIONS OF AI IN SMART CITY

Lacks artificial intelligence in intelligent cities as cities worldwide are more data-powered and connected, artificial intelligence is playing an important role in making urban areas "smart". AI technology is used widely in traffic management, energy efficiency, waste management, public safety, and healthcare. However, despite its benefits, the implementation of AI in Smart City Infrastructure comes with significant deficiencies and challenges. It is necessary to identify these boundaries to create more effective and inclusive urban systems.

1. **Availability and data quality:** AI depends on large amounts of high quality, accurate data. In smart cities, data comes from various sources such as monitoring systems, sensors, mobile phones and public databases. Poor or insufficient data quality can severely disrupt AI model performance (Bajaj & Sharma, 2018).
    - **Incomplete data:** Missing information may lead to wrong predictions or recommendations. For example, if some traffic sensors malfunction in specific areas, the AI system may wrongly optimize traffic flows.
    - **Data Silos:** In many cities, different departments collect and store separate data. Without proper data integration, AI cannot reach the entire picture, which reduces its effectiveness.
    - **Unarmed data:** Most of the data in urban settings is unarmed, such as social media posts or video feeds, which makes it difficult to process and analyze firmly without advanced natural language processing and computer vision technology.
    - **Infrastructure and technical intervals:** For AI applications to work effectively, a robust technological infrastructure is necessary. Many cities, particularly in developing nations, struggle with outdated infrastructure that isn't prepared for the digital age.
    - **Connectivity problems:** In some areas, the use of poor internet in some areas can obstruct real -time data collection and analysis, affecting AI's accountability.
    - **Sensor Quality:** Low quality or faulty sensors can provide incorrect data inputs affecting AI decision making.
    - **Insufficient hardware infrastructure:** Many urban hubs may have a shortage of AI processing or affordable devices and high-demonstration servers.
2. **Ethics and privacy issues:** Residents in smart cities provide a lot of personal information. While it can enhance services, it also raises significant ethical and privacy issues.
    - **Monitoring and Civil Freedom:** Continuous video monitoring and facial recognition may feel infiltration, causing the monitoring states to be apprehensive and reduced personal freedom.
    - **Data misuse:** Without effective control, personal information can be misused for commercial or political purposes.
    - **Lack of consent:** People often do not understand how their data is collected and used, violating the criteria of informed consent and transparency.

3. **Prejudice and discrimination:** In algorithms for AI The historical data used to train the AI algorithm may be inherently biased. This prejudice may result in some groups being treated unfairly in the setting of a smart city.
   - **Policing and justice:** AIs used in forecasting policing can inconsistently target specific communities if training data refers to biased law enforcement practices.
   - **Service allocation:** AI may recommend resource distribution like healthcare or education, based on flawed beliefs, keep the deprived groups more marginalized.
   - **Facial identification failures:** Facial identification technologies often have a high error rate for women and minorities.
4. **Economic and social inequality:** The use of AI frequently benefits regions with more developed infrastructure, which exacerbates the gap between wealthy and impoverished urban areas.
   - **Digital divide:** Most of the residents of smartphones or internet access remember smart city services.
   - **Job Displacement:** Automatic jobs in urban industries, such as traffic management or waste collection, can cause unemployment for low - income individuals.
   - **Uneven benefits:** Rich neighbourhoods can benefit from better data collection, leaving behind poor communities.
5. **High implementation and maintenance cost:** AI-based system installation might be highly costly. When building intelligent AI systems, the costs associated with software development, hardware configuration, and staff training can mount up rapidly.
   - **Initial investment:** AI projects in smart cities require significant upfront investment, which can struggle to bear small cities. Operations Expenditure: AI infrastructure, software updates, cloud services and cyber security can stress the city's budget.
   - **ROI Issues:** Cities can be difficult to measure withdrawal on investment for AI applications, especially in non-monetary terms such as better quality of life.
6. **Cyber-attack and security:** Risk with greater dependence on digital systems, smart cities attract cyber-attacks.
   - **Hacking and Malware:** Mailly actors can compromise the AI system for traffic, electricity, or water management, which can lead to widespread disruption.
   - **Data violations:** Sensitive personal information collected by the smart city system can be exposed, resulting in identity theft or other disadvantages.

- **AI weaknesses:** adverse attacks can manipulate the AI model, which can lead to wrong decisions, such as making items incorrectly in monitoring footage.

7. **Shortage of efficient workforce:** The operation and maintenance of the AI system requires special skills that are often in low supply.
    - **Lack of talent:** In many cities, there is a deficiency of data scientists, AI engineers and system integrators, which are necessary to design and manage complex AI systems.
    - **Training intervals:** City employees require continuous training to effectively use AI tools.
    - **Seller dependence:** City can rely on private vendors due to lack of internal skills, causing potentially high cost and low control.

CONCLUSION

Artificial intelligence is not a cure-all, despite the fact that it can transform city life. Smart cities will have to strike a balance between the hazards and rewards of artificial intelligence. To address issues with data privacy, bias, spending, and infrastructure, this is essential for fair and long-lasting development. In order for AI to improve city life without endangering the most vulnerable members of society, policymakers, engineers, and city planners should collaborate. In actuality, an intelligent city is not just technologically sophisticated but also ethically, socially, and environmentally (Ouyang & Jiao, 2021).

In the twenty-first century, urbanization has advanced significantly. Cities are growing at a never-before-seen rate, which brings pressure and opportunity. In the midst of these developments, artificial intelligence has become more than just a tool; it is a revolutionary force that can function, develop, and benefit its populace. When considering the full smart city development scenario, it is evident that AI-based data modeling is essential to creating cities that are not only more intelligent but also more resilient, skilful, and adaptable.

A New Era of Urban Intelligence: AI has fundamentally changed how urban issues are perceived and handled in a variety of fields, including public safety, energy systems, traffic management, and healthcare. AI has enabled cities to transition from reactive to active systems by processing large amounts of data in real time, identifying patterns, and offering insights into the future. The municipal administrator can now rely on the AI-powered real-time dashboard to comprehend what is occurring in their courts, as opposed to the manual comments or outdated data that were used to make judgments. Governments now have the ability to make better, quicker, and better-informed decisions, which will result in more responsible services, less waste, and a higher standard of living for all citizens (Singh & Gupta, 2012).

Smart Solutions, Real Impact: The effect of AI in smart cities is not theoretical - it is already visible in practical applications:

1. **Traffic and mobility:** The AI-operated traffic signal system in cities like Bengaluru and Kolkata has reduced congestion and travel time. With the help of adaptive signal control and predictable analytics, passengers spend less time in traffic, resulting in reduced emissions and improves productivity.
2. **Public safety and emergency response:** AI-operated monitoring system, such as used in Chandigarh, automatically identifying traffic violations and criminal activities. The emergency response system can predict the fastest routes for ambulances, improve survival rates and reduce nervousness in high-stool conditions.
3. **Energy Efficiency:** AI models predict energy usage and adjust the consumption patterns accordingly. City can reduce the use of electricity during peak hours using dynamic pricing, ensuring reliable supply through accurate forecasts.
4. **Healthcare and Public Health:** During health crisis such as Kovid-19 epidemic, AI proved necessary to track the spread of virus and manage hospital resources efficiently. Disease modeling devices helped authorities to plan vaccine rollouts and identify more rapidly weaker areas than ever.
5. **Waste management:** Cities like Indore and Bhopal have made major progress in cleanliness and cleanliness, with smart compartments and route adaptation. AI ensures that waste is collected efficiently, the overflow is reduced and hygiene is maintained in crowded areas.

These are not isolated cases-they represent a growing trend where AIs are being rapidly adopted to solve city-specific problems, often corresponding to unique challenges of each region.

The Vision of Sustainable Cities: AI also contributes to a large vision - which is permanent. As the cities consume most of the world's resources and are responsible for more than 70% of global carbon emissions, the creation of a permanent urban environment is no longer alternative. AI helps in running this mission:

- Reducing carbon footprints through efficient energy use and customized transport systems.
- Management of natural resources such as water and electricity with future stating analysis.
- Encouraging behavioural change in citizens through feedback systems and encouragement-based smart tools.

- Supporting green infrastructure development through data-supported urban planning.

For example, AI can simulate the environmental impact of new construction projects, which can help planners select environmentally friendly materials or optimal layouts. It may also recommend areas to install solar panels based on urban forests or install solar panels based on real -time climate data Mounce (Mounce et al., 2015).

Efficiency: Doing More with Less- Efficiency is one of AI's primary promises for smart cities; it uses fewer resources while producing better outcomes. This is especially crucial in emerging nations like India, where severe problems are caused by high population density and budget deficits. AI automates routine tasks, cuts down on physical labour, and gets rid of excesses. AI frees up human workers to concentrate on important tasks, such as digitizing public service forms or optimizing power plant operations. Furthermore, efficiency is not limited to technology; it also extends to government. City officials may assess the effectiveness of policies, change course as necessary, and continuously enhance service delivery via data-operated government.

Building Resilient Urban Systems: Flexibility refers to the ability to withstand a city's ability and recover from disintegration - as it is natural disasters, epidemic or economic crisis. Helps to create such flexibility through AI:

- **Disaster Preparation:** AI -run weather modeling and initial warning system can predict floods, heat and storms. This allows emergency teams to prepare in advance and reduce damage.
- **Real-time monitoring:** AI-operated sensor bridges, pipelines, and roads such as damage or stress signs monitor infrastructure, preventing frightening failures.
- **Healthcare flexibility:** AI hospital can predict overload, help manage emergency resources, and prevent the spread of infectious diseases.

Flexible cities can bounce back rapidly and save their weakened population more effectively. This capacity is especially important in countries suffering from climate related disasters or infrastructural challenges.

The Role of Data in Empowering AI: Without data, artificial intelligence cannot function. Numerous streams of data flow through a smart city, including citizen feedback apps, public transportation cards, traffic cameras, pollution sensors, and medical records. Artificial intelligence systems learn, adapt, and enhance this data.

However, there are drawbacks to managing this data as well: privacy, cyber security, and ethical issues need to be properly considered. It is crucial that cities become more accountable, transparent, and inclusive as they get smarter.

Challenges That Cannot Be Ignored: Despite all its capabilities, smart cities do not have AI limits.

- **Challenges include:** Prejudice in AI model: If the data used to train AI is biased, this decision may be inappropriate - to negatively affect some communities.
- **Data privacy:** Personal data of citizens must be preserved. Monitoring and tracking systems should not be converted into control or discrimination tools.
- **High cost and infrastructure needs:** not all cities can afford smart technologies. Rural areas and poor municipalities can be left behind.
- **Lack of skilled workforce:** Many cities still have a shortage of trained professionals required to manage, monitor and maintain the AI system.
- **More dependence on technology:** AI should help, not to replace, take human decisions. It is important to attack the balance.

These limitations highlight the need to adopt AI responsible with clear guidelines, human inspection and a strong legal framework.

Inclusive Smart Cities for All: In addition to being high-tech, a perfect smart city is also human-centered. AI should contribute to the development of inclusive, accessible, and just cities. Everyone, including the elderly, the impoverished, and those with varying levels of involvement, should have access to technology. The voices of all inhabitants should be reflected in urban planning, not those of companies or algorithms. To make sure AI benefits individuals as well as the system, community involvement, open data access, and transparent policies are crucial Letourneau (Letourneau et al., 1999).

The Road Ahead: Opportunities and Responsibility: As we look at the future, AI will continue to develop. With the development of 5G, age computing and quantum processing, the possibilities for smart cities will only expand.

Suppose:

- Autonomous vehicles that talk to each other to avoid traffic.
- Buildings that adjust their own energy use based on weather and occupancy.
- Virtual assistants help you navigate the services of the city immediately.

But this is the responsibility of this occasion. Governments, technical companies, urban planners and citizens should collaborate to ensure that AI is used morally and for public good.

Final Thoughts: Artificial intelligence based data modeling is not just another urban planning trend; it is the cornerstone upon which the future of urban living is being built. Cities are able to think more intelligently, work more quickly, and provide better services as a result. Action is uncertain in actionable ideas, and thus transforms difficult problems into doable functions. AI can tell us what cities can be not just places to live, but places to thrive as demonstrated by India's Smart Cities Mission and the Global Sustainability Goals. The objective is clear as we proceed along this path: Building future-focused, green, and prepared communities in addition to smart ones. With thoughtful implementation, continuous innovation and people-first policies, AI has the power to convert today's urban problems into permanent solutions tomorrow.

REFERENCES

Bajaj, R., & Sharma, V. (2018). Smart education with artificial intelligence based determination of learning styles. *Procedia Computer Science*, *132*, 834–842. DOI: 10.1016/j.procs.2018.05.095

Bokhari, S. A. A., & Myeong, S. (2022). Use of artificial intelligence in smart cities for smart decision-making: A social innovation perspective. *Sustainability (Basel)*, *14*(2), 620. DOI: 10.3390/su14020620

Caragliu, A., & Del Bo, C. F. (2019). Smart innovative cities: The impact of Smart City policies on urban innovation. *Technological Forecasting and Social Change*, *142*, 373–383. DOI: 10.1016/j.techfore.2018.07.022

Harrison, C., Eckman, B., Hamilton, R., Hartswick, P., Kalagnanam, J., Paraszczak, J., & Williams, P. (2010). Foundations for smarter cities. *IBM Journal of Research and Development*, *54*(4), 1–16. DOI: 10.1147/JRD.2010.2048257

Karthika, D. (2023). A study on artificial intelligence for monitoring smart environments. *Materials Today: Proceedings*, *80*, 2009–2013. DOI: 10.1016/j.matpr.2021.06.046

Khan, A. A., Arshad, S., & Mohsin, M. (2014). Population growth and its impact on urban expansion: A case study of Bahawalpur, Pakistan. *Universal Journal of Geoscience*, *2*(8), 229–241. DOI: 10.13189/ujg.2014.020801

Letourneau, S., Famili, F., & Matwin, S. (1999). Data mining to predict aircraft component replacement. *IEEE Intelligent Systems & their Applications*, *14*(6), 59–66. DOI: 10.1109/5254.809569

Li, Z., Zhang, Z., Shi, J., & Wu, D. (2019). Prediction of surface roughness in extrusion-based additive manufacturing with machine learning. *Robotics and Computer-integrated Manufacturing*, *57*, 488–495. DOI: 10.1016/j.rcim.2019.01.004

Lingitz, L., Gallina, V., Ansari, F., Gyulai, D., Pfeiffer, A., Sihn, W., & Monostori, L. (2018). Lead time prediction using machine learning algorithms: A case study by a semiconductor manufacturer. *Procedia CIRP*, *72*, 1051–1056. DOI: 10.1016/j.procir.2018.03.148

Liu, L., & Zhang, Y. (2021). Smart environment design planning for smart city based on deep learning. *Sustainable Energy Technologies and Assessments*, *47*, 101425. DOI: 10.1016/j.seta.2021.101425

Mounce, S., Pedraza, C., Jackson, T., Linford, P., & Boxall, J. (2015). Cloud-based machine learning approaches for leakage assessment and management in smart water networks. *Procedia Engineering*, *119*, 43–52. DOI: 10.1016/j.proeng.2015.08.851

Ouyang, F., & Jiao, P. (2021). Artificial intelligence in education: The three paradigms. *Computers and Education: Artificial Intelligence*, *2*, 100020. DOI: 10.1016/j.caeai.2021.100020

Reis, J., Santo, P. E., & Melão, N. (2019). Artificial intelligence in government services: A systematic literature review. In *Advances in intelligent systems and computing* (pp. 241–252). Springer., DOI: 10.1007/978-3-030-16181-1_23

Singh, K. P., & Gupta, S. (2012). Artificial intelligence based modeling for predicting disinfection by-products in water. *Chemometrics and Intelligent Laboratory Systems*, *114*, 122–131. DOI: 10.1016/j.chemolab.2012.03.014

Song, K., Yan, F., Ding, T., Gao, L., & Lu, S. (2019). A steel property optimization model based on the XGBoost algorithm and improved PSO. *Computational Materials Science*, *174*, 109472. DOI: 10.1016/j.commatsci.2019.109472

Yang, L., Elisa, N., & Eliot, N. (2018). Privacy and security aspects of e-government in smart cities. In *E-government, e-services and smart cities* (pp. 89–102). Elsevier., DOI: 10.1016/B978-0-12-815032-0.00007-X

Chapter 5
Foundations of Predictive Spatial Data Management in Smart Cities

Munir Ahmad
http://orcid.org/0000-0003-4836-6151
Survey of Pakistan, Pakistan

Arifur Rahman
http://orcid.org/0009-0006-4494-4558
International American University, USA

Bivash Ranjan Chowdhury
http://orcid.org/0009-0008-2081-5318
International American University, USA

Hossain Mohammad Dalim
http://orcid.org/0009-0009-2341-7750
International American University, USA

ABSTRACT

Predictive spatial data management is a transformative approach enabling smart cities to transition from reactive governance to proactive, data-driven decision-making. By integrating GIS, IoT, cloud computing, and machine learning, cities can optimize infrastructure, enhance mobility, improve environmental sustainability, and deliver citizen-centric services. Despite technical, ethical, and organizational challenges, emerging trends such as AI-driven autonomous services, advanced spatial analytics, and adaptive infrastructure promise more precise, resilient, and responsive urban

DOI: 10.4018/979-8-3373-4202-3.ch005

systems. This chapter outlines foundational principles, technological enablers, predictive modeling approaches, and practical strategies to operationalize spatial intelligence for sustainable urban development.

INTRODUCTION

The enormous growth of population in the world, especially in the urban areas, escalating infrastructure demands, traffic, and environmental pressures are pressuring and overwhelming urban centers around the world like never before. The conventional methods of managing cities that usually depend on reactive planning and data systems that are segregated can no longer support these challenges effectively. Predictive spatial data management is a shift in paradigm that unites geospatial information with the power of advanced analytics and machine learning to predict urban trends and use them as a basis of proactive decision-making (Andronie et al., 2023; Zhou et al., 2024).

Predictive spatial data management is fundamentally about utilizing the rich spatial datasets on many scales, including geographic coordinates and land-use maps, real-time sensor measurements, and socio-economic indicators. Multiplied by predictive algorithms, such datasets allow planners and city administrators to find patterns, predict the future state, and optimize resource distribution. In other words, traffic that is predicted can help de-congest traffic jams, the monitoring of infrastructure using the IoT can help avoid system breakdowns, and the spatially-informed environment analysis can inform a sustainable development of the urban environment (Abijith & Saravanan, 2022; Chang et al., 2022; Leta et al., 2021; Wang et al., 2022).

The integration between predictive analytics and spatial data continues to be supported by access to enabling technologies like Geographic Information Systems (GIS), cloud computing, and Internet of Things (IoT) networks. GIS offers the platform to represent and interpret the intricate urban patterns, sensors of the IoT constantly provide real-time data on mobility, energy, and environmental situations, and cloud environments can supply the required computational power to handle extreme volumes of data effectively. Collectively, the technologies have a networked and reactive urban environment that can help in dynamic decision-making (Chen et al., 2021; Selmy et al., 2023; Tahir et al., 2025; Truong et al., 2023).

The chapter discussed predictive spatial data management and the practical use of the subject with respect to smart cities. It examines data acquisition, integration, modeling, and real-time analytics capabilities and how cities can transition to proactive governance using them. This chapter also discusses how spatial intelligence could be used to increase urban resiliency, optimize infrastructure and services, and achieve citizen-focused solutions with the ability to optimize quality of life through an evaluation using the technological, methodological, and operational aspects.

CORE PRINCIPLES OF PREDICTIVE SPATIAL DATA MANAGEMENT

A body of core principles inherent in making predictive spatial data management successful is that data should not merely be collected, but managed, structured, and analyzed so as to present actionable outcomes that can be applied in urban decision-making. Primarily, spatial data covers various data, including geographic positions, land-use patterns, transport infrastructure, outputs of IoT sensors, and social and economic factors of population density, income, or access to services. Organization of these different data to be used in predictive modeling should be done with caution on points of quality, relevance, and compatibility. Table 1 shows a summary of a few principles applied in urban contexts.

Table 1. Principles applied in urban contexts

Principle	Urban Application Example	Benefit
Data Quality	High-resolution traffic data	Improves the accuracy of congestion forecasts
Interoperability	Integration of municipal + sensor data	Enables holistic city management
Temporal Relevance	Real-time pollution sensor updates	Supports timely public health alerts
Spatial Granularity	Neighborhood-level energy usage monitoring	Informs localized conservation measures
Alignment with Objectives	Linking mobility data to sustainability targets	Ensures data use supports policy goals

Data Quality

The cornerstone of successful predictive modeling in the context of urban governance is data quality therein the better the quality, lack of omissions and the consistency of data. Insufficient data, such as that generated with incorrect measurements, incompatible formats, and missing data points, may cause considerable bias, alter the analytical results, and eventually reduce the credibility of decisions and their performance as well. To deal with these risks, it is necessary to introduce comprehensive quality assurance procedures involving the use of automated and manual validation measures, systematic data cleaning, standardized formats, and ontologies so that the data can be interoperable with a variety of data repositories (Gupta et al., 2021; Jain et al., 2020; Moskolaï et al., 2021; Ren et al., 2021).

In real-time urban monitoring systems, i.e., environmental sensors operated by the Internet of Things, public transit monitors, traffic flows tracking, and data quality management also require consideration of such problems as sensor drift, network disruptions, and latency. The frameworks of continuous quality monitoring, in turn, assisted by the methods of anomaly detection algorithms, may allow locating and correcting them on the go, not accumulating false information into the predictive models. There should also be legible records of data origin and refresh rates, as far as accountability and transparency are concerned. Urban authorities can make sure that predictive models remain robust, credible, and actionable as data quality protection can be established at every step of the data lifecycle, thus maximising the accuracy and fairness of service delivery.

Interoperability

In the case of predictive analytics in urban situations, interoperability is a forte of successful analytics, where data across multiple sources can be merged, processed, and analyzed without problems. Urban data is heterogeneous by nature, based on urban administrative datasets, IoT-enabled infrastructure, remotely sensed platforms, crowd-sourced applications, and commercial services, each using potentially different formats, structures, or semantic standards. In the absence of interoperability, such streams of data are all siloed, which has a cascading effect on the scope and accuracy of predictive models (Fernandez & Pérez, 2024; Ranatunga et al., 2025; San Emeterio de la Parte et al., 2023).

They are needed in order to achieve meaningful interoperability: common data exchange standards and metadata schema, as well as open APIs enabling systems to speak the same language. Standardization schemes, e.g., promoted by international organizations (e.g., ISO, OGC), can be used to clarify a common method of data formatting, storage, and transfer. Metadata annotation is important to characterize content, origin, quality, and temporal attributes of datasets to provide automated systems with the accuracy to comprehend datasets and join them.

In addition to technical alignment, interoperability also requires institutional agreements and governance systems that allow data sharing across (and between) departments and sectors. This can entail implementation of legal structure, data-sharing, and privacy protection regulations that promote cooperation and prevent exposure of sensitive data. Incorporating interoperability concepts and practices in technical architecture and governance decisions offers the urban authorities an opportunity to establish a coherent data ecosystem, where predictive modeling could be more reflective, multidimensional data sets facilitated, and more actionable information could be supplied to plan and manage cities.

Temporal Relevance

Where fast-moving environments exist, e.g., transportation networks, energy grids, public health responses, or environmental surveillance, predictions can quickly become useless when founded on inaccurate or stale information. As an example, traffic congestion predicted based on the data gathered even several hours ago might not be able to reflect the effects of unexpected road closures, weather conditions, or demonstrations of people (R. Kumar et al., 2024).

The effective temporal relevance can be based on the incorporation of data feeds based on real-time or near-real-time into the predictive models. This is usually done through the use of IoT sensor networks, satellite, GPS tracking, and automated reporting systems that stream data constantly. These inputs can be quickly ingested, processed, and analyzed by advanced data architectures like event-driven processing and streaming analytics, which can provide time-sensitive insights to decision-makers.

Historical data would be useful, especially when seasonal trends, tendencies, and long-term correlations have to be identified. They, however, need to be put in context with present circumstances so that one does not end up with false assumptions. This necessitates temporal metadata tagging, time-window filters, and weighting processes that prioritize the most current data but are also able to maintain the overarching historic dimension. Incorporation of time relevance into the development of predictive systems means that the most up-to-date and contextually relevant information can be used to drive decisions, a model that allows a city planner and manager to intervene proactively, prevent response lag, and make city services more resilient.

Spatial Granularity

Spatial granularity transforms into the degree of geographic detail that is reflected within a set of data and yields an influential role in determining the accuracy of the applicability of predictive models (Yi et al., 2024). Fine-grained data, in an urban setting, can involve traffic at particular intersections, energy use at specific buildings, or air quality understood through sensors of a localized nature, which allows extremely local forecasting and actions that can be highly specific. This level of information can be used to back up efforts such as micro-level congestion control, energy-efficient retrofitting of buildings, or community-based or type-specific environmental mitigation programs.

Nonetheless, fine-grained data sets usually demand massive amounts of computational power to store, process, and analyse, and complex model architectures that can process high-dimensional inputs. Conversely, coarse-grained data (that is, data aggregated to neighborhood, district, or city-level) can be easier to work with and easier to plan with, long-term strategy. It, however, faces the danger of ignoring important spatial differences that might be essential in solving local problems or adapting services to local populations.

Balancing detail and efficiency at their optimal point comes down to paying close attention to the use case it is supposed to serve, what is accessible in terms of computing infrastructure, and the context of operations. A predictive system can be able to change the level of spatial details between levels by using techniques like hierarchical modeling, multi-resolution analysis, and on-demand data aggregation techniques. Urban data managers and predictive modelers can overcome this dilemma by tuning spatial granularity properly to ensure that the forecasts are capable of being meaningfully acted upon, and computationally feasible--with the heaviest emphasis placed on the relevance and decision support value of the spatial insights.

Last but not least, mapping predictive spatial systems to the goals of each city can be crucial in assuring that analytics can lead to actionable results. Cities can focus on sustainability, safety, operational efficiency, or the well-being of the citizens, and the predictive systems should be built according to those objectives. As an illustration, the predictive model developed to manage the traffic should not only think of minimizing the congestion but should also focus on environmental benefits and efficiency of the emergency measures. Investing in the objectives of a city into predictive frameworks can guarantee that the value-based strategic decisions enjoyed by the idea produced, instead of being mere technical results. Predictive spatial data management on the basis of such values as the quality of data, interoperability, the significance of the time, the granularity of space, and the alignment with city goals becomes the true credit of the smarter, responsive urban management.

DATA ACQUISITION AND INTEGRATION IN URBAN CONTEXTS

The key to effective planning and prediction with spatial data is to achieve a consistent approach toward constant integration and acquisition of different types of data capable of reflecting the complexity of town systems. The collection of assets of contemporary urban areas results in a tremendous flow of information of different sorts, with each of them offering a different lens on the urban process. IoT devices installed in streets in the form of sensors in roads and street lights, buildings, and even transport networks provide real-time information on parameters like traffic volume, air quality, energy usage, and infrastructure status that allow real-time situational awareness. Satellite imagery and remote sensing technologies have an even greater capability over this, due to the detection of land-use changes, urban sprawling, and facilitating environmental monitoring with continuity in time and space.

Predictive models are enhanced by crowdsourced datasets, based on mobile applications, social media, and citizen reporting systems, which can provide hyperlocal, timely dynamics of a given event, congestion, or service disruption that may not be represented in official monitoring systems. Meanwhile, records maintained by municipalities and administration, including zoning policy, utility grids, cadastral mapping, and demographic statistics, have been valuable, authoritative, and well-organized data, and in doing so, provide a spatial and legal foundation to real-time and crowdsourced information.

The difficulty is that it is hard to combine these heterogeneous data in a single analytical environment. This demands not merely technical interoperability and support of standard data formats, but also plans of quality assurance, long-term alignment, and spatial harmonization. When effectively integrated, high-frequency sensor streams, large geospatial imagery, participatory data, and official records can be used by urban managers to create a multi-layered and comprehensive information foundation that complements predictive modelling with enhancements in precision, reliability, and contextual richness.

Data Integration and Harmonization

Unification of heterogeneous urban data sources is one of the keys to predictive spatial data management, converting raw, disparate data into a logical basis upon which analyses may be defined. This takes the form of intense data cleaning to identify and fix errors, eliminate redundancies, and deal with missing data, then normalization and standardization to harmonise variability in units of measurement, type of classification, and file types. It is especially important when using datasets that have been recorded at different spatial or temporal frequencies (e.g., neighborhood-level census data and building-specific energy readings) or when a user has datasets recorded at the hourly level (sensor update) and much longer time periods (decadal satellite measurements) (Cheng et al., 2024; Loulaki et al., 2024; Noardo, 2022).

The alignment in space and in time is a key factor to be able to sustain predictive accuracy. The problems with the coordinate system, timestamp formats, or geographic boundaries can create systematic biases distorting outputs of the model. These risks are reduced with advanced data fusion. The reconciliation methods using machine learning may determine and fix the discrepancies among sources, and geostatistical methods (e.g., kriging or spatial interpolation) may interpolate lacking values or incompatible ones with spatially pertinent precision.

The growing amount and speed of urban data are driving the requirement to have scalable storage and processing infrastructures. Spatially enabled data lakes that may be built on cloud-based platforms provide such capabilities because they provide flexibility to ingest, store, and process multi-source data sets with near real-time capabilities, in turn allowing a dynamic update to predictive models as new data arrives. Together with strong metadata management and access control, these platforms not only enable analytics at scale but also make integrated datasets transparent, traceable, and operationally meaningful to be processed in the context of decision-making. Figure 1 illustrates IoT sensors, satellite imagery, crowdsourced datasets, and municipal records feeding into a unified spatial repository.

Figure 1. Urban data sources and integration workflow

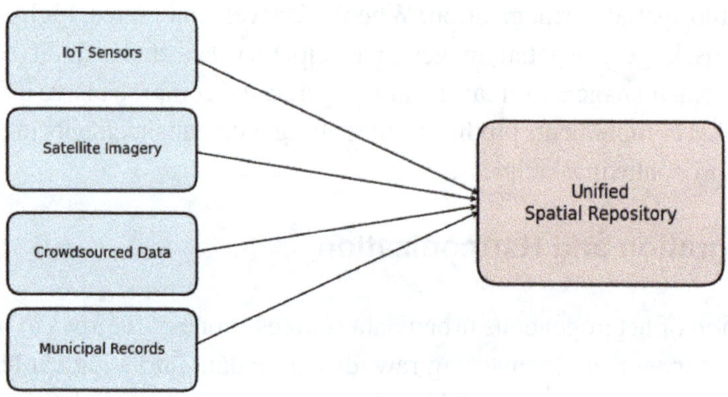

Challenges in Predictive Spatial Data Management

In spite of the high pace of development in the fields of sensing technologies, analytics platforms, and storage facilities, the area of predictive spatial data management is still burdened with some permanent challenges that may undermine the quality of models and their functioning. Missing data is still one of the most evident problems, which typically occurs due to unequal distribution of sensors in an urban environment, gaps in municipality records, or due to the time aspect of remote sensing systems, like poor satellite coverage at particular times, cloud-covered imagery, etc. Such gaps may produce blind spots in spatial analyses that result in the underrepresentation of specific geographic areas or socioeconomic classes.

Data latency is also a serious challenge, especially for time-sensitive requirements such as traffic control systems, response during a disaster, or monitoring of critical infrastructure. Small delays to the reporting of sensor results or network transfer may impact the timeliness and relevance of the results of the prediction output; therefore, the delays have to be mitigated as much as possible through the use of strategies like edge computing, local caches, and stream processing frameworks to alleviate downstream delays.

Possibly more important are the issues of privacy, confidentiality, and security. Urban data are frequently full of personally identifiable information (PII), traces of geolocation, or other sensitive information regarding infrastructure networks, which in the wrong hands can be dangerous to any given individual or even damage society. To mitigate these risks, custodians of data sets are increasingly using anonymization and pseudonymization techniques to eliminate or hide identifying attributes; they are using both encryption in use and at rest to prevent unauthorized access to data, and differentially private frameworks provide the ability to conduct statistical studies without revealing individual-level data. Such practices exist side by side with a strong governance framework, legal compliance underpinnings, and multi-level access restrictions that find the optimal balance between data usefulness and ethical care.

Quality Assurance and Governance in Spatial Data Integration

Lastly, the integration should not only be able to combine datasets that are heterogeneous but also implement a solid quality assurance and governance mechanism centralized in the process. This necessitates putting in place homogeneous procedures of constant checking, mistake identification, and automatic anomaly reporting, where the integrity of the data must always remain throughout its lifecycle of collection, storage, and utilization. Quality checks can include cross-checking sensor readings with past baselines, verifying the spatial coordinates with certificate geodatabases, and identifying outliers by means of statistical validation models or machine learning-based validation models (Ait Errami et al., 2023; Wagner & Henzen, 2022).

It is also vital to document thoroughly metadata, recording information on data provenance, on how it was collected, temporally, spatially, and on what preprocessing it has undergone. Such metadata facilitates both transparency and reproducibility and enables predictive modeling teams to evaluate the suitability of each dataset to suit particular analytic goals. Clear roles and responsibilities of data custodians ought to be defined in governance frameworks, version controls should be implemented so that the changes can be traced over time, and compliance monitoring should be completed so that the relevant law, ethical, and institutional standards are adhered to. Integration of strong administration with engineering-based protections can enable cities to shift past ad hoc data management to permanent, sustainable, and trust-based spatial data management.

Once cities can acquire, harmonize, secure, and govern their urban datasets, they are on the path to high-quality predictive models that can aid proactive management of the urban environment. This allows for smarter interventions in the 3 areas of mobility planning, environment monitoring, and infrastructure maintenance to create more resilient, responsive, and data-driven cities. Table 2 provides a summary of challenges in the context of data acquisition.

Table 2. Data acquisition challenges and mitigation strategies

Challenge	Description	Mitigation Strategy
Data Gaps	Missing or incomplete urban datasets	Use multi-source integration, interpolation
Latency	Delay in sensor or satellite data availability	Edge processing, near-real-time updates
Privacy Concerns	Sensitive personal mobility/health data	Anonymization, differential privacy methods
Heterogeneity	Diverse formats across IoT, imagery, and records	Standardized metadata, interoperability

PREDICTIVE MODELING AND MACHINE LEARNING APPLICATIONS

Predictive data modeling/machine learning is at the heart of predictive spatial data management, transforming big sets of raw urban data into usable information fed into decision-making. These models also allow cities to predict problems before they become out of control because they analyze patterns, trends, and correlations in both spatial and temporal data. Figure 2 shows the predictive modeling pipeline.

Figure 2. Predictive modeling pipeline

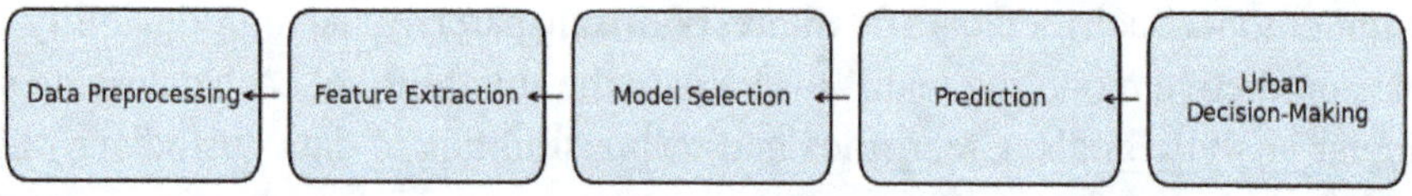

There are numerous machine learning Python algorithms and statistical methods used in urban environments. Some of the most common types of supervised learning are regression models, decision trees, and random forests used in extrapolating numerical results such as the amount of traffic, energy demands, and infrastructure wear. Techniques such as unsupervised learning, e.g., clustering and dimensionality reduction, can help identify hidden patterns, such as hotspots of congestion or areas prone to environmental hazards. Sophisticated deep learning methods, such as convolutional neural networks (CNNs) for spatial urban data and recurrent neural networks (RNNs) for time-series prediction, now enable more powerful processing on higher-order, more complex urban data. The fields of smart cities where predictive modeling can be used are wide and varied, as listed below and summarized in Table 3.

Table 3. Applications of predictive modeling in smart cities

Application Domain	Predictive Goal	Example Algorithms Used
Traffic Management	Forecast congestion & travel times	Random Forest, LSTM, ARIMA
Energy Optimization	Predict load fluctuations	Gradient Boosting, Neural Networks
Environmental Hazards	Early detection of floods/air pollution	CNN (imagery), Bayesian models
Infrastructure Maintenance	Predict structural degradation	Survival Analysis, Regression Trees

Traffic Congestion Forecasting

Predictive modeling is revolutionary in the context of controlling a city hood mobility as cities are capable of predicting traffic flows and mitigating bottlenecks. Using the historical and current data available because of the presence of road sensors, GPS systems, and connected cars, the models can be used to identify dynamic congestion conditions in time before reaching the critical stage. The dynamic traffic signal timing, strategic diversion of routes, and time-based specific public transit timetables are all possibilities via this foresight. The spin-off is less congested roads, less commuting time, and reduced vehicular emissions, resulting in more sustainable city transportation (Akhtar & Moridpour, 2021; K. Kumar et al., 2024).

Predictive Maintenance of Infrastructure

Internet of Things (IoT) Smart city infrastructure is progressively affixed with an array of sensors to continually track parameters of vibration, load, temperature, and humidity. The learning algorithms of these machine learning applications analyze these streams in order to identify early warning signs of machinery wear, material strain, or bridge, road, or utility-related structural limping. City authorities have an opportunity to be proactive, which is to schedule repairs in the most time-efficient periods, with minimal inconveniences to service availability and huge savings on repairs. This can prolong the life of assets and improve the safety of the people (Abbassi et al., 2022; Hallaji et al., 2022).

Environmental Hazard Alerts

Resilience of urban places is based on the ability to anticipate and respond to environmental threats. Predictive models combine different data sources, such as air quality indices, weather forecasts, hydrological variables, and land-use data, to understand situations in which hazards like air pollution events, floods, or heatwaves may occur (Neußner, 2021). The forecasts enable city agencies to give early warning, implement emergency response plans, and carry out specific interventions like opening cooling centers, cleaning out drainage systems, or varying the level of industrial activity. This way, cities can be able to safeguard vulnerable groups and reduce environmental health hazards.

Urban Energy Optimization

Predictive analytics is another useful feature of smart cities since it also assists with energy management in the community, predicting trends in consumption throughout residential areas, industrial centers, and public infrastructures (Klemm & Wiese, 2022). Knowing when loads are likely to be highest and when there are likely to be seasonal changes enables predictive models that can perform load balancing and demand-side management, putting less strain on the grid. Moreover, such predictions help to incorporate renewable energy by ensuring the harmonization of the supply of such resources as solar energy, wind energy, or other green sources with forecasted usage trends. This not only helps to streamline energy efficiency but also to achieve sustainability and decarbonization outcomes in this city.

Smart city predictive modeling can rely heavily on the control of accuracy, validation, and adaptability. The usefulness of any model is dictated by how well it can make credible predictions in actual practice, and serious testing and continual development are critical. The ability to convey the changing dynamics of a city, the variations between seasons, and the unanticipated outcomes of events like the collapse of the infrastructure or an extreme-weather event necessitates continuous updates with new databases made up of IoT sensors, municipal open records, or citizen-provided data. Various metrics of evaluation can also be utilised to ensure these models work as intended: root mean square error (RMSE) used to estimate average difference in magnitude between prediction and actual, precision and recall gauging performance in classification tasks, and F1 score, which is a compromise between the previous two. In addition to technical performance, the possibility of defining predictive models in a way that lets one understand them is an increasingly crucial issue. OPEN methodologies, transparently reported assumptions, and visual explanation tools contribute to establishing confidence in the urban planning community, policymakers, and other stakeholders so that predictions are not only more precisely true but also trustworthy and executable.

Predictive modeling and machine learning unveil raw spatial data and grant the urban authorities the ability to transform it into strategic foresight, which contributes to smarter, more resilient, and citizen-centered urban management.

TECHNOLOGICAL ENABLERS: GIS, IOT, AND CLOUD COMPUTING

The performance of the predictive spatial data management of smart cities is dependent on the overall use of sophisticated technologies that facilitate the real-time data capture, evaluation, and display. The backbone of the current predictive spatial technologies is built on three main technological enablers and includes Geographic Information Systems (GIS), the Internet of Things (IoT), and cloud computing. In combination, they enable vibrant, data-driven urban decision-making and management. Table 4 lists the technological enablers for predictive urban analytics.

Table 4. Technological functions in predictive spatial systems

Technology	Primary Function	Example Application
GIS	Spatial analysis, visualization	Heatmaps of congestion risk
IoT Sensors	Real-time monitoring	Air quality + traffic density sensing
Cloud Computing	Data storage + large-scale processing	Predictive modeling on big datasets
Edge Computing	Localized processing reduces latency	Instant alerts from traffic cameras

Geographic Information Systems (GIS)

GIS are the analysis backbone of spatially-informed choices in the smart city. These systems give the possibility to store, process, visualise, and interpret highly complex geospatial data, taking crude geospatial data and analysing it in a meaningful way. GIS allows urban planners to build layer maps through which various sets of information, such as infrastructure composition, demography, environmental variables, and socio-economic trends, may be interrelated into a multidimensional interpretation of the urban systems. In terms of predictive modeling, GIS has further abilities that include spatial queries to determine a certain feature or condition, geostatistical analysis to identify correlations and spatial trends, and hot spot detection to uncover a location of concern or opportunity. Incorporating predictive outputs in GIS-derived dashboards, the city authorities have the option of tracking emerging patterns, simulating future scenarios, and making location-based decisions, clearly, quickly, and accurately. This combination of spatial intelligence and predictive analytics changes GIS into an active decision support framework rather than a moldy mapping application.

Internet of Things (IoT)

IoT is the sensory nervous system of a smart city and enables the high-resolution, constant-data-stream that is essential to conduct real-time analysis and predictive modeling. Thousands of sensors located on almost every piece of transportation and utility infrastructure, buildings in the public estate, and environmental stations gather and relay life-saving information, no matter what the time. Traffic cameras monitor the flow and congestion of traffic, smart meters measure minute-by-minute energy consumption, weather stations measure local climatic phenomena, and structural health sensors sense the earliest signs of material failure or degradation in bridges and buildings (Ahmad et al., 2020). Real-time data is essential to predictive analytics since the models can always be trained and refreshed with the latest situation, making them more accurate and responsive to their forecasts. Feeding IoT data into machine learning algorithms, cities can then identify anomalies before they become full-blown and be proactive, forecasting service needs, making proactive interventions, meaning readjusting traffic light patterns, arranging preventative inspections of infrastructure, or deploying emergency assets before dangerous weather is expected to hit (Al-Atawi, 2024; Dhingra et al., 2021; Hussein et al., 2023).

Cloud computing

Cloud computing is the baseline infrastructure to manage, process, and distribute the huge heterogeneous data sets created within the environment of modern cities. The scalable computing resources that can ingest, store, and analyse large volumes of spatial and temporal data streams are what predictive spatial data management relies on. This ability to scale is available in cloud platforms, which include elastic storage that includes data lakes and high computing resources to execute complex predictive models. These platforms allow centralised data to be available to multiple stakeholders working within the municipality, cross-departmental collaboration, and eliminate data silos. Edge-cloud hybrid designs also support real-time analytics, with IoT devices doing some preprocessing before sending refined data to the cloud, where more intensive computing can be done. This distributed solution has the benefits of reducing latency, maximizing the utilization of available bandwidth, and ensuring critical decisions are made on the basis of the most recent and relevant information (Al-Atawi, 2024; Dhingra et al., 2021; Hussein et al., 2023; Uddin & Ahmad, 2020).

Moving beyond interoperability and integration becomes the connective tissue enabling GIS, the IoT, and cloud computing technologies to work as a coordinated predictive spatial ecosystem instead of only as siloed technology. To meet that goal, data as varied as geospatial databases, sensor networks, and more must be able to be shared across platforms and put in context in order to create any sense of predictive framework. The way to do it is to adopt a uniform set of protocols, a well-documented API, and common data formats so that different datasets can be heterogenized, analyzed, and presented freely. Real-time event streaming technologies like Apache Kafka or MQTT become communication channels through which the data is continuously consumed to make sure that each of the systems' updates comes in synchronization. Meanwhile, edge servers handle the actual computing locally, like noise-filtering, file-compression, or lightweight analytics, and send PP results to the cloud. Such an optimization of bandwidth consumption can also increase the response rate so that the city authorities may be able to quickly respond to any new tendencies, oddities, or disasters. The power of any predictive spatial data management system is not only in the complexity of individual parts, but in the harmoniously coordinated use of its capabilities.

With the integrated ecosystem powered up with GIS, IoT, and cloud computing, cities can be equipped with technology to have responsive, adaptive, and intelligent urban management in cities. Such enablers do not just enable predictive analytics but also enable scenario simulations, early warning systems, and interactive dashboards, allowing city authorities to be one step ahead, resource efficiently, and deliver citizen-centric services and so much more efficiently and resiliently.

OPERATIONALIZING PREDICTIVE SPATIAL DATA FOR SMART CITY PLANNING

The middle between predictive analytics and actionable urban governance is the cornerstone of smart cities' potential. The next challenge lies in operationalizing predictive spatial data; i.e., converting analytical results into practical tools, workflows, and decision-making frameworks that urban leaders can employ to make better planning, service delivery, and resilience decisions. The section examines actions, frameworks, and factors to consider in integrating predictive intelligence into the daily management of an urban environment. Figure 3 illustrates scenario simulation, real-time alerts, and predictive insights for planners.

Figure 3. Urban decision-support dashboard example

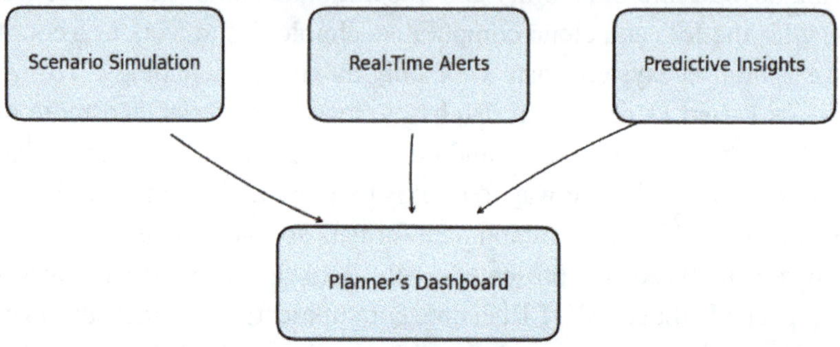

Decision-Support Systems (DSS)

DSS are the functional core of predictive spatial data management, and turn surges of raw data and model results into effective, comprehensible decision guidance to help decision-makers. DSS platforms can integrate predictive models, real-time sensor data feeds, and GIS-hosted displays into intuitive dashboards to allow city planners, municipal officials, and first responders to keep track of the changing conditions in cities, simulate possible choices of response, and decide on the most suitable course of action. Such systems can create early warnings of any upcoming issues, e.g., traffic leafs, air quality decline, or energy surges, and subsequently, match those warnings to immediate guidance, e.g., iterative tuning of traffic lights, specific community advice, or planned energy redistribution in power-heavy areas. In focusing on the complexity of analytics in understandable insights, DSS has the potential to solve the difference between the sophistication of analytics and the real-time needs of governance.

Scenario Simulations

The ability to simulate scenarios adds more strength to DSS as it allows one to develop theories of applying within the urban setting without incurring heavy costs. These simulations can, using predictive spatial models, anticipate the effects of a broad array of plans, including massive infrastructure improvements and zoning adjustments through to specific emergency responses, as well as climate adaptation projects. As an example, simulations can be used to understand how the redirection of vehicle flows might remove congestion hotspots, how open drainage systems could reduce risk during storm events with extreme weather, or how targeted energy efficiency services can smooth demand curves at peak demand. Comparing alternative policy options across outcomes helps planners to determine high-impact management strategies, measure the risks, and prioritize the interventions before investing in the public welfare. This futuristic ability not only gives greater accuracy in planning but also limits chances of price-intensive errors, developing cities with greater resilience and adaptiveness in the modern, busy city situation.

Urban Dashboards

The urban dashboards can be considered a nerve hub of smart city activities because they give a centralised, real-time look at the dynamics of a city in various areas (Bovkir & Aydinoglu, 2021). These dashboards help the authorities continuously monitor the activity going on in the city by visualising mobility flows, energy consumption, environmental conditions, metrics of public safety, and performance of infrastructures to be able to identify, at an early stage, anomalies that could develop into major problems. High-end dashboards contain interactive functionality, e.g., drill-down, geographic filtering, and predictive alerts, to enable analysts at the start of the decision-making process to investigate at a variety of levels of detail. As an example, a citywide energy pattern can be zoomed down to a building-level energy pattern, or 'maps of traffic congestion at the global scale' can be zoomed into the specific intersection of its focus, making the intervention highly specific. Moreover, scenarios, as well as what-if analyses, can be performed using dashboards to allow authorities to speculate on the effects of a policy change, a construction project, or an emergency response in a safe, virtual environment.

Stakeholder Collaboration and Participatory Governance

The application of predictive spatial insights needs to be well operationalized through sound stakeholder engagement and participatory governance. The actual worth of predictive modeling is achieved by sharing knowledge between departments and municipalities, utility companies, transportation agencies, emergency services, and community organizations. Common working platforms and data sharing policies help in strategizing together, coordinating actions, and prioritization agendas across industries. Citizens may also be directly engaged by the display of elements of the dashboard that are public-facing, such as informing them of transparency and building trust and a desire to be involved in urban planning. Cities can thus begin to geographically integrate participatory governance mechanisms with centralized visualization so that not only can decisions be made based upon predictive spatial data, but that action is encouraged and enabled by the data so that cities can transact spatial data and take action more rapidly, thus becoming more resilient, efficient, and altogether a better place to live. Table 5 provides the strategies and benefits of the operationalization.

Table 5. Operationalization strategies and benefits

Strategy	Description	Benefit
Decision Support Systems (DSS)	Interactive tools for planners	Informed urban planning
Scenario Simulations	Model future conditions (e.g., flooding)	Test resilience strategies
Dashboards	Real-time monitoring & alerts	Quick situational awareness
Stakeholder Collaboration	Shared spatial platforms	Improved governance and trust

Ethical Considerations and Data Governance

Smart cities cannot go without ethical considerations and effective data governance as a component of predictive spatial analytics. Urban data are often of a sensitive nature, and such data may include personally identifiable information of citizens, geographic footprint, and infrastructure-related data of critical importance. The collection, storage, and analysis of such information can be a big privacy and security risk without proper protection measures. To diminish these fears, cities must put in place a full governance regime that stipulates clear policies to access the information, anonymizing, encryption, and responsible use of data. The same can be said about transparency: people deserve to know which data is being gathered, how this data is going to be used, and what consequences this might have on privacy, and most importantly, part of the delivery of these services. Data practices are held accountable through accountability mechanisms such as frequent audits, access logs, oversight committees, and others, which ensure practices are within legal and ethical requirements. With a combination of technical properties (differential privacy, role-based access controls, secure data-sharing mechanisms) and ethical measures, municipalities can be able to retain that public trust as they continue to use predictive spatial analytics to make smart, equitable, and effective urban management decisions.

Iterative Policy Integration

Policy integration is also essential and must be iterative to convert the predictive insights on spatial issues to long-term, actionable enhancements in urban leadership (Noardo, 2022). The cities are dynamic systems, and mobility patterns, energy demand, environmental conditions, and demographic trends constantly change. Predictive models need to be constantly updated based on newfound information to keep them relevant, and they should be frequently used in the creation of effective policies, regulations, and standard operating procedures. This necessitates creating feedback loops where the model forecasts can be compared with material results and lessons learnt, and corrections can be made to the models as well as the policies they guide. Predictive insertion into the long-term planning cycles allows the urban authorities to foresee the upcoming challenges and allocate resources as well as design service delivery proactively, as opposed to reactively. Furthermore, iterative integration fosters adaptive governance: policies develop into living tools that adjust to new realities, influence, and inputs of stakeholders, as well as new technology possibilities. By doing this, cities can be able to ensure that predictive spatial analytics delivers the greatest value to society; in this way, urban areas continue to develop in resilient, equitable, and sustainable ways based on data-driven decisions.

Incorporating predictive spatial information into decision-support, scenario simulation, and urban dashboards coupled by collaboration, ethical governance, and iterative policy application can enable cities to transform reactive management to anticipatory, proactive planning. Such an operational strategy can improve the resilience of the city, streamline resource distribution, and create data-driven, citizen-centric services that can address not only current demands but also challenges into the future.

Table 6. Challenges vs. strategies and future solutions

Challenge Type	Example Issue	Strategy / Future Solution
Technical	Limited model accuracy in dynamic settings	Hybrid AI + physics-based models
Ethical	Risk of surveillance, biased predictions	Transparent algorithms, ethics boards
Organizational	Lack of capacity in municipal institutions	Training, public-private partnerships
Data Governance	Ownership & accessibility conflicts	Open-data policies, blockchain traceability

CONCLUSION

The management of predictive spatial data is a radical approach to streamlining the growth of smart cities that helps the urban authorities shift their paradigm towards proactive, data-directed governance, becoming no longer reactive to the issues that occur. In this chapter, we have considered the basic principles of such pre-emption, the technological means that enable pre-emption, pre-emptive models, and methods of operation that cannot form the basis of how cities can pre-empt the problems, optimize infrastructures to exploit pre-emption, maximize mobility, and maximize environmental sustainability. Introducing GIS, Internet of Things (IoT), cloud, and machine-learning capabilities, predictive spatial systems transform the heterogeneous data in the city into knowledge that can be used to guide citizen and resilient urban ecosystems.

Nonetheless, despite these opportunities, a major problem is present when it comes to implementing predictive spatial data management. Obstacles that are technical concerns include heterogeneity of data, real-time processing requirements, and computational constraints. Such considerations as ethical and legal, including the privacy of data, data security, and transparency of algorithms, demand effective governance systems to assert the faithfulness of the population. Organizational issues may block the successful implementation of predictive systems, not only due to interdepartmental coordination, collaboration with stakeholders, and building knowledgeable capacity, but also because of the heterogeneous urban environment with different socio-economic and infrastructural situations.

Moving ahead, some emerging trends are most likely to influence the course of predictive spatial data management. Autonomous city services fuelled by AI, like self-regulating traffic networks, predictive energy distribution, and automated environmental monitoring, can also likely improve efficiency and responsiveness. High-resolution data linked with complex machine learning models would be able to provide more specific and localized interventions in the future because of advances in spatial analytics. A dynamic changes approach (adaptive infrastructure) can enable resilient urban systems to deal with risks in real time based on predictive insights.

The future of predictive spatial management in smart cities lies in the further incorporation of technology, ethical policies, and strategic planning. Cities should aim to harness predictive spatial intelligence to turn a challenging and demanding present into one that is fruitful, sensible, and centered on people by acknowledging innovations as well as taking on challenges that face urban areas today. Such a whole-model approach addresses and solves the issue of smart cities ideals, such as the robust infrastructure, streamlined services, and quality of life, becoming fully feasible.

REFERENCES

Abbassi, R., Arzaghi, E., Yazdi, M., Aryai, V., Garaniya, V., & Rahnamayiezekavat, P. (2022). Risk-based and predictive maintenance planning of engineering infrastructure: Existing quantitative techniques and future directions. In *Process Safety and Environmental Protection* (Vol. 165). DOI: 10.1016/j.psep.2022.07.046

Abijith, D., & Saravanan, S. (2022). Assessment of land use and land cover change detection and prediction using remote sensing and CA Markov in the northern coastal districts of Tamil Nadu, India. *Environmental Science and Pollution Research International, 29*(57), 86055–86067. Advance online publication. DOI: 10.1007/s11356-021-15782-6 PMID: 34510357

Ahmad, S., Hasan, M., Shahabuddin, M., Tabassum, T., & Allvi, M. W. (2020). IoT based pill reminder and monitoring system. *International Journal of Computer Science and Network Security, 20*(7), 152–158.

Ait Errami, S., Hajji, H., Ait El Kadi, K., & Badir, H. (2023). Spatial big data architecture: From Data Warehouses and Data Lakes to the LakeHouse. In *Journal of Parallel and Distributed Computing* (Vol. 176). DOI: 10.1016/j.jpdc.2023.02.007

Akhtar, M., & Moridpour, S. (2021). A Review of Traffic Congestion Prediction Using Artificial Intelligence. In *Journal of Advanced Transportation* (Vol. 2021). DOI: 10.1155/2021/8878011

Al-Atawi, A. A. (2024). Enhancing data management and real-time decision making with IoT, cloud, and fog computing. *IET Wireless Sensor Systems, 14*(6), 539–562. DOI: 10.1049/wss2.12099

Andronie, M., Lăzăroiu, G., Iatagan, M., Hurloiu, I., tefănescu, R., Dijmărescu, A., & Dijmărescu, I. (2023). Big data management algorithms, deep learning-based object detection technologies, and geospatial simulation and sensor fusion tools in the internet of robotic things. *ISPRS International Journal of Geo-Information, 12*(2), 35. DOI: 10.3390/ijgi12020035

Bovkir, R., & Aydinoglu, A. C. (2021). Big urban data visualization approaches within the smart city: Gis-based open-source dashboard example. *International Archives of the Photogrammetry, Remote Sensing and Spatial Information Sciences - ISPRS Archives, 46*(4/W5-2021). DOI: 10.5194/isprs-archives-XLVI-4-W5-2021-125-2021

Chang, M., Ding, Z., Cai, Z., Zhao, Z., & Yuan, L. (2022). Prediction of Evolution Behaviors of Transportation Hubs Based on Spatiotemporal Neural Network. *IEEE Transactions on Intelligent Transportation Systems, 23*(7), 9171–9183. Advance online publication. DOI: 10.1109/TITS.2021.3091708

Chen, C., Liu, Y., Sun, X., Di Cairano-Gilfedder, C., & Titmus, S. (2021). An integrated deep learning-based approach for automobile maintenance prediction with GIS data. *Reliability Engineering & System Safety*, *216*, 107919. Advance online publication. DOI: 10.1016/j.ress.2021.107919

Cheng, C., Messerschmidt, L., Bravo, I., Waldbauer, M., Bhavikatti, R., Schenk, C., Grujic, V., Model, T., Kubinec, R., & Barceló, J. (2024). A General Primer for Data Harmonization. In *Scientific Data* (Vol. 11, Issue 1). DOI: 10.1038/s41597-024-02956-3

Dhingra, S., Madda, R. B., Patan, R., Jiao, P., Barri, K., & Alavi, A. H. (2021). Internet of things-based fog and cloud computing technology for smart traffic monitoring. *Internet of Things : Engineering Cyber Physical Human Systems*, *14*, 100175. DOI: 10.1016/j.iot.2020.100175

Fernandez, V., & Pérez, V. (2024). A new approach to interoperability within the smart city based on time series-embedded adaptive traffic prediction modelling. *Networks and Spatial Economics*, •••, 1–16. DOI: 10.1007/s11067-024-09662-y

Gupta, N., Mujumdar, S., Patel, H., Masuda, S., Panwar, N., Bandyopadhyay, S., Mehta, S., Guttula, S., Afzal, S., Sharma Mittal, R., & Munigala, V. (2021). Data Quality for Machine Learning Tasks. *Proceedings of the ACM SIGKDD International Conference on Knowledge Discovery and Data Mining*. DOI: 10.1145/3447548.3470817

Hallaji, S. M., Fang, Y., & Winfrey, B. K. (2022). Predictive maintenance of pumps in civil infrastructure: State-of-the-art, challenges and future directions. In *Automation in Construction* (Vol. 134). DOI: 10.1016/j.autcon.2021.104049

Hussein, W. N., Hussain, H. N., Hussain, H. N., & Mallah, A. Q. (2023). A Deployment Model for IoT Devices Based on Fog Computing for Data Management and Analysis. *Wireless Personal Communications*, •••, 1–13. DOI: 10.1007/s11277-023-10168-y

Jain, A., Patel, H., Nagalapatti, L., Gupta, N., Mehta, S., Guttula, S., Mujumdar, S., Afzal, S., Sharma Mittal, R., & Munigala, V. (2020). Overview and Importance of Data Quality for Machine Learning Tasks. *Proceedings of the ACM SIGKDD International Conference on Knowledge Discovery and Data Mining*. DOI: 10.1145/3394486.3406477

Klemm, C., & Wiese, F. (2022). Indicators for the optimization of sustainable urban energy systems based on energy system modeling. *Energy, Sustainability and Society*, *12*(1), 3. Advance online publication. DOI: 10.1186/s13705-021-00323-3

Kumar, K., Kumar, M., & Das, P. (2024). Traffic congestion forecasting using multilayered deep neural network. *Transportation Letters, 16*(6), 516–526. Advance online publication. DOI: 10.1080/19427867.2023.2207278

Kumar, R., Bhanu, M., Mendes-Moreira, J., & Chandra, J. (2024). Spatio-temporal predictive modeling techniques for different domains: A survey. *ACM Computing Surveys, 57*(2), 1–42. DOI: 10.1145/3696661

Leta, M. K., Demissie, T. A., & Tränckner, J. (2021). Modeling and prediction of land use land cover change dynamics based on land change modeler (Lcm) in nashe watershed, upper blue nile basin, Ethiopia. *Sustainability (Basel), 13*(7), 3740. Advance online publication. DOI: 10.3390/su13073740

Loulaki, M., Kiourtis, A., Mavrogiorgou, A., & Kyriazis, D. (2024). Addressing complex data integration and harmonization scenarios. *2024 E-Health and Bioengineering Conference (EHB)*, 1–4.

Moskolaï, W. R., Abdou, W., Dipanda, A., & Kolyang. (2021). Application of deep learning architectures for satellite image time series prediction: A review. In *Remote Sensing* (Vol. 13, Issue 23). DOI: 10.3390/rs13234822

Neußner, O. (2021). Early warning alerts for extreme natural hazard events: A review of worldwide practices. In *International Journal of Disaster Risk Reduction* (Vol. 60). DOI: 10.1016/j.ijdrr.2021.102295

Noardo, F. (2022). Multisource spatial data integration for use cases applications. *Transactions in GIS, 26*(7), 2874–2913. Advance online publication. DOI: 10.1111/tgis.12987

Ranatunga, S., Ødegård, R. S., Jetlund, K., & Onstein, E. (2025). Use of Semantic Web Technologies to Enhance the Integration and Interoperability of Environmental Geospatial Data: A Framework Based on Ontology-Based Data Access. *ISPRS International Journal of Geo-Information, 14*(2), 52. DOI: 10.3390/ijgi14020052

Ren, X., Li, X., Ren, K., Song, J., Xu, Z., Deng, K., & Wang, X. (2021). Deep Learning-Based Weather Prediction: A Survey. *Big Data Research, 23*, 100178. Advance online publication. DOI: 10.1016/j.bdr.2020.100178

San Emeterio de la Parte, M., Martínez-Ortega, J. F., Hernández Díaz, V., & Martínez, N. L. (2023). Big Data and precision agriculture: A novel spatio-temporal semantic IoT data management framework for improved interoperability. *Journal of Big Data, 10*(1), 52. Advance online publication. DOI: 10.1186/s40537-023-00729-0

Selmy, S. A. H., Kucher, D. E., Mozgeris, G., Moursy, A. R. A., Jimenez-Ballesta, R., Kucher, O. D., Fadl, M. E., & Mustafa, A. (2023). Detecting, Analyzing, and Predicting Land Use/Land Cover (LULC) Changes in Arid Regions Using Landsat Images, CA-Markov Hybrid Model, and GIS Techniques. *Remote Sensing (Basel), 15*(23), 5522. Advance online publication. DOI: 10.3390/rs15235522

Tahir, Z., Haseeb, M., Mahmood, S. A., Batool, S., Abdullah-Al-Wadud, M., Ullah, S., & Tariq, A. (2025). Predicting land use and land cover changes for sustainable land management using CA-Markov modelling and GIS techniques. *Scientific Reports, 15*(1), 3271. DOI: 10.1038/s41598-025-87796-w PMID: 39863826

Truong, T. X., Nhu, V. H., Phuong, D. T. N., Nghi, L. T., Hung, N. N., Hoa, P. V., & Bui, D. T. (2023). A New Approach Based on TensorFlow Deep Neural Networks with ADAM Optimizer and GIS for Spatial Prediction of Forest Fire Danger in Tropical Areas. *Remote Sensing (Basel), 15*(14), 3458. Advance online publication. DOI: 10.3390/rs15143458

Uddin, M. Y., & Ahmad, S. (2020). A Review on Edge to Cloud: Paradigm Shift from Large Data Centers to Small Centers of Data Everywhere. *Proceedings of the 5th International Conference on Inventive Computation Technologies, ICICT 2020.* DOI: 10.1109/ICICT48043.2020.9112457

Wagner, M., & Henzen, C. (2022). Quality Assurance for Spatial Research Data. *ISPRS International Journal of Geo-Information, 11*(6), 334. Advance online publication. DOI: 10.3390/ijgi11060334

Wang, H., Zhang, R., Cheng, X., & Yang, L. (2022). Hierarchical Traffic Flow Prediction Based on Spatial-Temporal Graph Convolutional Network. *IEEE Transactions on Intelligent Transportation Systems, 23*(9), 16137–16147. Advance online publication. DOI: 10.1109/TITS.2022.3148105

Yi, C., Zhao, X., Sun, Q., & Wang, Z. (2024). Assessing the accuracy of remote sensing data products: A multi-granular spatial sampling method. *Future Generation Computer Systems, 159*, 151–160. DOI: 10.1016/j.future.2024.04.062

Zhou, Y., Liu, Y., Ning, N., Wang, L., Zhang, Z., Gao, X., & Lu, N. (2024). Integrating knowledge representation into traffic prediction: A spatial–temporal graph neural network with adaptive fusion features. *Complex & Intelligent Systems, 10*(2), 2883–2900. Advance online publication. DOI: 10.1007/s40747-023-01299-7

Chapter 6
Integrating IoT and Artificial Intelligence:
A Comprehensive Framework for Creating Smart Environments

G. Swetha
http://orcid.org/0000-0001-8620-7149
RR Institute of Technology, Bangalore, India

M. S. Veena
http://orcid.org/0000-0001-5554-6295
RR Institute of Technology, Bangalore, India

Tejaswini Krishnamurthy
RR Institute of Technology, Bangalore, India

S. Druva Kumar
http://orcid.org/0000-0001-5157-6294
Dayananda Sagar College of Engineering, India

M. Shruthi
Dayananda Sagar College of Engineering, India

S. Vishwanatha
http://orcid.org/0000-0002-0485-6194
Jain University, India

D. Rajeshwari
http://orcid.org/0000-0001-9952-3183
The National Institute of Engineering, India

N. Raghu
http://orcid.org/0000-0002-2091-8922
Jain University, India

Kamal Narayanan
University of Nevada, Las Vegas, USA

G. B. Arjun Kumar
Dayananda Sagar College of Engineering, India

DOI: 10.4018/979-8-3373-4202-3.ch006

ABSTRACT

The convergence of Artificial Intelligence and the Internet of Things has paved the way for the development of smart environments that are efficient, adaptive and sustainable. The IoT enables vast networks of interconnected sensors, device and actuators to capture real-time data, AI provides the intelligence needed to process, analyze and act upon this information. Together, they transform reactive systems into proactive context-aware ecosystems capable of decision-making and automation. This chapter presents a comprehensive framework for integrating AI with IoT highlighting its architecture, component and applicability across domains such as smart cities, healthcare, agriculture, transportation and industrial automation. The framework emphasizes enabling technologies including edge, fog and cloud computing, communication protocols and open platforms like FIWARE, which collectively support scalability, interoperability and real-time analytics. the study also critically addresses challenges related to data management, security, privacy, ethical considerations and workforce readiness.

INTRODUCTION

The technological progress of Internet of Things (IoT) and Artificial Intelligence (AI) has created a total revolution in contemporary technological advancement. The first IoT idea required linking physical devices to internet protocols which allowed them to transfer and receive information. The first applications operated at a basic level to track and control operations. The growth of IoT networks brought forth three main technological solutions which included smart homes and connected health systems and industrial automation systems. AI experienced a transformation from basic rule-based systems and expert programs into complex machine learning and deep learning models which recognized patterns and made predictions and autonomous decisions. AI provided the intelligence to process real-time data that IoT delivered in vast quantities. AI and IoT joined forces to establish modern adaptable systems which will remain relevant in the future. These technologies enable both connected systems and environments which adapt while learning from their operations. The integration between AI and IoT enables cities to operate efficiently and produces wearable devices that track health data instantly thus reshaping modern living. AI-powered IoT stands at the innovation forefront to lead the world toward a more intelligent and sustainable future.

The Internet of Things and Machine learning have fundamentally altered our interactions with the environment and data utilization. The Internet of Things (IoT) is a network including interconnected devices, sensors, and actuators that gather data. Artificial intelligence (AI) refers to a collection of advancement that allows machines to replicate the abilities of humans. Technologies such as computer vision and machine learning, natural language processing exemplify this category. Nonetheless, the endeavor of deriving substantive insights from this enormous volume of data is far more arduous (Nagaraj, 2023). Help from AI in such a scenario with AI algorithms and techniques, which are derived from the data generated by Internet of Things new insights are gained that enable efficient decision-making processes and can also be used for automation. IoT + AI — The Two Are Inseparable Creating smart environments requires these two technologies combination of the Internet of Things (IoT) and Machine learning, gives rise to numerous opportunities for smarter, more energy-efficient, sustainable settings aware of human requirements.

The sensing capability of the Internet of Things gagdets along with the cognitive capabilities of artificial intelligence systems together can reinvent entire cities, healthcare, transportation, and agriculture. The Internet of Things combined with Machine learning, for example, can enable smart cities to deploy intelligent transportation systems that increase public safety and ease traffic congestion by better controlling the flow of cars. Internet of Things (IoT) gadgets used to screen patients remotely combined with artificial intelligence analytics which can identify patterns, forecast maladies and personalize treatments may help the healthcare business. One of the ways in which the transportation sector can simplify this daunting task and enhance efficiency at machines is through implementation connected automobiles as well as AI-driven traffic management.This research study's primary goal is to provide a comprehensive framework for combining machine learning with the internet of things to build intelligent settings. The framework will include the key technology, components, and design needed to successfully integrate artificial intelligence with the Internet of Things. The purpose of this essay is to analyze the issues, prospects, and possible applications of this all-encompassing framework in a number of industries.

The agricultural field is using Internet of Things sensors to keep track of the growing conditions of the crops. In addition, irrigation systems and crop management structures can be improved further by the use of approaches based on artificial intelligence. This research paper, therefore, is aimed at describing a comprehensive framework employing artificial intelligence with the internet of things. This framework will also cover design, components and appropriate technologies necessary for successful integration of internet of things with artificial intelligence. Furthermore, this article aims to examine the challenges, opportunities or possible domains of relevance of this hybrid framework in practice.

Figure 1 presents a succinct summary of the FIWARE Internet of Things architectural components. FIWARE, as an open-source platform, facilitates compatibility across the world of technology of Things, huge amounts of data, cloud computing, and the future Internet. It offers throughout a large number of the business sectors a full range of standards and technology which may be applied for the creation of modern and scalable applications and services. The following is an illustration of how FIWARE can easily fit in with these technologies: Assume that a network of several sorts of IoT devices is deployed in different areas of the city, collecting information in real time. Some of these devices might have sensors that track Air quality, traffic patterns, trash management, and energy consumption.

Figure 1. FIWARE framework for an integrated future of Internet of Things

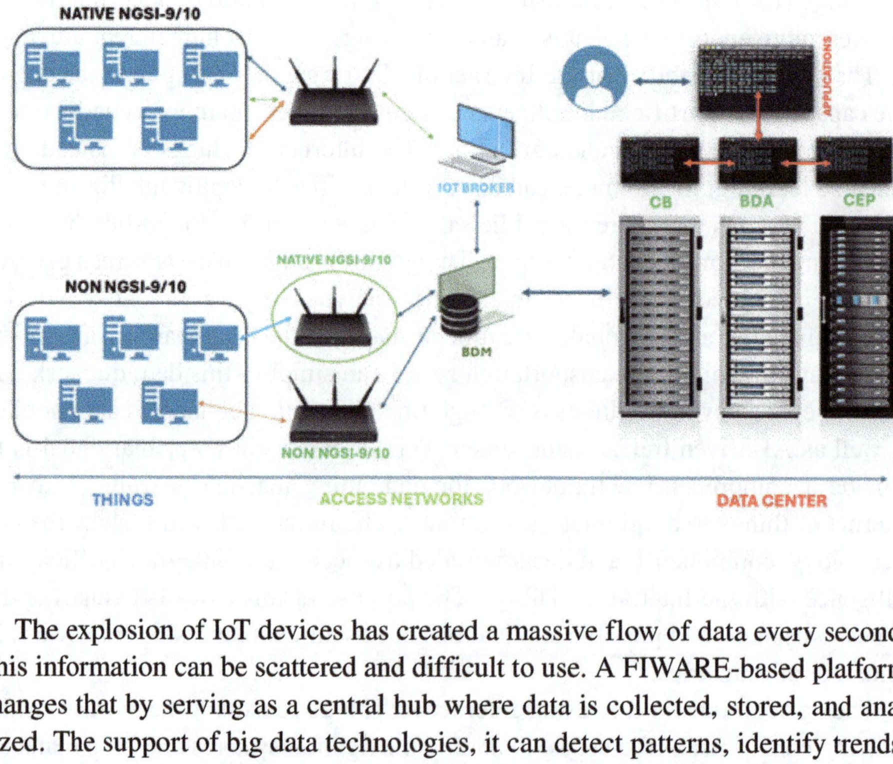

The explosion of IoT devices has created a massive flow of data every second. This information can be scattered and difficult to use. A FIWARE-based platform changes that by serving as a central hub where data is collected, stored, and analyzed. The support of big data technologies, it can detect patterns, identify trends, and spot unusual events turning raw numbers into insights that help guide better decisions. One of FIWARE's biggest strengths is its cloud-native design. As more devices connect and data volumes grow, the platform automatically scales to handle demand while staying reliable. Thanks to open APIs and standardized data models, developers can tap into this data with ease, reusing it to build useful applications. In smart mobility, for instance, real-time traffic updates can be transformed into accurate directions and optimized routes, cutting travel time and reducing congestion.

The impact goes far beyond traffic management. In waste collection, FIWARE can help cities create efficient schedules, reducing unnecessary trips and saving resources. Its open ecosystem of developers, service providers, and innovators ensures that raw IoT data is turned into practical tools that improve everyday services and urban living. The FIWARE stand out is the way it brings together IoT, cloud computing, and big data analytics in one platform. From gathering information to analyzing it and building applications, the platform supports the full journey through open standards. While often associated with smart cities, FIWARE is also driving progress in healthcare, agriculture, transportation, and industrial automation.

The merging IoT with Artificial Intelligence also raises challenges. Protecting data privacy, ensuring security, addressing ethical concerns, managing costs, and developing the right skills are all critical. Yet these challenges also bring opportunities. By finding solutions, platforms like FIWARE can unlock intelligent, adaptive environments that are not only smarter but also more sustainable. FIWARE is more than just infrastructure, it is a bridge between raw data and meaningful action, enabling the creation of digital ecosystems that make communities more connected, efficient, and responsive to real-world needs.

LITERATURE SURVEY

The author illustrated review how data aggregation can simultaneously improve energy efficiency and data security in IoT-driven healthcare systems by reducing redundant transmissions and by enabling lightweight, privacy-preserving aggregation at intermediate nodes. The authors survey aggregation strategies cluster-based collection, compressed sensing, edge/ fog aggregation, and secure aggregation protocols and discuss trade-offs between reduced communication and the increased computation or trust requirements at aggregators. They highlight practical techniques such as adaptive sampling, duty-cycling, and lightweight cryptographic schemes that balance energy use with confidentiality and integrity requirements, and note emerging enablers like edge computing and blockchain for tamper-resistant, decentralized aggregation. The paper concludes by identifying open challenges: heterogeneity of medical devices, strict privacy/regulatory demands, and the need for low-overhead secure aggregation methods that preserve data utility for clinical use (Shetty & N, 2024). Overall, the study is a useful roadmap for researchers aiming to design energy-aware, privacy-conscious data aggregation in healthcare IoT.

The authors present an in-depth exploration of advanced encryption strategies aimed at safeguarding patient data within Internet of Medical Things (IoMT) applications. The study emphasizes that as healthcare devices become increasingly interconnected, traditional encryption techniques often fail to balance security, efficiency, and real-time data accessibility. To address this, the authors analyze modern encryption mechanisms such as homomorphic encryption, lightweight cryptography, and hybrid key management systems, which enable secure data transmission without imposing heavy computational loads on medical sensors. They also discuss how integrating AI-based threat detection and blockchain frameworks can enhance data integrity and traceability across healthcare networks (V. M. S. et al., 2025). The paper concludes that achieving strong protection for patient information requires a multi-layered encryption approach tailored to IoMT constraints, combining both robust security algorithms and low-power implementation techniques.

The Authors state that the dynamic junction of three cutting-edge technologies, namely Artificial Intelligence (AI), Internet of Things (IoT), and Cloud Computing, as well as the tremendous impact that these technologies have had on a variety of different fields. After providing an overview of artificial intelligence and the issues it faces, it then goes into Internet of Things applications in several industries, including transportation, industry 4.0, healthcare, and agriculture. Artificial intelligence (AI) on the cloud is investigated, with topics including finance, e-commerce, smart cities, healthcare, and robotics being covered. This section examines several applications, such as intelligent meters, smart buildings, intelligent farming, smart hospitals, and smart industry, focusing on the combination of artificial intelligence and cloud-based Internet of Things technologies (Nagaraj, 2023). Challenges such as security and confidentiality of information are examined, along with the prospective future trajectories of technologies like quantum computer technology and fog computing. A collection of use cases that illustrate the practical implementations of these transformational technologies across a variety of industries is presented in the final section of the book. In addition, a list of scholarly references is included at the end of each chapter which is intended for more advanced readers.

The authors illustrate that the Internet of Things (IoT) has established a globally interconnected environment, driven by wired and wireless sensors that generate heterogeneous data across various industries and applications. Advanced artificial intelligent (AI) technology, particularly deep learning (DL), has enhanced data analysis, prediction, and decision-making by extracting comprehensive information from data (Chang et al., 2021). The merging of AI with IoT has expedited the advancement of AI-of-Things (AIoT) technologies that intelligently evaluate and react to environmental stimuli autonomously. The amount, speed, and validity of data, along with catastrophic transmission delay, impair networking infrastructures, rendering cloud data processing challenging or unfeasible. Edge computing can address these significant challenges. This essay thoroughly examines an end-edge-cloud coordinated architecture for flexible AIoT systems. Fundamental concepts such as IoT, AI, and computing at the edge are presented. These principles direct its examination of the comprehensive AIoT architecture, a practical AIoT instance to illustrate AI's applications in real life, and prospective AIoT applications. The subsequent review focuses on the advancement of AI model interpretation and training techniques at the network edge. The remaining challenges and potential trajectories of this promising sector are examined.

The Authors have claimed that Internet of Things (IoT), billions of gadgets and sensors gather data from their surroundings, process it, send it to data centers in the cloud, and then receive feedback through the Internet to assist with perception and connection. However, it is challenging to communicate large quantities of diverse data, to perceive complicated situations from this data, and to make timely smart decisions. Deep learning and other forms of artificial intelligence (AI) have recently achieved notable success in a number of domains, such as computer vision, voice recognition, and NLP. When AI is integrated into the IoT, it signals the beginning of the AI of things (AIoT) era (Zhang & Tao, 2020). This article provides an extensive overview of AIoT to demonstrate how AI may enhance the IoT by making it more efficient, intelligent, environmentally friendly, and secure. In particular, we provide a high-level overview of the AIoT architecture within the framework of cloud, fog, and edge computing. Then, we cover the state of AI research for the Internet of Things (IoT) from four angles: 1) perception, 2) learning, 3) reasoning, and 4) behavior. We will now go on to a brief overview of several exciting AIoT uses that have the potential to change our world in a significant way. Lastly, we discuss the difficulties of AIoT and point out several areas where future study could be fruitful.

The authors states that the IoT is a network that allows physical objects to communicate with one another and share data through the use of applications, sensors, and actuators (Kankanhalli et al., 2019). Artificial intelligence (AI) might help governments build apps and learn from big data sets collected from individuals, sensors, and networks. Services in the areas of energy, transportation, medical care, schooling, and public safety can be developed with the use of the Internet of Things (IoT) and artificial intelligence (AI). The Internet of Things (IoT) and artificial intelligence (AI) are set to revolutionize smart government. In this special edition guest editorial, the authors address the obstacles that have so far prevented their widespread adoption and implementation in the public sector, and they present a thorough research methodology to address these issues. Following a brief editorial introduction, this special issue contains six papers. Using the suggested framework and existing literature gaps, the articles included in this special issue advocate for further study of the Internet of Things (IoT) and artificial intelligence (AI) in the context of smart government. On the agenda are studies that are specialized to certain domains, studies that go above adoption studies to look at how these technologies are put into practice and evaluated, studies that zero in on particular problems and successes quickly, and studies that broaden the scope of research methodologies and theoretical underpinnings.

The Authors have reported that Internet of Things (IoT) is industrializing in numerous real-world applications such as smart transportation, smart city to improve human life trustworthy (Singh et al., 2020). Industrial IoT sensors are producing too much sensing data as industrialization increases. AI is a powerful analytic tool for big data processing, providing scalable and accurate results in real time. However, centralized architecture, security, privacy, resource limits, and a lack of training data make AI-based big data analysis tool design and development difficult. Blockchain, a new technology, fosters decentralization. It encourages secure data and resource sharing between IoT network nodes to reduce centralized control and solve AI problems. The research aims to create a blockchain-AI IoT infrastructure for big data processing. To effectively integrate blockchain technology with AI for the Internet of Things, this article suggests a Blockchain-enabled Smart IoT Architecture that makes use of state-of-the-art methods and tools. As part of its quantitative and qualitative assessments, the suggested design defines "AI-driven Blockchain" and "Blockchain-driven AI" as they pertain to Internet of Things (IoT) systems. In order to compare studies on precision, delay, privacy and security, complexity of computation, and cost of energy in IoT applications, we objectively analyze the Block IoT Intelligent architecture. This evaluation covers gadget, fog, edge, and cloud intelligence. When compared to current Internet of Things (IoT) systems, the results show that the suggested architecture performs better and solves more problems.

The growth of intelligent services and apps driven by AI is expanding the reach of the Internet of Things (IoT) into a variety of people's everyday lives, as stated by the authors in (Baccour et al., 2021). Because of issues with data privacy and the scalability of IoT networks, traditional AI methods that rely on centralised data collection and processing may not function in practical application settings. The federated learning (FL) approach enables a plethora of intelligent IoT applications by facilitating AI training at dispersed IoT devices without requiring data transmission. From the inception of FL and the Internet of Things (IoT) to their integration, we cover every angle of FL's expanding use in IoT networks. Data sharing, unloading, caching, identifying attacks, location, mobile crowdsensing, privacy and security are some of the Internet of Things (IoT) services that we highlight as having the ability to be enabled by FL. Next, we take a close look at FL's applications in smart industries, smart cities, UAVs, smart transportation, and smart healthcare. The crucial lessons learned about FL-IoT services and applications are highlighted. At the end of this study, we will go over some of the present problems and future directions for research in this exciting new area (Rawat, 2023).

IOT AND AI: FUNDAMENTALS AND PRINCIPLES

This class of gadgets can include the now commonplace product such as telephones, wearables, household apparatuses, automobiles, manufacturing tools, transport and structures. The architecture of the Internet of Things typically consists of four primary layers: Sensors, actuators, and embedded systems are the make-up of the perception layer; it is an extensive range. These are functional components which enable the organization to collect information from the external environment (Nguyen et al., 2021). These devices are equally able to monitor many aspects such as location, motion, temperature, and Humidity. Information that is collected by the devices can be transferred to the cloud or a central hub easier due to the network layer that facilitates. Some of the communication that it uses include cellular networks, Bluetooth, Zigbee, and Wi-Fi among others (Hao et al., 2021).

Data that is accumulated by gadgets can be communicated to the centralized system or cloud more easily thanks to the network layer, which is responsible for making this possible (Campolo et al., 2021). Technology like as cellular networks, Bluetooth, Zigbee, and Wi-Fi are among the communication tools and protocols that it employs.

The cloud and platform layer is responsible for storing, processing, and analyzing the massive amounts of data that are produced by Internet of Things devices (Muhammad & Alhussein, 2021). For the purposes of data management, analytics, and application development, cloud platforms provide scalable infrastructure as well as tools. Applications and services that utilize data make up the application layer from the Internet of Things for specific objectives such as the automation of smart homes, the monitoring of businesses, or the sensory analysis of the environment (Yu et al., 2022).

Figure 2. The overall order of conventional artificial intelligence

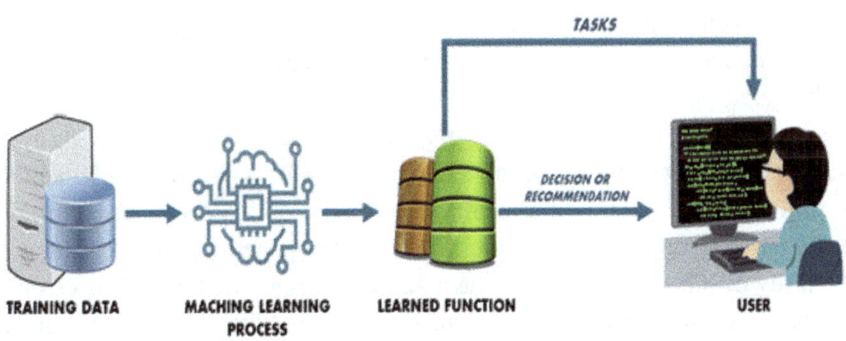

The phases involved in traditional artificial intelligence, also known as classical or symbolic AI, are depicted in Figure 2. These steps include general processing and problem-solving. An outline of the typical flow is shown in the following context:

Defining the problem that the artificial intelligence system is supposed to solve is the first step in the process. To accomplish this, it is necessary to have an understanding of the problem area, to determine the specific goal or objective, and to establish the criteria for success. Knowledge Acquisition: In traditional artificial intelligence, the process of acquiring knowledge involves collecting information about the topic matter from many sources, such as books, databases, experts, or other sources. In most cases, this information is communicated in the form of facts, rules, or claims that are logical. In the process of knowledge representation, the artificial intelligence system is able to process the information that it has learned after it has been organized and represented. In this context, formal languages may be utilized (Alberti et al., 2019). Some examples of these include logic-based representations, frames, semantic networks, and ontologies.

INFERENCE AND REASONING

The artificial intelligence system use reasoning and inference in order to arrive at make or conclusions the judgments by utilizing the information that it has learnt and the problem that it has identified. The use of logical principles, deductive reasoning, or probabilistic reasoning is often required in order to arrive at conclusions that are logical or probabilistic. Planning and Problem-Solving: In certain applications of artificial intelligence, planning and problem-solving are essential initial steps. To accomplish this, one must devise a sequence of steps or actions that will allow one to finish a certain work while taking into account the resources available, the constraints, and the outcomes that are sought. AI system generates a plan or a solution, it can either put it into action by actually implementing it or by simulating the process to ensure accuracy (Mastorakis et al., 2020).

At this stage, physical systems, robotics, or the modeling of actions and their consequences are all on the table. In terms of assessment and comments, the AI system looks at the outcomes or the accomplished objectives following the implementation of the strategy or solution. Using feedback, we can examine performance, find out how efficient the method is, and perhaps enhance the approach for further iterations. The efficacy of the system can also be assessed through feedback. It is critical to realize that this general order is not exhaustive and may differ based on the specific issue domain, AI techniques, and methodologies applied. Because conventional AI can't handle complex and unpredictable real-world problems, new paradigms in AI like machine learning and deep machine learning have arisen. Among the most important topics related to the Internet of Things are features include interoperability, scalability, real-time analytics, and device management. Through the Internet of Things (IoT), it is possible to connect physical items into the digital world in a seamless manner, which ultimately results in a networked environment that can be utilized for a range of reasons (Shafique et al., 2020).

The Many AI Approaches, and Methods Computers that can mimic human intelligence are able to carry out a wide variety of jobs hitherto reserved for people, a phenomenon known as "artificial intelligence" (AI). It is possible for artificial intelligence systems to perceive, reason, learn, and make decisions on their own. Artificial intelligence can be divided into numerous categories, including the following:

- AI that is limited: Artificial intelligence systems that are too narrow, often called weak AI, are created to solve very specific issues or execute very specialized jobs. Such technologies include methods for image identification, voice assistants, and engines for recommendations, among others.
- AI in general: Across a wide range of activities and topics, the goal of general artificial intelligence is to display intellect that is comparable to that of humans. It has capacities that are comparable of humans in terms of comprehension, learning, and knowledge application. Achieving comprehensive general artificial intelligence remains a challenging task.
- The primary goal of machine learning, a subfield of artificial intelligence, is to automate the collection of data and the improvement of computer systems' performance through the application of specific algorithms and approaches. All three types of learning supervised, reinforced, and unsupervised fall under one umbrella.
- "Deep learning" is shorthand for a branch of machine learning that uses multilayered artificial neural networks to find hidden meaning in data. Artificial intelligence and the processing of natural languages are only two of the many fields that have benefited from its use.

IoT AND AI SYNERGIES AND COMPLEMENTARITIES

The combination of AI and the Internet of Things (IoT) produces a smart environment because of the complementary and synergistic effects of the two technologies: Through their connections to the IoT, everyday objects collect and generate vast amounts of data about their environments. This information can be analyzed and processed using machine intelligence and other AI approaches to provide useful predictions, discoveries, and patterns for the future (Alankar et al., 2021). It is now feasible to make decisions all at once with the aid of machine learning algorithms that can send information from the Internet of Things in real time. By evaluating data collected by Internet of Things sensors and making real-time adjustments to the timing of traffic lights, machine learning can be utilized to enhance traffic flow in a smart city environment. Internet of Things devices may automate operations and adapt to varying conditions simultaneously by employing intelligent automation (AI) (Yarlagadda, 2018). As an example, AI-driven systems in industrial automation can optimize machine performance by monitoring it using IoT sensors and then adjusting operations automatically. This means is that AI systems may use IoT data to essentially predict what the future holds. This capability is useful for such ideas as sales forecast or break-down prediction as artificial intelligence can use the sensor data expect issues in equipment and arrange for measures so as to be prepared for the maintenance. Adaptability and Personalization: The information provided by the internet of things AI algorithms can adjust interface to different user and make systems more perceptive about individual preferences (Balas et al., 2020). As an example, there are already AI-controlled smart home systems that can monitor IoT device use patterns and adjust settings like temperature and lighting accordingly. The combination of the IoT's data-creation powers with AI's decision-making and analysis capabilities can lead to new opportunity and effective advances in many different fields.

INTEGRATED FRAMEWORK FOR SMART ENVIRONMENTS

The development of smart grids that maximize efficiency in power generation, distribution, and consumption is made feasible by the Internet of Things. This is referred to as smart energy management. Additionally, it enhances the management of inventory, enhances the efficiency of logistics and transportation, and reduces inefficiencies such as delays and losses. The effective monitoring of energy consumption, the implementation of demand response plans, and the incorporation of renewable energy sources are all made possible this way (Ma et al., 2020). Monitoring of the Environment: Internet of Things sensors have the capability to monitor and collect data on a variety of environmental elements, such as the levels of noise, the quality of the water, and the qualities of the air (Haseeb et al., 2025).

The data analysis and analytics layers are responsible for processing and examining the collected data. Machine learning and deep learning are two methodologies within artificial intelligence utilized for deriving insights, recognizing patterns, and generating predictions. Big processing methods can be employed to manage the substantial volumes of data produced through Internet of Things gadgets (De Micco et al., 2025). Using the data that has been examined and the AI models, the decision-making and automation layer is responsible for making judgments and initiating certain automated activities (Alberti et al., 2019). While automation mechanisms are responsible for directing actuators and Internet of Things devices to carry out the operations that are required, real-time decision-making algorithms make it possible to respond quickly to changing situations (Alberti et al., 2018).

The layer of integration: The integration layer is responsible for ensuring that the integrated framework is able to operate in conjunction with and is connected to the infrastructure and systems that are already in place. Integration with business software, cloud platforms, legacy systems, and other data sources is required for this purpose in order to make more efficient use of the resources that are easily accessible.

Figure 3. Crucial Internet of Things

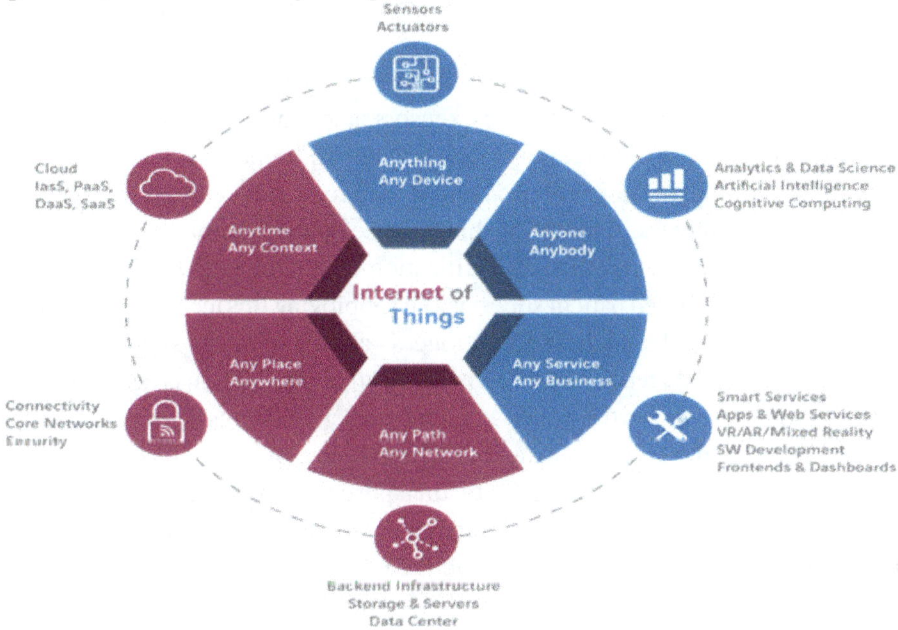

The Internet of Things (IoT) possesses extensive potential applications across various domains. Numerous significant Internet of Things applications are augmenting overall productivity and instigating transformations across various industries, as illustrated in Figure 3. The Internet of Things (IoT) now enables the automation and control of many home devices and systems. Homeowners can remotely control various domestic gadgets, such as lights, thermostats, and security systems, through linked devices or cellphones. The Industrial Network of Things is referred to as IIoT.

The IIoT links several pieces of industrial machinery, sensors, and other devices in order to gather and analyze data in real-time (Santana et al., 2018). This makes it possible to perform predictive maintenance, which in turn optimizes operations and boosts productivity in industries such as manufacturing, energy, transportation, and agriculture. Applications of the Internet of Things in healthcare make it possible to collect health data in real time, to use intelligent medical devices, and to monitor patients remotely. Patient care is improved as a result of this technology since it enables medical practitioners to monitor vital signs, provide tailored treatment, and ensure rapid actions. The some of major IoT applications that are creating revolutions in various sectors and enhancing productivity in large are: It is now possible to automate and control most of the household equipments and systems with the help of Internet of Things or IoT. Homeowners are capable of working remote lighting, thermostats, security systems, appliances, and other home appliances with the use of linked devices or smartphone. IIoT is the acronym that denotes Industrial Internet of Things (Sezer et al., 2018).

Real-time data collection and analysis is the concept of the Industrial Internet of Things (IIoT) which connects machines, sensors, and other industrial equipment's. This makes it possible to do predictive maintenance and hence leading to efficiency in operations and enhancement of productivity in production, energy, transport and even agricultural sectors. Internet of things in healthcare enable the acquisitions of health information in real time, the utilization of smart health products and services, and the home-based patient supervision. Employing this technology enhances patient care because; the medical practitioner is able to track the signs, administer treatment from the data received and know what immediate action is necessary to be taken. Smart cities refer to the notions of the Internet of Things acting as the major factor for boosting the productivity and eco-friendliness of large cities (Veena et al., 2025). Smart grids, intelligent modes of transportation, waste and environmental management, and security-supporting apps are just a few of the innovations that emerge from the interconnected infrastructure (Rehman et al., 2020).

In precision agriculture, IoT sensors in conjunction with drones along with data collection and analytics are used for purposes of health condition of crops, irrigation, animal tracking, and conditions of the soil. Because of this, farmers are able to increase the yields in their farms by practicing modern farming techniques which do not require a lot of resources and end up wasting them while making sound decisions based on facts that are made available to them by technology. Due to IOT it has become feasible to trace and track product in real time starting from supply chain to consumption. Thanks to Internet of Things, it can be created smart grid systems in order to generate, transmit and use electricity more efficiently. This is referred to as smart energy management (Liu et al., 2020). Further, it optimizes stocks, optimizes the processes of shipping and receiving, and minimizes such wastes as downtimes and damages. This way, it is possible to monitor the energy consumption, propose and manage demand response plans and integrate renewable energy sources. Monitoring of the Environment: Various Internet of Things sensors have the ability to detect as well as record. gather information on different elements of the environment including the levels of noise, quality of the water and even the quality of the air (Guo et al., 2018). The application of this information helps decision making on policies easier and ways and means on how to handle the risks that come with the environment. Wearable technology and individual fitness: Various gadgets for example fitness trackers and smart watches that are IoT enabled can also capture details of a person's health and fitness. They give data about such things as a pulse, sleep pattern, and exercises, which in turn assists people in tailoring their mode of living. Connected shelf and smart inventory are Internet of Things gadgets which are integrated into smart retail (Wang et al., 2021).

Deep Learning: Deep learning is highly effective for analyzing extensive quantities of complex data due to its utilization of neural network architectures (Chelloug & El-Zawawy, 2017). Consequently, deep learning is exceptionally efficacious. Unlike convolutional artificial neural networks (CNNs), which are primarily employed for the analysis of images, recurrent neural network networks (RNNs) are appropriate for processing time-series data.

Figure 4. The real-time decision-making and automation

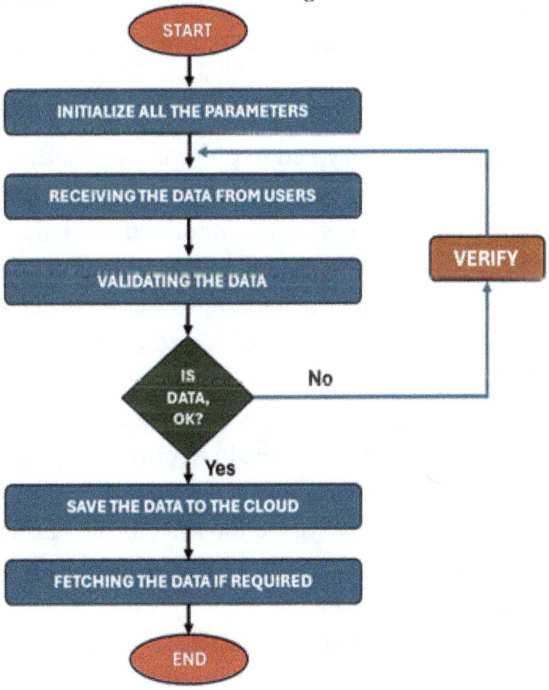

Decision-Making and Automation in Real-time Smart settings are able to respond rapidly to shifting situations as a result of real-time decision-making and automation as shown in Figure 4. Using artificial intelligence algorithms and rule-based systems, judgments are formed based on the data that has been processed, and these decisions then begin automatic activities. There are numerous scenarios in which real-time decision-making and automation can be utilized (Balachandra Achar & R. N, 2025). One example of this is the implementation of traffic management systems that dynamically alter traffic lights based on information regarding the flow of traffic at the moment. equipment for industrial automation that can make adjustments to the parameters of machines in real time in order to improve production lines. regulating energy use in smart homes based on occupancy patterns and the time of day are systems that are included in smart homes.

Scalability, handling of information, and analytical capabilities are all made easier with the combination of cloud-based services with contemporary data processing and storage technologies. Through this integration, data created by the Internet of Things can be processed and stored using resources and services available in the cloud (Mohanty et al., 2016).

Important Things to Keep in Mind Regarding Security and Privacy: Concerns around security and privacy arise as a result of smart environments' use of machine learning and the Internet of Things. To ensure the protection of private data and the reliability of the system, the following elements must be considered:

- Encryption of Data: It is advised that data sent between IoT gadgets and the combined structure should be encrypted to prevent unauthorized access.
- Access Control: It is essential to provide robust access control methods in order to guarantee that only appropriately authorized individuals are able to interact with the Internet of Things devices and the integrated framework.
- Authentication of devices: Before connecting to the integrated framework, Internet of Things devices should be authenticated in order to prevent unauthorized devices from gaining access to the system. When it comes to data privacy, it is important to establish policies and procedures that regulate the collection, storage, and exploitation of sensitive or personal data that is produced by Internet of Things devices.
- Network Security: The firewalls and detection systems for intrusions are two types of security measures for networks that must be established to safeguard against unauthorized access and attacks.
- Security Patches and Regular Updates: In order to address any vulnerabilities that may exist, it is important that the integrated framework and Internet of Things devices receive regular updates and security patches. When these considerations are taken into account, the integrated framework is able to ensure that smart environments are operated in a secure and confidential manner, which in turn increases the level of user confidence in the system.

APPLICATIONS OF THE INTEGRATED FRAMEWORK

Smart cities utilize the Internet of Things and the artificial intelligence framework to improve several facets of urban life.

Integration with Existing Infrastructure and Infrastructure Systems: There is a need for IoT and Artificial Intelligence to be integrated and merged with the present environment so that it will be harmonious and compatible. Through such approaches as system integration of the existing infrastructure, applications and data sources, functionality and efficiency can be enhanced. The following are important factors that determine integration: Others have taken its meaning to mean "Application Programming Interface". Through application programming interface or API the connected framework and other current systems can communicate and share information. This is especially useful since both the coordination or integration of the operations of different platforms and the integration of information from a number of sources are enabled by this. Integration with Legacy Systems: The term Integration with Legacy Systems means enrichment involved in the current systems or addition of new systems to support the Internet of Things devices as well as the utilization of artificial intelligence. This may entail making modifications or adding new modules to allow the connection and exchange of information depending on the situation that prevails. The adoption of cloud platforms with the traditional storage and computing systems makes it possible to have scalability, data management and analysis at a given instance. We have seen that through integration, the data that is produced by the Internet of Things is processed and stored utilizing cloud services. Some Things to Think About Regarding Privacy and Safety: Smart space and Internet of Things with the help of AI and integration generates the issues of privacy and security.

The integrated framework contributes to the advancement of agriculture by maximizing the utilization of resources, crop management, and livestock monitoring. In order to facilitate precision farming, Internet of Things sensors are installed in the fields. These sensors collect data on the moisture content, nutrient content, and environmental aspects of the soil. AI algorithms are used to process this data in order to improve pest management, fertilization, and irrigation, which ultimately results in an increase in crop yield and a reduction in the amount of resources that are wasted.

Smart irrigation utilizes Internet of Things devices and artificial intelligence analyses to optimize watering schedules according to plant specifications, forecasts for the weather, and soil moisture levels. As a consequence of this, water is utilized more efficiently, there is less water that is wasted, and the health of the crop is enhanced. Through the use of Internet of Things devices such as GPS trackers and wearable sensors, it is feasible to do real-time monitoring of the health, behavior, and position of cattle. The data are then examined by artificial intelligence systems in order to identify anomalies, forecast disease outbreaks, and enhance strategies for livestock management.

The following considerations are essential for the aim of protecting private data and the system's integrity: Security of Data: Therefore, it is suggested that data transferred between the Internet of Things gadgets and integrated framework must be encrypted.

Access Control: To ensure that only the right people get to engage the IoT systems and the integrated architecture it is necessary to incorporate strong signal access control mechanism that shall only allow: Authentication of devices: In other words, every device to be connected to the integrated framework should go through an identification process to ensure only right devices access the system. Concerning data privacy, there exists a need to formulate measures on how the information generated and managed by the Internet of Things devices should be collected, stored, or exploited. Network Security: Security and access control mechanisms such as firewalls and IDS are two of the methods that should be put in place in a network to prevent web accesses and other forms of unlawful attacks. Security Patches and Regular Updates: To eliminate any possibilities of risk, it is invariably relevant that the integrated framework and the IoT devices should be updated frequently and should be guarded with updated security layers. When these factors are considered, the integrated framework is effective in making certain that smart environments are run in a secure and an optimal way of keeping the data a secret from other people hence enhancing the level of confidence in the system among the users.

Therefore, smart cities extend the use of IOT and AI in order to improve a number of affairs of urban life. Internet of Things (IoT) sensors and artificial intelligence algorithms can perform smart transportation activities which include constant traffic surveillance, estimation of traffic densities and optimal signal timing. It increases throughput and reduces congestion, enhances the environment for driving a car and optimizes public transport. Intelligent Energy Management: The IoT grows energy use reduction capabilities, pattern recognition, and the integration of renewable energy through AI analytics to energy management. This helps to mitigate inefficient use of energy and hence improve on the energy efficiency. Smart garbage can IoT sensors can view fill levels and improve waste collection routes so that the garbage cans are not full. This contributes to lowering of the number of avoidable trips and costs of operation. Besides, many AI systems can analyze data and recognize pattern for suggestion of the possible ways and means for minimizing the generation of wastes. Systems that involve the use of Internet of Things and artificial intelligence could improve public safety in different ways such as predictive policing, video surveillance and in optimizing disaster response. Thus, by applying real-time data analysis, there is a possibility to assess the possible security threats as well as to take appropriate and adequate measures in due time. Some of the uses of the integrated framework in the area of the healthcare business permit the practice of differential management of patients, deliver high-quality care to the patients, and improve the overall effectiveness of the operations in healthcare business.

ETHICS AND GOVERNANCE

As Artificial Intelligence (AI) and the Internet of Things (IoT) become more deeply embedded in everyday life, ethical concerns and governance have taken on increasing importance. These technologies can collect and process enormous amounts of information, which allows for powerful insights and automated decision-making but also introduces risks. Privacy, security, and accountability are central challenges. Without careful oversight, personal data can be misused, and decisions made by automated systems can unintentionally harm individuals or communities.

Addressing these challenges requires more than technical fixes. Ethical considerations must be integrated into every stage of design and development. Developers need to prioritize transparency, fairness, and inclusivity, ensuring that systems do not reinforce social biases or exclude certain groups. In the case of IoT, this means implementing strong protections for personal data and giving users clear control over how their information is used. For AI, it involves creating systems whose decisions can be traced and understood, so that they can be reviewed or corrected when necessary.

Governance frameworks complement these efforts by providing standards, regulations, and best practices. Governments, industry bodies, and international organizations play an important role in shaping policies that encourage innovation while protecting society. Public engagement and accountability mechanisms are equally essential, giving communities a voice in determining how these systems are used. Embedding ethical principles and strong governance into AI-powered IoT systems, it is possible to harness their potential safely. This approach ensures that future smart technologies are not only effective and intelligent but also responsible, trustworthy, and aligned with human values.

CHALLENGES AND OPPORTUNITIES

Problems with Scalability and Interoperability: Scaling Internet of Things installations and integrating a wide variety of devices, protocols, and platforms can be challenging because of the vast volume of data and devices involved. It is possible for interoperability issues to arise when equipment built by different manufacturers or utilizing different communication protocols are required to work together without any interruptions.

It is possible that scalability and interoperability can be improved by the standardization of frameworks and their protocols. The use of open standards and protocols makes it possible to integrate and communicate with devices, systems, and platforms without any interruptions. It is possible that attempts to achieve interoperability could be bolstered by collaboration between Standards organizations and industry stakeholders.

Challenges Related to Data Management and Processing: The Internet of Things (IoT) devices provide issues in storing, analyzing, and analysis due to the substantial volume, rapid generation, and diverse nature of the data they produce. Traditional data management solutions may be inadequate for managing the substantial volume and complexity of data created by the Internet of Things.

The application of big data, cloud computing technology, and edge computing to address data management challenges is a potential prospect. Cloud platforms provide scalable storage as well as processing capabilities, whereas edge computing minimizes latency and bandwidth requirements by relocating computation nearer to the data source. Enhancements can be implemented in AI algorithms to satisfy the processing demands of data generated by the Internet of Things (IoT). Security and privacy are important issues with IoT devices because of the cyber dangers they face, such as hacking, data leaks, and hostile attacks. Privacy issues are also brought up by the gathering and processing of personal or sensitive data.

Opportunity: The implementation of stringent security measures, including encryption, authentication techniques, and secure protocols, is essential to safeguard the data and devices linked to the Internet of Things (IoT). Technologies that enhance privacy, such as data anonymization and consent management, can be utilized to address concerns regarding security and confidentiality. An ongoing commitment to research and collaboration is essential in order to stay one step ahead of the ever-evolving cybersecurity threats.

Concerns Regarding Ethical Behavior and Responsible Artificial Intelligence the AI algorithms that are a part of the integrated framework ought to adhere to ethical standards of fairness and transparency. The potential for employment displacement, algorithmic bias, and accountability in decision-making are all challenges that need to be addressed.

Creating ethical paradigms, standards, and guidelines for the application of artificial intelligence in smart environments is an opportunity that must be taken advantage of Ethical AI principles such as explainability, justice, and human oversight should be incorporated into the process of developing and utilizing artificial intelligence technologies. It is absolutely necessary for politicians, industry, and academic institutions to work together in order to address these ethical concerns.

Deficiencies in Skills and Training Requirements within the Workforce: Confrontation: Successful integration of the Internet of Things with artificial intelligence necessitates a workforce proficient in developing, implementing, and maintaining these intricate systems. There can be a dearth of experts in fields related to artificial intelligence (AI), the Internet of Things (IoT), and their overlaps. Possibility: Learning programs, training initiatives, and interdisciplinary collaboration can effectively bridge skill gaps, allowing us to address this issue. Learners can acquire the information and skills needed to interact with IoT and AI technologies through educational programs offered by universities, corporate collaborations, and online learning platforms.

Business Models and Their Economic Implications: Degree of difficulty: The integration of the Internet of Things and machine learning may necessitate significant expenditures in gadgets, platforms, and infrastructure. Assessing the viability of business models, return on investments, and income streams in emerging applications and industries can be tough.

It is possible to discover new economic opportunities by conducting research on innovative business models. Some examples of these models are data monetization, service-oriented approaches, and value-added services. It is possible for stakeholders, such as technology companies, business leaders, and legislators, to contribute to the promotion of sustainable economic growth through collaborative efforts.

In order to completely realize the possibilities of smart circumstances, it is vital to overcome these challenges and exploit the opportunities given by the interconnected Internet of Things and intelligent machines framework. The only way to make smart surroundings deliver on their potential. Integration of the Internet of Things (IoT) and machine learning (AI) will need ongoing research, new ideas, and cooperation across many fields and stakeholders if it is to overcome these obstacles and make the most of AI's benefits.

FUTURE DIRECTIONS AND RESEARCH AGENDA

Explainable AI, 5G connectivity, learning federation, the Internet of Medical Things (IoMT), and edge computing are a few examples of the new developments in AI and the Internet of Things. Research can investigate and make use of these technologies. In order to stimulate further growth, it is necessary to investigate the possibilities offered by these technologies and to appreciate the implications that they have for intelligent settings.

Creating and implementing standardized protocols, frameworks, and architectures for the integration of artificial intelligence with the internet of things can be facilitated by research. Both regulatory frameworks and efforts to standardize are being made. Collaborative efforts with standardization bodies, industry participants, and legislators can assure the preservation of compatibility, confidentiality of data, and security in smart environments. Studying enables a comprehensive understanding of the rules and regulations governing the legal and responsible implementation of machine learning in Internet of Things platforms. When thinking about the challenges and limitations of integrating AI with the IoT, many issues must be considered. These include issues of scaling, connectivity, organizing data, security, and privacy. Smart environments can run smoothly and safely with the help of newly researched methods, tools, and algorithms that solve these problems. To maximize the integrated framework's potential, researchers can also look into state-of-the-art processing of data, resource utilization, and system reliability technologies.

Research has the ability to highlight the need of designing smart settings with users in mind, with a focus on usability, accessibility, and user experience, as well as human-centered approaches. Research on user habits, views, and opinions has the potential to pave the way for more personalized services, adaptable systems, and interfaces that are easy to use. The advancement of people-centric methodologies in intelligent surroundings may also be shaped by investigations into the social, cultural, and ethical implications of integrating IoT with AI.

Collaborative research across several fields: Interdisciplinary studies involving the integration of AI and the internet of things can be facilitated by the development of collaboration among investigators, practitioners, and industry experts from other domains. Smart environment innovation can be enhanced by interdisciplinary collaboration in engineering, computer science, social science, healthcare, and design. This can also provide you a full picture of what the integrated framework can do. Research that incorporates different fields of study can tackle intricate problems and provide all-encompassing solutions.

The integration of the Internet of Things (IoT) and machine learning (AI) for smart settings can continue to advance if researchers follow these future paths and prioritize these areas of study. Application efficiency, environmental impact, and user experience will all see significant improvements as a result. Ongoing research and collaboration are crucial for managing the ever-changing demands and obstacles of smart environments and making full use of the combined structure of the Internet of Things and machine learning.

RESULTS AND DISCUSSION

This research presents an integrated framework leveraging IoT and AI to create smart environments, significantly enhancing efficiency, productivity, and quality of life in smart cities, healthcare, transportation, and agriculture. Key findings include a comprehensive understanding of IoT and AI, detailed framework architecture, and practical applications across sectors. The study addresses critical challenges like scalability, interoperability, data management, security, privacy, ethics, and economic implications, proposing solutions and emphasizing standardization and interdisciplinary collaboration.

Implications:

- Efficiency: Intelligent automation and resource optimization boost productivity.
- Quality of Life: Enhanced public services and urban living.
- Sustainability: Improved energy use, waste reduction, and resource management.
- Personalization: AI-driven customization enhances user experiences.
- Economic Growth: Data-driven innovations and new business models foster development.

CONCLUSION

Intelligent machines (AI) and the Internet of Things (IoT) collaborate in the comprehensive smart environment framework to create responsive, efficient, and smart spaces. All of this is made possible by the framework's design and components. It connects to existing infrastructure, automates operations, makes choices in real time, processes data, and does analytics. It is talked about that worries about privacy, security, scaling, and interoperability are all taken care of. Intelligent cities, health care, agriculture, transportation, and industry automation are just some of the fields that can use the framework.

Important Findings and Contributions from the Research: This research paper offers a thorough analysis of the topic by combining the Internet of Things (IoT) with artificial intelligence (AI). The following are examples of important findings and contributions: Familiarity with the terminology, classifications, and processes of artificial intelligence and the Internet of Things has been achieved, by analyzing the similarities and differences between the two technologies demonstrates the combination of IoT and AI can improve automation, judgment, and efficiency. The practical applications in various industries will be investigated in this chapter, with a focus on smart cities, medical care, agriculture, transportation, and industrial automation.

REFERENCES

Alankar, B., Kaur, H., & Chauhan, R. (Eds.). (2021). *Transforming the Internet of Things for Next-generation Smart Systems*. IGI Global. DOI: 10.4018/978-1-7998-7541-3

Alberti, A. M., Santos, M. A., Souza, R., Da Silva, H. D. L., Carneiro, J. R., Figueiredo, V. A. C., & Rodrigues, J. J. (2019). Platforms for smart environments and future internet design: A survey. *IEEE Access : Practical Innovations, Open Solutions*, *7*, 165748–165778. DOI: 10.1109/ACCESS.2019.2950656

Alberti, M., Bontempo, M. M., Santos, J. R. D., Sodré, A. C., & Righi, R. da R. (2018). NovaGenesis applied to information-centric service-defined trustable IoT/WSAN control plane and spectrum management. *Sensors (Basel)*, *18*(9), 3160. https://www.mdpi.com/1424-8220/18/9/3160. DOI: 10.3390/s18093160 PMID: 30235844

Baccour, E., Mhaisen, N., Abdellatif, A., Erbad, A., Mohamed, A., Hamdi, M., & Guizani, M. (2021). Pervasive AI for IoT Applications: A Survey on Resource-Efficient Distributed Artificial Intelligence. *IEEE Communications Surveys and Tutorials*, *24*(4), 2366–2418. DOI: 10.1109/COMST.2022.3200740

Balachandra Achar, H. V., & R. N. (2025). The impact and future of artificial intelligence in learning. In *2025 3rd International Conference on Integrated Circuits and Communication Systems (ICICACS)* (pp. 1–6). IEEE. DOI: 10.1109/ICICACS65178.2025.10967881

Balas, V. E., Kumar, R., & Srivastava, R. (Eds.). (2020). *Recent trends and advances in artificial intelligence and internet of things*. Springer International Publishing. DOI: 10.1007/978-3-030-32644-9

Campolo, C., Genovese, G., Iera, A., & Molinaro, A. (2021). Virtualizing AI at the Distributed Edge towards Intelligent IoT Applications. *J. Sens. Actuator Networks*, *10*(1), 13. DOI: 10.3390/jsan10010013

Chang, Z., Liu, S., Xiong, X., Cai, Z., & Tu, G. (2021). A Survey of Recent Advances in Edge-Computing-Powered Artificial Intelligence of Things. *IEEE Internet of Things Journal*, *8*(18), 13849–13875. DOI: 10.1109/JIOT.2021.3088875

Chelloug, S. A., & El-Zawawy, M. A. (2017). Middleware for internet of things: Survey and challenges. *Intelligent Automation & Soft Computing*, •••, 1–9. DOI: 10.1080/10798587.2017.1290328

De Micco, F., Di Palma, G., Ferorelli, D., De Benedictis, A., Tomassini, L., Tambone, V., Cingolani, M., & Scendoni, R. (2025). Artificial intelligence in healthcare: Transforming patient safety with intelligent systems—A systematic review. *Frontiers in Medicine*, *11*, 1522554. DOI: 10.3389/fmed.2024.1522554 PMID: 39845830

Guo, H., Ren, J., Zhang, D., Zhang, Y., & Hu, J. (2018). A scalable and manageable IoT architecture based on transparent computing. *Journal of Parallel and Distributed Computing*, *118*, 5–13. DOI: 10.1016/j.jpdc.2017.07.003

Hao, Q., Nazir, S., Li, M., Khan, H., Wang, L., & Ahmad, S. (2021). AI-Enabled Sensing and Decision-Making for IoT Systems. Complex., 2021, 6616279:1-6616279:9. DOI: 10.1155/2021/6616279

Haseeb, K., Saba, T., Rehman, A., Abbas, N., & Kim, P. W. (2025). AI-driven IoT-fog analytics interactive smart system with data protection. *Expert Systems: International Journal of Knowledge Engineering and Neural Networks*, *42*(1), e13573. DOI: 10.1111/exsy.13573

Kankanhalli, A., Charalabidis, Y., & Mellouli, S. (2019). IoT and AI for Smart Government: A Research Agenda. *Government Information Quarterly*, *36*(2), 304–309. DOI: 10.1016/j.giq.2019.02.003

Liu, Y., Yuan, X., Xiong, Z., Kang, J., Wang, X., & Niyato, D. (2020). Federated learning for 6G communications: Challenges, methods, and future directions. *China Communications*, *17*(9), 105–118. DOI: 10.23919/JCC.2020.09.009

Ma, Y., Ping, K., Wu, C., Chen, L., Shi, H., & Chong, D. (2020). Artificial Intelligence powered Internet of Things and smart public service. *Library Hi Tech*, *38*(1), 165–179. DOI: 10.1108/LHT-12-2017-0274

Mastorakis, G., Mavromoustakis, C. X., Batalla, J. M., & Pallis, E. (Eds.). (2020). *Convergence of artificial intelligence and the Internet of Things*. Springer International Publishing. DOI: 10.1007/978-3-030-44907-0

Mohanty, S. P., Choppali, U., & Kougianos, E. (2016). Everything you wanted to know about smart cities: The Internet of things is the backbone. *IEEE Consumer Electronics Magazine*, *5*(3), 60–70. DOI: 10.1109/MCE.2016.2556879

Muhammad, G., & Alhussein, M. (2021). Convergence of artificial intelligence and Internet of Things in smart healthcare: A case study of voice pathology detection. *IEEE Access : Practical Innovations, Open Solutions*, *9*, 89198–89209. DOI: 10.1109/ACCESS.2021.3090317

Nagaraj, A. (2023). *The Role of AI in Enhancing IoT-Cloud Applications*. Bentham Science Publishers., DOI: 10.2174/97898151657081230101

Nguyen, D., Ding, M., Pathirana, P., Seneviratne, A., Li, J., & Poor, F. (2021). Federated Learning for Internet of Things: A Comprehensive Survey. *IEEE Communications Surveys and Tutorials*, *23*(3), 1622–1658. DOI: 10.1109/COMST.2021.3075439

Rawat, R. (2023). Harnessing the Power of IoT and AI for Human Evolution. *International Journal of Research In Science & Engineering*, *33*(33), 58–68. DOI: 10.55529/ijrise.33.58.68

Rehman, M. H. ur, Salah, K., Damiani, E., & Svetinovic, D. (2020). Towards blockchain-based reputation-aware federated learning. In *Proceedings of the IEEE INFOCOM Conference on Computer Communications Workshops (INFOCOM WKSHPS)* (pp. 183–188). IEEE.

Santana, E. F. Z., Chaves, A. P., Gerosa, M. A., Kon, F., & Milojicic, D. S. (2018). Software platforms for smart cities: Concepts, requirements, challenges, and a unified reference architecture. *ACM Computing Surveys, 50*(6), 78:1–78:37.

Sezer, O. B., Dogdu, E., & Ozbayoglu, A. M. (2018). Context-aware computing, learning, and big data in Internet of Things: A survey. *IEEE Internet of Things Journal*, *5*(1), 1–27. DOI: 10.1109/JIOT.2017.2773600

Shafique, K., Khawaja, B. A., Sabir, F., Qazi, S., & Mustaqim, M. (2020). Internet of things (IoT) for next-generation smart systems: A review of current challenges, future trends and prospects for emerging 5G-IoT scenarios. *IEEE Access : Practical Innovations, Open Solutions*, *8*, 23022–23040. DOI: 10.1109/ACCESS.2020.2970118

Shetty, G. S., & N, R. (2024). Strategies for Achieving energy efficiency and data Security through data aggregation in IoT Healthcare Applications: A Comprehensive study. *International Journal of Computer Networks and Applications*, *11*(2), 191. DOI: 10.22247/ijcna/2024/224440

Singh, S., Rathore, S., & Park, J. (2020). BlockIoTIntelligence: A Blockchain-enabled Intelligent IoT Architecture with Artificial Intelligence. *Future Generation Computer Systems*, *110*, 721–743. DOI: 10.1016/j.future.2019.09.002

V. M. S. S. S., Krishnamurthy, T., Chaithra, R. N., & S. G. (2025). Advanced encryption strategies for protecting patient data in IoMT applications. In *2025 6th International Conference for Emerging Technology (INCET)* (pp. 1–6). IEEE. DOI: 10.1109/INCET64471.2025.11139937

Veena, V., Manjunath, R., Asha, V., Gangaraju, V., Swetha, G., & Raghu, N. (2025). Smart Homes with IoT Technology Provide Safety and Security for Elderly Individuals. In Mallik, S., Mathivanan, S., Sangeetha, S., & Soufiene, B. (Eds.), *Real-World Applications of AI Innovation* (pp. 425–450). IGI Global Scientific Publishing., DOI: 10.4018/979-8-3693-4252-7.ch020

Wang, Z., Qiu, J., Zhou, Y., Shi, Y., Fu, L., Chen, W., & Letaief, K. B. (2021). Federated learning via intelligent reflecting surface. *IEEE Transactions on Wireless Communications*, *21*(2), 808–822. DOI: 10.1109/TWC.2021.3099505

Yarlagadda, R. T. (2018). Internet of things & artificial intelligence in modern society. *International journal of creative research thoughts (IJCRT), ISSN*, 2320-2882.

Yu, W., Liu, Y., Dillon, T., Rahayu, W., & Mostafa, F. (2022). An integrated framework for health state monitoring in a smart factory employing IoT and big data techniques. *IEEE Internet of Things Journal*, *9*(3), 2443–2454. DOI: 10.1109/JIOT.2021.3096637

Zhang, J., & Tao, D. (2020). Empowering Things With Intelligence: A Survey of the Progress, Challenges, and Opportunities in Artificial Intelligence of Things. *IEEE Internet of Things Journal*, *8*(10), 7789–7817. DOI: 10.1109/JIOT.2020.3039359

Chapter 7
Optimized Approaches in Data Security and Privacy Using IoT Devices for Smart Cities

Sonu Sharma
http://orcid.org/0009-0004-3836-590X
Poornima University, Jaipur, India

Nikhil Kumar Goyal
http://orcid.org/0009-0007-4532-8033
Poornima University, Jaipur, India

ABSTRACT

This chapter presents a comprehensive analysis of methods for securing IoT networks in smart cities, addressing the technical, operational, and regulatory challenges unique to these environments. It begins by exploring the foundational role of IoT in enhancing urban efficiency, connectivity, and sustainability. Security threats, including device vulnerabilities, unauthorized access, and privacy breaches, are examined, highlighting the limitations of traditional models. Advanced solutions such as cryptographic techniques, secure connectivity, and distributed authentication are discussed. Privacy-preserving data processing approaches like differential privacy, federated learning, and secure multi-party computation are explored, ensuring a balance between data protection and system performance. The chapter also emphasizes the significance of AI and ML in threat detection, anomaly analysis, and proactive response. Edge and fog computing are highlighted for their potential in reducing latency and enhancing data privacy.

DOI: 10.4018/979-8-3373-4202-3.ch007

INTRODUCTION TO SMART CITIES AND IOT SYSTEMS

Defining Smart Cities

The idea of smart cities provides an incredibly new perspective on urban development and implementation of modern digital technology and data analytics to enhance well-being of resident; manage their resources effectively; and, promote environmental sustainability. (Albino et al., 2015) As a result, the concept of smart cities empowers a radical model for urban development Smart cities leverage their network of systems to increase operational efficiency, reduce the environmental footprints and support economic growth in your face of the urbanization challenges. The base for smart city achievement is the use of various sensors, communication networks and advanced analytics platforms for collecting and interpreting data in real time from areas ranging from traffic control, waste systems, distribution of power. By analyzing real-time information, cities obtain insights which facilitate agile decision making, enhanced service management and more robust urban resilience in change.

What makes smart cities different at its heart is not just a question of technological advancement itself. With the element of social, economic, and environmental factors it aims at constructuring societies that are inclusive, networked, and resilient to change. To realize this comprehensive vision, concerted effort in the country's public authorities, leaders in industry, educational interests and the locals themselves is needed, which in turn promotes an atmosphere of constant innovation and advancement.

Role of IoT in Smart City Infrastructure

The IoT is the backbone of smart cities allowing easy gathering, distribution, and analysis of data across the urban fabric. Subordinated streetlights, smart meters and other IoT also comprise the core systems that enable smart city networks (Alomair & Poovendran, 2014). The output of the data streams of these devices provide immediate and detailed analysis for urban environments ranging from the environmental factors such as air quality and cost of energy to operational facts such as traffic flows and emergency management.

When IoT connections are integrated into smart city infrastructure, cities get to benefit from a variety of benefits which include better public safety, effective energy consumption, superior transportation networks, and less environmental load. To provide another example, IoT empowered traffic lights can adjust their time-sequences in a instant based on the actual traffic situation that eases the traffic jam condition, thereby reducing the environmental footprint. Moreover, smart water networks which apply IoT can allow automatic detection of leaks, optimization of water supply itself, and reduction of waste, which increases sustainability in managing water resources. In spite of all the benefits, such progressive connectivity implies severe challenges regarding the data security, privacy, and systems' interoperability. The proliferation of connected devices is flashing the urgent requirement of ensuring the secure confidentiality, integrity, and availability of sensitive urban information. This requires the establishment of secure structures for cybersecurity, state of the art encryption systems, as well as networks infrastructure that can successfully withstand advanced threats (Anthopoulos, 2015).

Data-Driven Urban Management

Data driven urban management is at the heart of the smart city concept. It is a process which entails the application of live data analytics to make sound decisions, analyze future risk, and refine the delivery of public services. Through transforming raw data into usable insights, cities can increase operational efficiency, and solve the changing needs of its citizens effectively.

Smarter traffic systems can crunch relevant data from road sensors, CCTV cameras and connected vehicles in real time so as to be able to adjust signal timings, reduce congestion and promote road safety (Bartoli, 2011). Smart energy grids are able to adjust their power outputs dynamically at any given moment in time to zone in on the current demand drifting energy wastage and greenhouse gases to a minimum. Similarly, systems that tie into the waste management infrastructure give real time insights into fullness of the refuse bin, which makes possible more efficient route planning and reduces costs and ecological burdens (Centenaro et al., 2016). Moreover, not only does this strategy improve the activities going on, but it also becomes a foundation for long-term provisions for urban sustainability and expansion. Urban planners gain from such data; they are able to identify shifting patterns and forecast future resource needs and assess the impact of actions of policy. In other words, data-driven management promotes cities' ability to be nimble, bounce back from misfortune, and stay in the long term.

SECURITY CHALLENGES IN IOT-ENABLED SMART CITIES

IoT, forming the basis of smart city infrastructure, has created unimaginable possibilities to effect efficient city administration and improve interaction with citizens. However the wide-ranging nature of IoT makes it an overly weak security network that exposes urban systems to cyber attacks with the potential scale to threaten the safety and privacy of the public as well as the very integrity of cities. The complexities and the scale of the smart city infrastructures have led to the identification of high-level cybersecurity as a critical goal for city officials, tech developers as well as policy makers (Chourabi et al., 2012).

Data Breaches and Cyber Threats

In the era where almost everything connects to the Internet, even the possibility of cyberattacks on critical infrastructures and data leaks is an important security issue for IoT society-based security of smart cities. Be it small opportunistic cyberattacks and theft of data or highly advanced and meaningful intrusions of essential city services, compromising private details, the threats are numerous and substantial. Smart city breaches can wreak havoc on all components of life, including individual security, general safety, financial stability, and nation's security posture. For instance, having illicit insights into live traffic watching, such as emergency service platforms, or electrical installations control systems, may cause substantial disrupting which will compromise public safety and cause a great chaos (Dandotiya et al., 2026).

Moreover, the distributed and wide nature of the IoT network architecture means a robust and complex attack space is created that creates challenges to efficient monitoring, security and defense. Because of this complexity, cyberattackers often use such means as DDoS, MITM, and ransomware to gain access to IoT networks (Dandotiya et al., 2024).

A good example is the Mirai botnet attack of 2016 that demonstrated the enormous damage that can be caused if IoT devices are under threat. By compromising millions of insecure cameras, routers, and sensors, the perpetrators released a deadly DDoS attack that severed giant internet infrastructure for a brief period highlighting the need for more restrive measures to secure. Managing these challenges will require smart city developers to come up with complex security plan, which will highlight, strong encryption, a continuous network monitoring and advanced intrusion detection technology. Continuous assessments of security, prompt installations of patches, and an all around incident response process should also be given priority with a view of the damages that might occur from the security incidents.

Device Vulnerabilities and Attack Vectors

In addition to threats on network scales, IoT devices also put forward significant security issues on an individual scale. The built-in deficiencies in processing, memory as well as energy make it difficult to apply conventional cybersecurity practices to these devices. Developers who concentrate on producing IoT sensors and actuators at a low cost and quickly often forget security issues, resulting in outdated firmware and poor defenses for sensor/actuator attacks.

The risks at the levels of devices tend to arise from simple or default passwords, absence of secure protocols for communication and low levels of authentication mechanisms for connected devices. Such weaknesses allow attackers to get into devices, take control of their functions, or run malicious scripts, which may lead to total compromise of IoT networks. (Department of Economic and Social Affairs, 2014) In addition, the absence of remote update capacities in many IoT devices makes them vulnerable to relentless security problems because new vulnerabilities are unexpected from time to time. This concern is particularly difficult in critical infrastructures systems, as device disruption capability can compromise public safety and service reliability. The attack vectors, through which attackers target IoT devices, depend on the device's purposes and the environment that it operates on as well as the typical threat vectors, which often recur:

- **Physical Attacks:** Tampering with devices or intercepting signals through direct physical access.
- **Firmware Exploits:** Manipulating outdated or unpatched device firmware to gain control or extract sensitive data.
- **Side-Channel Attacks:** Exploiting unintended data leaks, such as electromagnetic radiation or power consumption patterns, to bypass security measures.
- **Supply Chain Attacks:** Introducing vulnerabilities during the manufacturing or distribution stages of IoT devices.
- **Botnet Infections:** Co-opting vulnerable devices into large-scale botnets for use in coordinated cyberattacks.

Addressing these challenges require synergizing robust device engineering, responsible production procedures and continued lifecycle management efforts. In order to prevent these threats it is highly essential to put in place a series of robust methods of authentication, secure transmission of data and secure boot mode to ensure devices are not compromised. In addition, manufacturers and city planners should work together in establishing concrete guidelines on security, promoting top security practices and incorporating security at all stages of the IoT development process, from planning through to the end-state of build and ongoing maintenance.

SECURITY CHALLENGES IN IOT-ENABLED SMART CITIES

As smart devices connect cities, the confidentiality and integrity of citizens' data is now an important matter. Despite improving such factors as safety, efficiency, and transportation by IoT technologies, IoT technologies also create critical problems regarding data privacy and ability to maintain safe systems which grow bigger. Overcoming these challenges is important for the development of public reception and continuation of the development of the smart city programs.

Privacy Concerns in Citizen Data

The ability to consistently operate large volumes of personal data is an ever present aspect in smart cities functionality. As a result, smart city developments, from immediate traffic control to the public health and emergency response, make intensive use of strong and sometimes individual data aggregations. However, the way smart cities work based on data leaves a number of serious threats to privacy, which can include unnecessary surveillance, theft of personal identities, and abuse of sensitive information. Smart cities have a significant problem since huge amounts of different personal data are gathered by interconnect devices. The private data that is gathered all includes such as location, biometric info, communication habits, daily routines each able to disclose sensitive personal information if mishandled. For example, sensors for smart cameras, wearable devices, and mobile phone applications continually gather information, and if such information gets leaked, people may become targets for specific attacks or manipulative strategies (Eckhoff & Wagner, 2017).

To alleviate such concerns, smart city engineers must opt for an all-embracing, multi-stage strategy in terms of privacy assurance. This approach puts high-quality encryption and safe storage systems of data, and advanced controls for access to information, to use. To this end, developers should take advantage of available tools such as differential privacy and secure multi-party computation, and federated learning in order to protect personal data during the analysis and the machine learning activities. If anything, developing thorough and transparent rules regarding data governance is as important when deciding how to responsibly handle personal data. It requires giving informed permission from citizens, minimising data collection, and providing the individual with practical oversight into what personal data is shared. It is important that throughout all stages of development of smart city, privacy is a priority ensuring stakeholders gain public trust and secure a safer digital environment (Farooqi et al., 2026).

Scalability and Interoperability Issues

As smart cities expand and become more complex, the need for networks which can expand and integrate with ease becomes more pronounced. Practical scalability implies that an IoT system can handle the increasing amount of connected devices and volumes of data while providing a high performance and the strong security. In contrast, interoperability ensures that IoT devices, platforms, and different communication protocols can interoperate successfully, wherever made by whom and with what technology beneath.

However, it goes without saying that bringing scalable and interoperable networks to smart cities is complicated by numerous challenges. One of the significant problems is that there are numerous vendors who produce IoT devices, as each vendor uses its proprietary specifications for communication processes, data processing, and security. The lack of standardization leads to rupture networks, increased integration costs, and increased security weaknesses. The subsequent explosive growth of the IoT networks exerts significant pressure on the available communication systems leading, among other issues, to congestions in the networks, high data latency, and reduced reliability. In particular, such challenges represent serious threats in mission-critical environments – emergency response systems, where time as well as efficiency are of crucial importance (ForeScout Technologies, 2016).

As networks expand, the complexity of maintaining security becomes greater along with more points of vulnerability that expose them to increased cyber threats. respond<|/system|>Americans think that this is a surefire way to get them higher pay. answer Americans believe that this is a surefire way of earning them higher pay. These challenges demand cities to adopt flexible and interoperable IoT infrastructures that make it easy for technologies to seamlessly cooperate with each other. To do this, they must incorporate the edge and fog computing capabilities to address data overload, reduce reaction times, and reinforce their network security. Moreover, adoption of standardized standard protocols including MQTT, CoAP, and 5G can successfully raise interoperability and reduce the risk of security breaches (Gharaibeh et al., 2017).

DATA SECURITY STRATEGIES FOR IOT NETWORKS

The management of sensitive data in vast volumes is key to the operation of smart city infrastructure; a task that IoT networks perform. Since the IoT networks play an essential role in the infrastructure, they are often targeted by cyber threats with incidents such as data breaches, identity theft, and elaborate network breach common. Intensive security measures needed to ensure that this critical information is secured on a security integrity, confidentiality and availability standpoint. Fundamentally, strategies depend on cryptographic techniques and protected communication programs to protect data against threats in storage, while in transit, and while in processing.

Cryptographic Techniques for IoT

Data security in IoT data networks heavily depends on the cryptography, which provides the necessary mathematical methods to ensure data's privacy, ensure data's integrity, and ensure user identification. Yet, traditional cryptographic techniques are quite prone to fail in the case of the security requirements of IoT devices in particular, due to the fact that they are resource-limited, energy-efficient and distributed across a wide network. Consequently, targeted cryptographic solutions have been put in place to deliver robust and low-resource compatible security for IoT devices (Goyal et al., 2026a).

1. **Lightweight Cryptography**

Although the standard encryption routines usually require massive amounts of processing power and memory for their operations, the lightweight cryptography has been designed to function best in devices that have limited computational powers, memory and energy capabilities. Over Dedicated algorithms like AES-128, PRESENT, and SPECK protect encryption in a more efficient way.These algorithms are perfectly applicable for implementing in battery-powered sensors wearable devices and smart meters which valuably require power efficiency.

2. **Elliptic Curve Cryptography (ECC)**

ECC is another widely used approach for securing IoT communications. It provides a high level of security with relatively small key sizes, making it ideal for constrained IoT devices. For example, a 256-bit ECC key offers comparable security to a 3072-bit RSA key, significantly reducing the computational burden without compromising security. ECC is commonly used in secure messaging, digital signatures, and public key infrastructures (PKIs) for IoT devices.

3. **Hash Functions and Message Authentication Codes (MACs)**

The most important role performed by Hash functions is that of ensuring data integrity and authenticity, which has them turn different input sizes into unassailable fixed outputs. In the context of IoT networks, widely used hash functions such as

SHA-256 and SHA-3 are crucial to the integrity of transmitted data and to detecting any accidental tampering. In addition, such mechanisms as HMAC – Hashed Message Authentication Code – effortlessly authenticate messages and verify the integrity of data (Goyal et al., 2026b).

4. **Quantum-Resistant Algorithms**

With quantum computing developing, many traditional cryptosystems could lose the resilience to significant quantum assaults. Therefore experts are researching post-quantum cryptography (PQC) algorithms such as Lattice-based and Hash-based cryptography aimed at future proofing long-term IoT security from a quantum threat (He et al., 2017). Although these techniques are still not at full maturity, they are fundamental in enabling the securing of IoT devices with reliable solutions from now on.

Putting these cryptographic solutions into effect in IoT contexts is of utmost importance as regards the safeguarding of sensitive information from interception, corruption and unwanted entry of various kinds. In order to provide strong cryptographic security, robust key management procedures need to be installed as well as software is requiring constant updating and attention to postulated security weakspots.

Secure Communication Protocols

Cryptographic techniques, vitally important in IoT security, merit to be supplemented by the use of safe communication protocols for securing data over possibly insecure networks. Through these protocols, data transports are protected because unauthorized access is eschewed and sensitive information is not modified.

1. **Transport Layer Security (TLS) and Datagram TLS (DTLS)**

TLS is an important standard for protecting internet communications, whereby data is encrypted, its source is authenticated, and its content is unchanged. However, the standard version of the protocol might burn too much power and memory for neeti devices. In vie of achieving a connectionless, datagram-oriented structure, a lightweight cousin to TLS, DTLS, overcomes the resource limitations commonly experienced in the IoT realm, suiting.

2. **Message Queuing Telemetry Transport (MQTT) with SSL/TLS**

MQTT has been designed particularly for devices under hardware restrictions and those where bandwidth and response time are mostly suboptimal. Joining IoT systems with an SSL/TLS-encrypted MQTT provides systems with a secure and reliable way to share messages which support publish-subscribe and request-response models (Jing et al., 2014). This combination is well used in industrial IoT, smart agriculture and energy management applications.

3. **CoAP (Constrained Application Protocol) with DTLS**

CoAP is widely used in IoT deployments and is the best choice to be used in the context of resource constrains. Supported by its multicast capability, the low demand for processing, and fast responses, CoAP is best suited in smart cities, such as street lighting, environmental monitoring, and traffic control. The combination of CoAP with DTLS improves security with little deterioration in performance for resource constrained scenarios.

4. **LoRaWAN (Long Range Wide Area Network)**

LoRaWAN is a power friendly, long range network protcol tailored for the Internet of Things (IoT). LoRaWAN is protected with robust encryption and device validation and therefore safeguards its transit data with applications like asset tracking, water usage monitoring and residential metering being considered excellent for wide smart city usage.

Device Authentication and Access Control

The mechanism of making IoT networks secure largely relies on the implementation of a reliable device authentication and access control system. Adoption of these protocols means that networks can retain the integrity of their data, secure their connection, and reduces intrusion and data breaches risks. With such a large number of devices on the work in a smart city infrastructure, establishing strong authentication and access controls systems is a difficult, but crucial task.

Device Authentication

Their identity should be validated before IoT devices can push data in a network. The authenticating of devices is crucial in stopping the malign intent of un-legitimate users to masquerade as real nodes and mess with the exchanges or even flaw data. Successful device authentication calls for the exploitation of diverse trustworthy skills:

1. **Public Key Infrastructure (PKI)**

PKI emerges as an all too familiar method of authenticating devices within IoT settings. It distributes through the use of digital certificates, public-private key pairs, and trusted certificate authorities (CAs) to verify device identities (Kitchin, 2016). In most PKIs configurations, every IoT appliance obtains a unique digital certificate it uses to verify its identity when interacting with networks. Even though PKI provides strong security, it needs strong computational power, making it suitable for devices that will have a lot of processing power and memory such as smart meters, connected cars, and industrial sensors.

2. **Mutual Authentication**

In mutual authentication both the device and the server need to authenticate the other's identity before a secure connection is made, thus, enhancing the framework of the PKI model. This form of mutual authentication minimizes the risk of MITM occurrences, and resists unauthorized data access, and thus only ensures trustworthy devices and services communicate securely within the network.

3. **Hardware-Based Authentication**

Sometimes, the security of IoT networks is enhanced through the use of hardware-based techniques of authentication, including Trusted Platform Modules (TPMs), Hardware Security Modules (HSMs) as well as Physical Unclonable Functions (PUFs) (Kumari & Goyal, 2026). Using hardware specific and non replicable attributes, these methods allow generation of cryptographic keys that greatly undermine the ability of potential attackers to copy or alter the device.

4. **Lightweight Authentication Protocols**

Because the resources in most IoT devices are scarce, light-weight authentication protocol was developed and implemented such as EAP-TLS, DICE, and OAuth to secure access with minimal computational need. Such protocols are perfect for allowing secure authentication on battery equipped-sensors, wearables, etc.; low-power IoT devices.

Access Control

Once a device is authenticated, it is critical to set which resources on the network it can see. Strategic access control policies enable maintaining a balance by giving the proper amount of access to user and device allowing them to carry out their tasks, thus reducing risks associated with data insecurity, excessive rights, and internal misconduct.

1. **Role-Based Access Control (RBAC)**

RBAC is one of the most popular access control models in the IoT settings. Roles are set up in a way that determines permissions, ensuring that any device or user only has access to things relevant to what they are intended to use it for. Take a smart thermostat, for example: it should be given rights to conduct temperature data and operation controls management but denied for viewing or dealing with financial data or user identity.

2. **Attribute-Based Access Control (ABAC)**

Apart from RBAC, extra attributes are added in terms of device type, location, activity time, and user clearance level in ABAC. With the adoption of a more refined perspective, it is easier for network administrators to move from the flexibility and security provided. It makes it easier to create very contextual access policies (Kumari et al., 2026).

3. **Least Privilege Principle**

At the heart of access control is the principle of least privilege, one which denies each device or user any more privileges than he absolutely needs for his purpose. The implementation of the least privilege principle goes a long way in containing the ramifications of a breach of the device, limits the curtains of attacks and increases protection against a spread of an attacker into the network.

4. **Network Segmentation and Micro-Segmentation**

Growing network security is possible by isolating IoT environments into individual, isolated spaces, where security components and access restrictions can be handled autonomously. After deploying micro-segmentation, lateral movement of malicious software is constrained and the process of network intrusion recovery is streamlined.

Figure 1. IOT features for smart city

Real-Time Threat Detection and Response

Good authentication and access controls are an essential part of the blocking of unauthorized entry, but not sufficient to offset a continuous cyber-attack environment. Timely threat detection and immediate response are key toward timely identification, remediation, and recovery from security incidents before they become significant problems.

1. **Anomaly Detection Systems**

A vast number of IoT networks use machine learning methods and AI to detect abnormalities in device operations, network traffic, or data flows. These systems can detect threaten emergence immediately either through activation of automated countermeasures or by alerting of security personnel to take charge. Strategies involve: network behavior analysis (NBA); signature of malicious activity; and heuristic analysis **(Li et al., 2015)**.

2. **Intrusion Detection and Prevention Systems (IDPS)**

Through a combination of on-the-fly observation and network analysis with self-initiated movement to capture and neutralize harmful activity, IDPS technologies converge. This strategy includes both signature and anomaly detection aspects with the purpose of protecting networks against existing and future threats.

3. **Security Information and Event Management (SIEM)**

SIEM platforms gather and analyze data flowing from numerous security sources and provide real-time understanding of potential threats and incidences. response.

4. **Automated Threat Mitigation**

In order to allow for quick responses, a variety of IoT networks have automated threat mitigation measures ranging from SDP, firewalls to DDoS defenses. These solutions help remove access from impacted devices, filter away malicious network traffic, and control the incidents to keep them from getting any bigger.

AI AND MACHINE LEARNING FOR ENHANCED SECURITY IN IOT NETWORKS

The dramatic growth of IoT has significantly altered the way that smart cities manage urban economies as well as the public services. However, the MD (internet-of-things) infrastructure's need to utilize connected devices makes cities vulnerable to significant security concerns, such as data weaknesses and breach attempts. The past security methods cannot cope up with this dynamic security challenges in inter-connected IoT network. In turn, integration of the AI and ML into the IoT security frameworks has become a vital approach. AI and ML will enable IoT network to notice the threats more proactively, automate risk handling, provide dynamic security, and become indispensable for modern IoT systems (Zhang et al., 2017).

Predictive Threat Analysis

This analysis uses AI and ML to predict future cyber attacks, that aid in the formulation of effective preemptive countermeasures. Where traditional security measures concern well known attack patterns, predictive analysis is more forward looking, noting data trends and system activity and incorporating threat intelligence, namely, looking for risks to precede events.

1. **Data-Driven Threat Modeling**

Predictive models emerge after researching huge volumes of data from Internet of Things, network activity and previous security breaches. With information processed by AI algorithms, correlations as well as strategies are identified which may anticipate future cyberattacks. Using techniques such as supervised learning, including regression and classification, and unsupervised learning such as clustering and anomaly detection, such models are able to detect odd behavior in a short amount of time.

2. **Threat Intelligence Integration**

Harnessing external threat data, such as information from cyber security groups and high frequency breach notifications, greatly enhances predictive analysis. Artificial intelligence analyzes hypothetical vulnerabilities in relation to the current device configurations, assesses which are easily exploitable, and discharges the resources for remediation according to the analytics.

3. **Behavioral Analysis**

One great merit of AI-enabled predictive analysis is its ability to make identification of unusual activity patterns. For example, if a smart lighting system sends large amounts of data at times that are normally dormant for the system, this may be detected as a potential security problem. The constant exposure to new information allows for A I models to better recognize the normal fluctuations and identify real threats.

4. **Risk Scoring and Prioritization**

In predictive analysis, risk scores for such threats identified are computed using criteria such as impact potential, the significance of the device, and ease of execution for the attack. By doing so, security teams get to use these scores to spend their energies appropriately, focusing on the most serious vulnerabilities and monitoring the less intense threats.

Anomaly Detection in IoT Networks

Anomaly detection is one of the key components of AI in IoT security that means outliers that might indicate cyber breaches are detected. Since there is extensive variety and size of the IoT networks, highly advanced anomaly detection systems are very important in order to properly distinguish between genuine behavior and actual security threats.

1. **Statistical and Machine Learning Techniques**

Statistical approaches such as Z-score analysis or Principal Component Analysis (PCA) have historically been combined with the machine learning capabilities of Autoencoders and Gaussian Mixture Models (GMM) to identify outliers (Midi et al., 2017). Using these techniques, data flow rates, device communication frequency, and patterns in sensor data are examined. Consider for instance when a smart camera when starts sending data in a much faster rate than it would normally do, anomaly detection technologies can very quickly detect and examine the device (Tang et al., 2015).

2. **Deep Learning Models**

Sophisticated anomaly detection systems utilise the power of the Convolutional Neural Networks (CNNs) and Recurrent Neural Networks (RNNs) within deep learning to detect such subtle temporal patterns within IoT data feeds. Such models are good at detecting subtle or gradual cyber harm that tend to be unrecognized by more basic cyber threat detection mechanisms.

3. **Context-Aware Anomaly Detection**

Contextuality of IoT environments implies that what appears odd in one context might be completely normal in another. Context-aware technology uses information about device type, location, and operation time to reinforce the reliability of anomaly detection. The end result is that there are fewer false alarms, and ensuring that only genuine threats are raised.

4. **Adaptive Learning**

Gender model changes in IoT networks demand anomaly detection models to adapt to device pattern changes and network organization. Adaptive learning techniques update frequently with new data and therefore ensuring the models keep up with the times and stay accurate. Such constant refreshing is essential for maintaining high capabilities of threat detection in an ever-changing network environment.

Automated Risk Assessment and Mitigation

When AI and ML are used, real time security risk analysis can be carried out and mitigation procedures can be automated. Such pre-emptive steps focus on reducing the gap between noticing threats and implement measures, thereby reducing the risk factor for severe damage from the cyber incidents.

1. **Real-Time Risk Scoring**

Allen AI-powered systems, with continuous examination of network information, should be able to produce real-time risk scores for each device on the network. Systematic risk computations factor in weakness, possible threats, and the device's relevance in the network. If the devices are rated as having high risk, they get tighter monitoring, or they get taken from the network to prevent lateral movement.

2. **Automated Incident Response**

When a likelihood of threat is identified, AI systems will be able to respond fast, i.e., blocking unauthorized connections to systems, isolating impacted devices, or reverting back to less dangerous previous configurations of software. Automated response deployment significantly reduces human interaction hence reducing the impact of cyberattacks.

3. **Predictive Maintenance for Security**

In addition to responding to problems, AI can forecast at what point devices will be vulnerable to obsolete firmware or poor performance. By implementing and scheduling consistent updates and maintenance, organizations can ensure the maintainance of the security integrity and reduce vulnerabilities thus giving attackers a few targets to exploit.

EDGE AND FOG COMPUTING FOR PRIVACY PROTECTION IN IOT NETWORKS

The spread of IoT networks in smart cities necessitates closer attention to the privacy and security of data produced by the devices. Traditional cloud infrastructures are effective for data management but are often plagued by latency, bandwidth, and, especially, privacy issues that are characteristic of centralized data-processing systems. Consequently, Edge and Fog Computing have become critical infrastructures for protecting privacy through mitigating the distance that such sensitive information must traverse through the network by processing data infancy. Decentralized computing has effective means of addressing IoT privacy conflicts, fosters real-time data analysis, reduces vulnerability to hacking, and improves overall management of resources.

Data Processing Near the Source

Standard-based benefit of Edge and Fog Computing is the fact that it can perform data processing at the source which is possible from the source device (edge computing) to a middleman platform such as a gateway or the local. By decentralizing the data processing process, there is a significant reduction in dependency of the decentralized cloud data centers and a reduced risk of sensitive data exposure over less secure connection routes.

1. **Edge Computing**

Edge computing refers to the practice of processing data on or near IoT devices, such as smart sensors, wearables, and cameras. By performing computation directly on these devices or local edge nodes, sensitive data can be analyzed in real-time without being transmitted to a distant cloud server. For example, in a smart home setting, a security camera may process images locally, only sending alerts if an anomaly is detected, rather than streaming the entire video feed to the cloud. This reduces the risk of exposing sensitive data and ensures faster, more efficient operations.

2. **Fog Computing**

To enhance capability of edge computing, fog computing extends it by inserting an intermediate layer to identify more seamlessly IoT devices at an interface with the cloud. As reported in [20], fog nodes that are installed near the edge of the network are programmed to process and store the data near the devices – a function that does not necessarily have to occur on the devices themselves. This helps to scale fog more effectively, able to handle vast amounts of data while offloading the processing off central cloud resources. Fog computing is essential if real-time data processing is not possible through existing local resources. In urbanised environments fog nodes could manage operations like traffic monitoring and the creation of environmental data, thus speeding through responses without weighing the cloud with too much workload.

By processing data at the edge or fog layer, IoT networks can significantly reduce the amount of sensitive data that needs to be transmitted, thereby minimizing the attack surface and lowering the risks associated with data breaches.

Figure 2. Measures for smart city

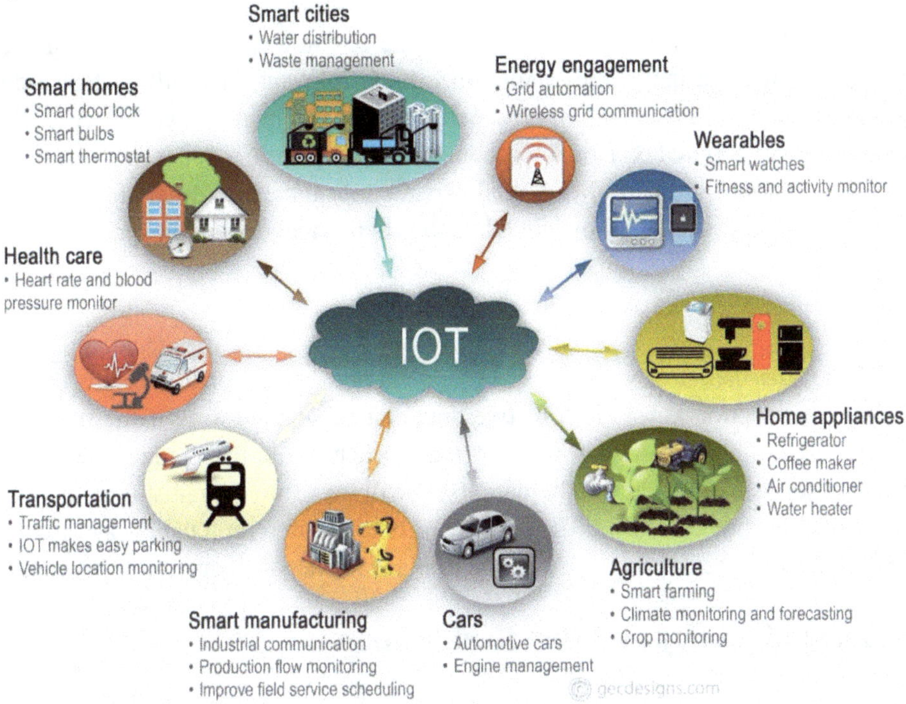

Secure Data Aggregation and Filtering

The combination of edge and fog computing enables efficient data aggregation and filtering, a key component in enhancing privacy protection. In many IoT applications, not all the data generated by devices is valuable or necessary for analysis. Secure aggregation and filtering procedures limit the cloud to allowing only relevant or non privacy implicative data to enter, thus securing sensitive data in private regional networks.

1. **Data Aggregation at the Edge**

Real-time data aggregation on IoT things at the edge can be done to aggregate, aggregate, and anonymize data to optimize transfer to the public cloud or centralized systems. Using this process, valuable, little data is transmitted through networks, thus helping reduce traffic and preserve the privacy of the users. For example, a smart grid could compile and analyze energy consumption information locally and then transfer summarized statistics to the cloud, rather than uploading individual-use statistics.

2. **Data Filtering and Anonymization**

The application of data filtering in an edge or fog level facilitates extraneous information discard hence only anonymized and relevant data will be forwarded for more analysis. Since location and health metrics can be sensitive user data; this describes a method that is critical for reducing risk exposure in related applications. Edge-based applications of differential privacy introduce statistical noise to data, maintaining aggregate analysis while keeping individual data points private.

3. **Edge-Based Encryption**

IoT devices can locally encrypt data that can be sent out for further processing. Through encryption keys, the sensitive data can be secured even if the transmission is compromised, as when the information leaves the IoT device, even if the transmission is compromised, the data itself will be secured against unauthorized reading. This way, not only is the privacy of data secured on local encryption but works well with other privacy measures in place for network operation purposes.

Reducing Latency and Bandwidth Costs

Edge and fog computing is a great plus since it reduces latency and bandwidth costs that are the key aspects of time-sensitive IoT systems. Traditional cloud systems frequently suffer high latency for data transmissions especially when data units in large amounts are exchanged over long networks. On the other hand, by handling data at the edge of the network these challenges are countered which enables quick decisions and putting less pressure on bandwidth resources.

1. **Low-Latency Decision Making**

Edge technology reduces time needed to analyze and respond to incoming data significantly. Efficiency in autonomous driving, traffic control, and healthcare technologies requires a close approximation for minimum latency to achieve correct real-time decision-making. An autonomous car that is equipped with the IoT sensors and edge computing capabilities so as to illustrate the point—this car can quickly compute the inputs from the sensors, discern any type of barrier, and come to immediate driving decisions in real time without having to contend with latency issues caused by communicating with the cloud.

2. **Reduced Bandwidth Consumption**

By making the use of edge or fog computing to process and filter data locally, IoT networks can dramatically reduce the volume of data that is to be sent to the central cloud. Consequently, more data has to be transferred through a network, which leads to the reduced bandwidth consumption and the cost of data transmission. For instance, envision cities that are provided with sensors that monitor, at all times, air quality, ambient temperatures, and traffic flows. IoT networks have the ability to prevent network congestion and reduce operational costs by filtering and analyzing sensor data before sending to the cloud.

3. **Efficient Use of Cloud Resources**

Offloading computing to local nodes with edge and fog computing leads to effective utilisation of resources in the cloud. This leaves the central cloud servers less stressed, and thus IoT systems can also grow and evolve more easily. Therefore, cloud infrastructure is able to focus on the complex analytics and perpetual data archiving whereas edge and fog nodes carry out current processing functions.

Use Cases in Smart Transportation and Smart Grid

Edge and fog are vital in supporting major smart city applications (transport and grid networks) with better privacy and working efficiency by merging local data processing.

1. **Smart Transportation**

Since the introduction of edge computing in smart transportation, real time analysis of traffic, vehicle data, and road conditions becomes possible, hence, prompt responses on congestion, accidents, or traffic signal changes can be undertaken in time. For instance, smart traffic lights use local edge computing to improve traffic by tailoring themselves to current vehicle density data, hence reducing delays and emissions. In addition, thanks to local edge computing, connected vehicles can analyze the sensor data and identify risks or change the driving course right away, ensuring the immediate safety of all passengers, not depending on the distant cloud processing.

2. **Smart Grid**

Edge and fog computing in the smart grid are capable of continuous monitoring of energy usage, flow of energy, in urban infrastructures. Local computers in devices such as smart meters and the transformers that are part of the energy grid process information to detect equipment problems or changes in energy demand, as described in (Mohanty et al., 2016). By aggregating and filtering data at the edge, utilities are able to send less sensitive information to central servers, thus increasing privacy and eliminating expectations of data breaches. Edge and fog nodes also support load balancing, favour energy efficiency and enable predictive maintenance, which ultimately enhances and optimises the performance of the energy grid (Zanella et al., 2014).

Through strong privacy protection, improving the performance of networks, and facilitating immediate decision making, edge and fog computing are essential for IoT networks. These models significantly enhance the security, and privacy of IoT applications through local data processing and lowered data transmission of sensitive information, even in urban environments such as smart cities. Due to the constant progress of IoT technologies, edge and fog computations will play a major role in defining the privacy benchmarks for the digital future (Mosenia & Jha, 2017).

CASE STUDIES AND REAL-WORLD IMPLEMENTATIONS OF IOT IN SMART CITIES

While being incorporated into cities their urban infrastructure is coming to be more efficient, sustainable and more appropriate to improve the quality of life of people residing in them. The practical benefits and difficulties associated with these technological innovations are demonstrated in various areas of application, including transportation, healthcare, energy, and utility. We explore some IoT projects in city settings, showing the advancement, lessons learned, and altered IoT impact towards building smarter cities.

Figure 3. IOT security adoptions for smart city

Smart Traffic Management Systems

IoT has identified one of its ways to apply in urban areas, that is, through smart traffic management systems. Using IoT sensors and cameras and networking infrastructure, these systems create real-time monitoring and control of traffic flow, increasing efficiency and protecting road users. With the introduction of AI and machine learning algorithms into its traffic management designs, the system becomes more effective in the prediction of traffic patterns and corresponding quasi-immediate responses to emergencies, while being able to adjust traffic signal timing in real time (Sehgal et al., 2012).

Case Study: Barcelona's Smart Traffic System

Innovative IoT technologies have been deployed in Barcelona in order to put the city in the forefront when it comes to urban management. Over 200 000 sensors distributed on roads and intersections allow the city of Barcelona, with its smart traffic system, to monitor speeds of vehicles, volumes of traffic and air quality (Sharma et al., 2026a). Process based upon real-time data which is collected at sensor level, automatically adapts traffic light sequence to reduce stoppage of congestion and consequently emissions are reduced. Real-time information is given to such other urban services as public transportation, which can be used for better coordination and operational efficiency.

Impact on Traffic Efficiency and Environmental Friendly Practices

Barcelona's smart traffic technology has addressed traffic jam efficiently and even supported superior environmental conditions. The system optimizes the dynamics of traffic and redirects resources to public transit, providing faster trips and more efficient urban mobility footprint. The real-time data also supports evidence based planning whereby the city planners could wisely shape the future infrastructure investments.

Connected Healthcare and Patient Data Security

IoT has greatly enhanced patients' outcome with the use of remote monitoring, customization of therapies and real time data as regards to the state of health. However, such developments have significant questions over privacy and security as now the personal health information is shared and saved electronically. The advent of connected healthcare systems brings out the gap between technology and patient privacy.

Case Study: There is an NHS (United Kingdom) digital health initiative.

NHS has introduced an IoT driven digital health platform for remote patient monitoring using connected medical devices (Sharma et al., 2026b). Chronic patients such as diabetes or heart disease can get wearable devices that will keep tabs of signs such as blood pressure, heart rate and blood glucose. The devices report to healthcare providers for constant monitoring and ability to identify any health risks in good time.

Data Security Challenges and Solutions

Given the extreme sensitivity of health information, any effort to secure privacy and security is paramount in the NHS initiative. As a reaction to this, the NHS adopts end-to-end encryption throughout all health data being sent from devices to healthcare professionals. To secure access, the initiative incorporates a robust authentication network that confirms for users to only access patient information who are authorized to do so. It goes without saying that regular audits and compliance with such data protection rule as GDPR significantly improve the security in patient data.

Improvement of Care and Efficiency in Workflow

Healthcare outcomes have received favor from the IoT system owing to early diagnosis, personalised treatment plan and smarter allocation of medical resources. The decrease in in-person consultations implies that the healthcare facility encounters less stress, which reduces the waiting times as well as expenses for operation. It is continually a top priority of the NHS to guarantee patient confidentiality is maintained and data is secure..

Smart Energy Grids and Utility Management

The transition to smart energy grids also presents a rather massive opportunity for IoT to change urban infrastructure. IoT-enabled sensors and smart meters and platforms for communication in smart grids support energy monitoring, optimized distribution, and the minimization of energy consumption and losses. Smart grid technologies can help support a more efficient grid connection of renewable sources of energy such as solar and wind.

Benefits of Smart Grids

The interventions for the integration of smart grids have made energy consumption and waste less in the system. With real time monitoring utilities will be able to make their operation fit demand patterns, avoid over generating at peak times and reduce both operational cost and the environmental impact by reducing the amount of carbon emitted. Furthermore, the smart grids allow effortless integration of renewable energy, meaning more sustainable electricity consumption.

Privacy and Security Concerns

Just as IoT devices face privacy and security risk challenges, smart grids also face such challenges. Information concerning the usage of energy might reveal private activities in homes and, under improper access, may threaten the privacy. As a result, the inclusion of secure encryption, user permission to use data, and adherence of privacy standards to be indispensable components in deploying the smart grid systems..

Lessons from Smart City Pilot Projects

Smart city pilots are significant testbeds for technical innovations and possible challenges, before IoT solutions could be applied on a wider scale. Through pilot projects, organizations can come up with the recognition and solution of the numerous technical, social, and ethical complexities that may result when applying IoT technologies within the existing urban infrastructure.

Case Study: Amsterdam Smart City Project

Amsterdam's smart city project is one of the most comprehensive pilot programs in Europe. The city has embraced a wide range of IoT applications ranging from energy efficient street lighting to smart waste management solutions that automate disposal route scheduling. From the project, valuable insights have been gained about involving citizens, protecting consumer information from misuse, and systematic integration of the system.

Key Lessons Learned

1. **Citizen Engagement and Trust**: Successful smart city projects must involve citizens at every stage, from planning to implementation. Public consultations

and transparent communication about data usage and privacy policies are critical for gaining public trust.

2. **Interoperability**: Many smart city projects struggle with the integration of diverse technologies and platforms. Ensuring that different systems can communicate and share data seamlessly is essential for the success of large-scale deployments.
3. **Scalability**: Pilot projects often face challenges when scaling up from a limited deployment to a city-wide system. Ensuring that IoT infrastructure can handle the increased load and complexity of a larger population is crucial for long-term success.

Figure 4. IOT applications for smart cities

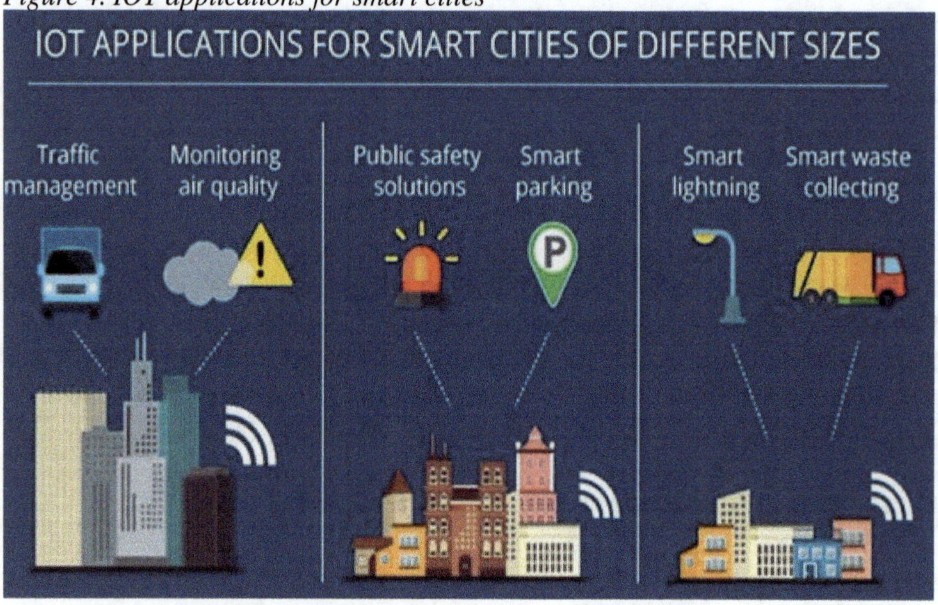

The examples in this chapter show how IoT technologies are transforming the face of smart cities in their very basic mechanism. IoT influences many developments in the smart city, from urban mobility and healthcare up to energy and utilities. However, every case highlights the crucial necessity to address privacy, security, and scalability problems in order to successfully incorporate these technologies into cities. In the future as cities develop, the findings of such trail-blazing efforts will enumerate good practices for addressing newly arising issues and optimising IoT use in developing next-generation smart cities.

REGULATORY AND ETHICAL CONSIDERATIONS IN IOT-ENABLED SMART CITIES

Whereas as smart cities deploy IoT, much debates occur on how to regulate and morally handle data with regards to privacy, security, and governance challenges. The amount of personal and sensitive information being collected increases dramatically as urban connectivity continues to deepen and more IoT devices are deployed. Information collected from such sources as trends in traffic and records accessed in regard to health can play a pivotal role in improving urban infrastructure and improving the quality of life between chiefly urban citizens. Still, along with this surge goes increased vulnerability to privacy slips, threats to security, and thorny questions around the use of citizens' data.

It is necessary, first of all, to understand the IoT governance applied now and thoughtfully navigate the ethical dilemmas that go along with large-scale data gathering to create smart cities that respect individual liberties and enable space for further advancement in technology. This component examines transitioning, or actualized, forms of influential privacy laws from the world, discusses the ethical questions arising from mass-data collection, and explores fundamental trade-offs that the companies have to make, between security and privacy on the one hand, technological development on the other.

GDPR, CCPA, and Global Data Privacy Laws

Given the rising numbers of personal information flowing out of the IoT networks in smart cities, countries across the world have established sever regulations to protect privacy and ensure proper ethical management of data. The primary data protection laws are; General Data Protection Regulation (GDPR), which applies in the Europe and The California Consumer Privacy Act (CCPA) in the US. These guidelines are important in determining the management of data in IoT scenario.

1. **General Data Protection Regulation (GDPR)**

The GDPR, which was applied in 2018, represents one of the greatest ever-wide-ranging data privacy standards in the world (Sharma et al., 2026c). Its primary goal is to protect the personal data of EU citizens and residents from any kind of unauthorized access regardless of the country in which the processing tkes place. Organizations are required under GDPR to ensure people whose data ought to be collected have given informed consent and explain the reason behind the collection of data and mandatory strict cybersecurity measures to ensure they are secure from unauthorized access. Further, people have the right to request copies of their own data, to have discrepancies corrected and to delete their personal details, giving them a great deal of control over their personal detail.

The execution of the GDPR has significant challenges to IoT networks in smart city infrastructures. Smart-city technologies, as an example, such as smart cameras, personal health, wearables, and vehicle connectivity systems collect personal information about people's activities and well-being. Systems working with data—matters of IoT—in smart cities should comply with the GDPR requirements, with special attention given to data minimization and genetic privacy measures to the architecture itself from the initial stage (Sheikh et al., 2026).

2. **California Consumer Privacy Act (CCPA)**

commonly, one of the next ordinary adverbs used in this sentence is Reply. Adverbs are words used to define how an action is done. In this example, by implementing the word 'Reply' the speaker always maintains his or her audience with the recipient's position. Reply is While sharing a great deal of the principles of data privacy with the GDPR, the CCPA is specifically designed for those business both in California and those who operate nationally and in California as well with major implications for the organizations (Sicari et al., 2015).

Organisations implementing IoT solutions, like those building smart city infrastructure, are also subject to CCPA. IoT data collection systems have a legal responsibility to disclose to consumers any data collected, what will be done with it and whether it will be passed on or sold. Because of these regulations, the use of smart infrastructure by cities requires them to be open about their data practices in order to adhere to local laws and build trust with their citizens (Yaqoob et al., 2017)..

3. **Other Global Data Privacy Laws**

The global data protection rules go way beyond the GDPR and CCPA and countries are developing their own frameworks too. Similar to the GDPR, Brazil's LGPD provides strong data protection rights for all citizens of Brazil, according . Adjacent to Asia, Japan, and South Korea have had their own frameworks put in place to protect their collection and processing of personal information (Townsend, 2013).

The expansion of smart city infrastructures requires a thoughtful blend through various legal environments to continue being compliant with local data privacy laws. Adoption of adaptable data management techniques, and understanding that IoT networks are frequently globally connected are necessary.

CONCLUSION

Eventually, the adoption of IoT in smart city settings is going to drastically improve public services, increase the efficiency of urban systems, and move things forward on environmental sustainability. With continued growth of IoT, it is now clear that effective deployment and public trust of IoT technologies depend on strong data protection, privacy preservation, and adherence to ethical guidelines. This chapter dissects complex challenges and their respective solutions related to IoT network protection, data security, and the cautious management of innovation in the face of privacy concerns. Adoption of evidence-based data protection solution such as encryption, authenticated communication mechanisms and proactive security tools, can minimize the incidence of cyber-attacks and protect privacy of citizens and credence of urban operations. Furthermore, the use of AI and machine learning for sophisticated security improvements together with edge and fog computing for personal data privacy demonstrates how the most innovative technologies can be used to avert the issues in contemporary dynamic urban ecosystems.

In view of the emergence of such regulations as the GDPR, CCPA etc. which stand on personal data, smart city projects should also be alert to their legal responsibilities, when they promote transparency and enhance public confidence. Equally important, regulators and city leaders need to be closely attentive to ethical aspects of data gathering and use of surveillance, and risks of algorithmic bias, to prevent harm and preserve impartiality in urban data systems. The final yardstick on the success of infiltrated smart cities will be the ability to balance among security, privacy and innovation. With the development of IoT technologies, it will be important that the government, players in the industry, and the public collaborate to design policies and systems that respect privacy, promote innovation, and help build sustainable, smart cities. If cities address these issues they will promote resilience, amplify inclusivity and build digital power creating the conditions for a future in which innovation serves all people without undermining individual rights.

REFERENCES

Albino, V., Berardi, U., & Dangelico, R. M. (2015). Smart cities: Definitions, dimensions, performance, and initiatives. *Journal of Urban Technology*, *22*(1), 3–21. DOI: 10.1080/10630732.2014.942092

Alomair, B., & Poovendran, R. (2014). Efficient authentication for mobile and pervasive computing. *IEEE Transactions on Mobile Computing*, *13*(3), 469–481. DOI: 10.1109/TMC.2012.252

Anthopoulos, L. G. (2015). Understanding the smart city domain: A literature review. In Rodríguez-Bolívar, M. P. (Ed.), *Transforming city governments for successful smart cities* (pp. 9–21). Springer., DOI: 10.1007/978-3-319-03167-5_2

Bartoli, A. (2011). Security and privacy in your smart city. In *Proceedings of the Barcelona Smart Cities Congress* (pp. 1–6).

Centenaro, M., Vangelista, L., Zanella, A., & Zorzi, M. (2016). Long-range communications in unlicensed bands: The rising stars in the IoT and smart city scenarios. *IEEE Wireless Communications*, *23*(5), 60–67. DOI: 10.1109/MWC.2016.7721743

Chourabi, H., Nam, T., Walker, S., Gil-Garcia, J. R., Mellouli, S., Nahon, K., Pardo, T. A., & Scholl, H. J. (2012). Understanding smart cities: An integrative framework. In *Proceedings of the 45th Hawaii International Conference on System Sciences* (pp. 2289–2297). IEEE. https://doi.org/DOI: 10.1109/HICSS.2012.615

Dandotiya, M., Goyal, N. K., Yadav, P., Sankhla, K., & Kumari, M. (2026). GENALYTICS: A deep dive into real-time insights via generative AI. In Elbaghazaoui, B., Amnai, M., & Gherabi, N. (Eds.), *Generative AI-powered data architectures: From governance to autonomous analytics* (pp. 105–134). IGI Global Scientific Publishing., DOI: 10.4018/979-8-3373-5616-7.ch005

Dandotiya, M., Sharma, S., & Goyal, N. K. (2024). ANFIS and LSTM techniques based model for time-series forecasting. In *Proceedings of the 2024 International Conference on Augmented Reality, Intelligent Systems, and Industrial Automation (ARIIA)* (pp. 1–6). IEEE. DOI: 10.1109/ARIIA63345.2024.11051439

Department of Economic and Social Affairs. (2014). *World urbanization prospects: The 2014 revision, highlights*. United Nations.

Eckhoff, D., & Wagner, I. (2017). Privacy in the smart city—Applications, technologies, challenges, and solutions. *IEEE Communications Surveys and Tutorials*, *20*(1), 489–516. DOI: 10.1109/COMST.2017.2748998

Farooqi, A., Das, D., Tripathi, M., Dutta, R., Goyal, N. K., & Sheikh, M. F. (2026). Navigating ethical frontiers in AI-driven geriatric care. In Stolakis, K., & Efthymiou, I. (Eds.), *AI for geriatric care in an aging society: Ethical, clinical, and policy challenges* (pp. 243–266). IGI Global Scientific Publishing., DOI: 10.4018/979-8-3373-5462-0.ch009

ForeScout Technologies. (2016). Smart cities require smarter cybersecurity.

Gharaibeh, A., Salahuddin, M. A., Hussini, S. J., Khreishah, A., Khalil, I., Guizani, M., & Al-Fuqaha, A. (2017). Smart cities: A survey on data management, security, and enabling technologies. *IEEE Communications Surveys and Tutorials*, *19*(4), 2456–2501. DOI: 10.1109/COMST.2017.2736886

Goyal, N. K., Goyal, N. K., Kumar, S., Dutta, R., Kothari, V., & Sinha, S. K. (2026). Empowering elderly care through artificial intelligence. In Stolakis, K., & Efthymiou, I. (Eds.), *AI for geriatric care in an aging society* (pp. 1–28). IGI Global Scientific Publishing., DOI: 10.4018/979-8-3373-5462-0.ch001

Goyal, N. K., Gulati, K., Avasthi, A., Kapoor, S. K., Keswani, B., & Gupta, S. (2026). Securing the imagination: Security and privacy challenges in generative AI systems. In Elbaghazaoui, B., Amnai, M., & Gherabi, N. (Eds.), *Generative AI-powered data architectures* (pp. 239–266). IGI Global Scientific Publishing., DOI: 10.4018/979-8-3373-5616-7.ch010

He, D., Zeadally, S., Kumar, N., & Lee, J.-H. (2017). Anonymous authentication for wireless body area networks with provable security. *IEEE Systems Journal*, *11*(4), 2590–2601. DOI: 10.1109/JSYST.2016.2544805

Jing, Q., Vasilakos, A. V., Wan, J., Lu, J., & Qiu, D. (2014). Security of the Internet of Things: Perspectives and challenges. *Wireless Networks*, *20*(8), 2481–2501. DOI: 10.1007/s11276-014-0761-7

Kitchin, R. (2016). *Getting smarter about smart cities: Improving data privacy and data security*. Maynooth University.

Kumari, M., & Goyal, N. K. (2026). Artificial intelligence in the cryptocurrency ecosystem: Crime mechanisms, detection strategies, and legal perspectives. In Lin, L. (Ed.), *Innovations in cryptocrime and financial fraud* (pp. 245–282). IGI Global Scientific Publishing., DOI: 10.4018/979-8-3373-0675-9.ch007

Kumari, M., Kumawat, R., Tripathi, M., Suganthi, K., Goyal, N. K., & Sheikh, M. F. (2026). Blockchain for transparency and traceability in supply chains. In Khan, M., Mehmood, T., & Rao, M. (Eds.), *Maximizing operational efficiency at the intersection of supply chain management and marketing* (pp. 35–64). IGI Global Scientific Publishing., DOI: 10.4018/979-8-3373-3109-6.ch002

Li, Y., Lin, Y., & Geertman, S. (2015). The development of smart cities in China. In *Proceedings of the 14th International Conference on Computer Urban Planning and Urban Management* (pp. 7–10).

Midi, D., Rullo, A., Mudgerikar, A., & Bertino, E. (2017). Kalis—A system for knowledge-driven adaptable intrusion detection for the Internet of Things. In *Proceedings of the IEEE International Conference on Distributed Computing Systems* (pp. 656–666). https://doi.org/DOI: 10.1109/ICDCS.2017.104

Mohanty, S. P., Choppali, U., & Kougianos, E. (2016). Everything you wanted to know about smart cities: The Internet of Things is the backbone. *IEEE Consumer Electronics Magazine*, *5*(3), 60–70. DOI: 10.1109/MCE.2016.2556879

Mosenia, A., & Jha, N. K. (2017). A comprehensive study of security of Internet-of-Things. *IEEE Transactions on Emerging Topics in Computing*, *5*(4), 586–602. DOI: 10.1109/TETC.2016.2606384

Sehgal, A., Perelman, V., Kuryla, S., & Schonwalder, J. (2012). Management of resource constrained devices in the Internet of Things. *IEEE Communications Magazine*, *50*(12), 144–149. DOI: 10.1109/MCOM.2012.6384464

Sharma, P., Kumawat, R., Jain, G., Keswani, G., Goyal, N. K., & Sheikh, M. F. (2026). Machine Learning-Based Intrusion Detection Systems for Healthcare Internet of Things (HIoT). In Shieh, C., Goel, P., Horng, M., & Pandey, H. (Eds.), *Resilient Healthcare Internet of Things Systems: Architectures, Security, and Trust* (pp. 185–214). IGI Global Scientific Publishing., DOI: 10.4018/979-8-3373-7610-3.ch007

Sharma, S., Goyal, N. K., Dandotiya, M., & Sharma, V. P. (2026). An Intelligent Data Encryption Approach with the Entangled Mermin-Based Qkd Protocol for Secure Quantum Communications. In Nayak, R., Mittal, N., Khunteta, A., & Kumar, M. (Eds.), *Recent Advancements in Artificial Intelligence. ICRAAI 2025. Lecture Notes in Networks and Systems* (Vol. 1468). Springer., DOI: 10.1007/978-981-96-7760-3_9

Sharma, S., Singh, V., Goyal, N. K., & Farooqi, A. (2026). Blockchain for Secure and Context-Aware Edge-Fog Computing. In Ramírez, J., Le, D., Kumar, A., Batta, P., & Rathore, P. (Eds.), *Semantic Annotation of Edge/Fog Services With Blockchain Integration* (pp. 263–292). IGI Global Scientific Publishing., DOI: 10.4018/979-8-3373-4387-7.ch009

Sheikh, M. F. A., Kapoor, S. K., Goyal, N. K., Tuba, E., & Rao, A. (2026). Enhancing healthcare security with fog computing and blockchain technology. In Nayak, R. (Eds.), *Recent advancements in artificial intelligence* (pp. 467–478). Springer., DOI: 10.1007/978-981-96-7760-3_38

Sicari, S., Rizzardi, A., Grieco, L. A., & Coen-Porisini, A. (2015). Security, privacy and trust in Internet of Things: The road ahead. *Computer Networks*, *76*, 146–164. DOI: 10.1016/j.comnet.2014.11.008

Tang, B., Chen, Z., Hefferman, G., Wei, T., He, H., & Yang, Q. (2015). A hierarchical distributed fog computing architecture for big data analysis in smart cities. In *Proceedings of ASE BigData Social Informatics*. DOI: 10.1145/2818869.2818898

Townsend, A. M. (2013). *Smart cities: Big data, civic hackers, and the quest for a new utopia*. W. W. Norton.

Yaqoob, I., Hashem, I. A. T., Mehmood, Y., Gani, A., Mokhtar, S., & Guizani, M. (2017). Internet of Things architecture: Recent advances, taxonomy, requirements, and open challenges. *IEEE Wireless Communications*, *24*(3), 10–16. DOI: 10.1109/MWC.2017.1600421

Zanella, A., Bui, N., Castellani, A., Vangelista, L., & Zorzi, M. (2014). Internet of Things for smart cities. *IEEE Internet of Things Journal*, *1*(1), 22–32. DOI: 10.1109/JIOT.2014.2306328

Zhang, K., Ni, J., Yang, K., Liang, X., Ren, J., & Shen, X. S. (2017). Security and privacy in smart city applications: Challenges and solutions. *IEEE Communications Magazine*, *55*(1), 122–129. DOI: 10.1109/MCOM.2017.1600267CM

Chapter 8
Privacy Preserving Federated Learning Frameworks for Secure, Scalable, and Citizen-Centric AI in Smart City Ecosystems

Anorgul Ashirova
Mamun University, Khiva, Uzbekistan

Nosir Khurramov
Termez University of Economics and Service, Termez, Uzbekistan

R. N. Ravikumar
http://orcid.org/0009-0009-3705-1681
Marwadi University, Rajkot, India

S. Aarthi
http://orcid.org/0009-0006-9064-2091
Marwadi University, Rajkot, India

S. Sindhuja
Kongu Engineering College, Erode, India

DOI: 10.4018/979-8-3373-4202-3.ch008

ABSTRACT

Smart cities increasingly depend on data-driven intelligence to enhance healthcare, mobility, energy, and citizen services, yet centralized AI models raise concerns over privacy, security, and regulatory compliance. Federated Learning (FL) emerges as a privacy-preserving paradigm that trains models locally on IoT devices, smart meters, and mobile systems while sharing only model updates. This chapter explores FL's architecture, core components, and integration with enabling technologies such as Blockchain, Differential Privacy, and Edge AI. Real-world use cases, including Traffic Management, Healthcare Diagnostics, and Energy Optimization, demonstrate its transformative potential. Challenges like device heterogeneity, communication overhead, and data imbalance are critically analyzed, with solutions and future policy directions proposed to ensure ethical, secure, and sustainable adoption of FL in citizen-centric smart cities.

1. INTRODUCTION

The accelerated transformation of smart city also alters the way cities function by relying on data-driven systems, thereby making them the most efficient, sustainable, and healthy to the inhabitants. Nevertheless, this dependency on big data full of impersonal information to healthcare and energy consumption raises the privacy and security issue on a large scale. Conventional centralized machine learning models have higher chances of requiring that sensitive information be kept in a single database, which can become the source of a breach, intrusion, or violation of data protection regulations like GDPR (Jiang et al., 2020). Federated Learning (FL) has proven to resolve them as a consistent step in this direction, empowering the abilities of multi-participant intelligence without infringing on the privacy of one specific person. Fl will also enable the models to train on the distributed devices and then secure aggregate rather than transmitting raw data to them. The plan does not only promote trust and resilience but as per the vision of smart cities that are people-centered. With complementary technologies such as a blockchain, differential privacy, and edge AI to FL, urban systems will have privacy-preserving, sustainable, and scalable intelligence.

1.1 The Need For Privacy In Urban AI Systems

By implementing artificial intelligence to enhance mobility, health care, energy management, and citizen services, cities have created challenges over privacy of urban AI systems and tools. Smart cities rely on the real-time information gathered by a variety of devices and systems including traffic cameras, IoT, wearables, smart meters and mobile apps to make important decisions (Belcastro et al., 2023). The main issue, however, is that the security of such a sensitive data is concentrated, which only increases the likelihood of its violation, the possibility of unauthorized monitoring, and the loss of trust of citizens. Moreover, the articulation of strict laws, including GDPR, and the increasing interest in social concern, to the problem of digital rights is compelling logical management of data. One can progressively understand that the citizens doubt AI systems, which could access confidential information, or encourage abuse, behavioral tendencies. This is why privacy is not only a technical factor but also a minimum when it comes to sustainable and a moral smart city planning (Deep and Verma, 2023). Not only will the respect of individual data guarantee the potential compliance with the law, but also popular confidence, integration, and transparency, which will guarantee that urban AI ecosystems can flourish without being irresponsible.

1.2 Federated Learning: Concept, Evolution, And Relevance

FL belongs to the new technologies that suggest the change of the training paradigm of AI models, in particular, sensitive environments, such as smart cities. Unlike the traditional network-scale centralized designs where raw data is accumulated at a central point, in a network of edge devices with FL only communicates model changes to a centralized server. Such a decentralized system reduces the disclosure of sensitive information and it solves privacy, security concerns and the regulations. First suggested by Google in 2016 to be applicable on mobile devices, FL has since developed into a powerful platform in the field of healthcare, finance, IoT and urban management (Tanwar et al., 2023). Its value to smart cities consists in the fact it can enable the use of intelligence contained in such distributed systems as connected cars, smart meters, and medical equipment without breaching confidentiality. Further advances, such as secure aggregation, differential privacy, blockchain integration, and more, make FL more tamper-resistant against cyber-attacks and adversarial attacks (Yang et al., 2019). Having linked data sovereignty to collaborative intelligence, FL enables smart cities to embrace AI solutions that are not only resourceful but also trustworthy, ethical and citizen-focused. The Figure 1 shows the iterative cycle of FL starts with the training of models on devices then sends updates to a central server where these updates are aggregated to make a global model and preserve privacy using privacy-preserving techniques, and to remain compliant with regulation.

Figure 1. Federated learning workflow cycle

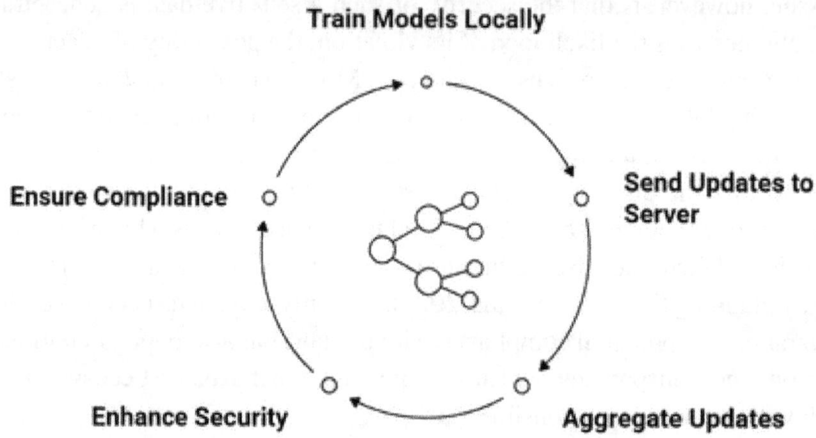

1.3 Objectives Of The Chapter

This chapter discusses the importance of FL as a privacy-aware AI model of smart cities. It identifies the basis of architecture, technologies in support and application in real world cases and issues surrounding challenges and policy implications. These goals are set to inform readers about the transformative capabilities that FL has in urban ecology centered on citizens.

- To explain the fundamental concepts, architecture, and types of Federated Learning.
- To examine the integration of FL with supporting technologies such as Edge AI, IoT, blockchain, and differential privacy.
- To analyze real-world use cases in traffic management, healthcare, and energy optimization.
- To identify key challenges, limitations, and technical barriers in FL deployment.
- To propose future directions and policy frameworks that promote ethical, secure, and sustainable smart city applications.

2. BACKGROUND STUDY

Smart cities implementation can be motivated by the necessity to create effective and sustainable, people-oriented cities in the physical infrastructure, governance and management using analytical intelligence based on data. The traditional distributed machine learning algorithms are effective at generating insights, but they need to collect massive amounts of sensitive data in centralized data sets. There are serious questions related to privacy, security, and the compliance with GDPR in such direction (Li et al., 2020). Digital citizens are feeling more rightful to their rights, and this fear of falling prey to being data mined can reduce confidence in smart city projects. FL is a new paradigm which can assist in solving such issues. The fact that the models are trained on the devices and the updates are summarized rather than being stored in raw form permits the collaborative training without the threat to privacy. The situation of the study justifies why FL shall become a key force of success in attaining Innovation-Ethics- Secure- Sustainable Urban Environment balance (S. Ahmad et al., 2020).

2.1 Literature Review

The growing idea behind smart cities has generated numerous studies on the topic of data-driven city management, in which AI plays a crucial role in streamlining an assortment of city services such as healthcare, traffic, energy and citizen connectivity (Shaheed, 2023). Traditional machine learning systems have largely relied on centralized systems where data based on different sources is pooled in a centralized point and is subsequently trained on models. Even though it will guarantee effectiveness and centralized management, this approach will be vulnerable to several important threats, such as data privacy risks, misuse, and regulatory breaches (Priyadarshini, 2024). Growing attention to global data rights and systemic safeguards such as GDPR is raising awareness of the shortcomings of centralized learning, and researchers are seeking out distributed alternatives. FL has arisen as a potentially effective approach towards these concerns. In contrast to centralized models, FL allows distributing training across a variety of devices or organizations without any need to send raw data (Tan et al., 2021). This method limits the risk of privacy and still has capability of creating strong global models. Initial research showed how FL would apply to mobile applications, where a collection of devices could be used to improve a predictive-text model without any parties uploading any personal content.

Further studies applied the principles to other aspects including healthcare, transportation and energy showing its use in urban ecosystems. Another common thing in the literature is the flexibility of FL to diverse settings (Rahman et al., 2019). Smart cities consist of diverse sources of data with a number of sources of data ranging between sensors, smartphones, smart meters and cars with data generated in large and often uneven data. This type of diversity is enabled by the fact that FL enables local models to be trained independently before updating a global model. It has been established that through this decentralization not only privacy is available, but also the bandwidth usage and communication expenses are minimized under the assumption that optimized protocols may be applied. Besides, secure aggregation, differential privacy, and integration into blockchains have become particularly popular in terms of their roles to bring trust and resilience to the federated ecosystems. It has also been suggested by the literature that there are significant issues that showcase the issues related to the concern. The inter-machine communication non-homogeneity, time latency of the network, and information asymmetry are commonly mentioned as obstacles on the way to achieving high model accuracy.

There are also communication overhead and convergence delays that are another bottleneck in large scale deployment due to the high number of nodes. To solve these obstacles, researchers suggest the adaptive federated optimization transitions, gradient compression, and clustering-based aggregations (Bi et al., 2022). However, the trade-offs among accuracy, efficiency and privacy remain a recurrent issue that shows how challenging real-life implementations can be. The contemporary academic debate draws a greater focus on the contribution of FL to construct ethical and sustainable smart cities. Due to its ability to secure individual privacy at the same time as allowing the collective intelligence, FL is one of the most important factors that will contribute towards citizen trust. integration with edge computing, IoT, open-source frameworks are viewed as essential to scale (Rasheed et al., 2021). Furthermore, the literature refers to policy-making, regulatory systems, and the cooperation between the state and business as the major contributor to responsible adoption protection. The cumulative resource presents direction of definite tendencies: FL is no longer a tool and technical device but a socio-technical structure that reinvents the city in the manner of using data to provide intelligent and citizen centred services. The Table 1 summarizes the main findings of the literature on FL in smart cities and connects key themes with an implication of them to the urban governance and technology adoption.

Table 1. Summary of literature review on federated learning in smart cities

Theme	Key Insights	Implications for Smart Cities
Centralized vs FL	Centralized models raise privacy and compliance risks. FL keeps data local.	Enables GDPR-compliant, privacy-preserving AI solutions.
Device & Data Diversity	FL adapts to heterogeneous devices and non-uniform datasets.	Supports wide-scale urban IoT and mobile ecosystems.
Supporting Technologies	Secure aggregation, blockchain, and differential privacy strengthen FL.	Builds trust, resilience, and transparency in multi-stakeholder environments.
Technical Challenges	Heterogeneity, latency, imbalance, and communication overhead slow performance.	Necessitates adaptive optimization, gradient compression, and efficient protocols.
Governance & Trust	Ethical oversight, fairness, and citizen trust are critical.	Promotes inclusive, transparent, and socially sustainable smart city applications.

2.2 Research Gap

The use of FL in smart cities has not been exploited yet, and it has demonstrated great potential and utility in healthcare, finance, and mobile applications. The present research scenario is more of technical possibility excluding issues of large scale implementation, models of governance, and the notion of policy frameworks of citizen trust and regulatory adherence. In addition, the heterogeneity of devices, the effectiveness of the communication, and the impartiality of Non-IID information are not examined in-depth in terms of different urban infrastructure. This omission suggests that there is a need to adopt citizen-led, cross-cutting FL mechanisms to be used in smart cities.

2.3 Problem Statement

Real-time data plays a major role in smart cities with services including healthcare, transportation, and energy management improvements. Early centralized machine learning systems and their models present substantial risks, as well, associated with such factors as privacy violations, abuse of surveillance, and regulatory non-compliance. Citizens do not usually trust centralized AI systems because of their doubts regarding the ownership and misuse of data. A decentralized and privacy preserving learning Framework is urgently needed with scalability, fairness, and efficiency. FL can provide a potential, and the effective implementation of this approach in heterogeneous, large-scale urban ecosystems is one of the current problems.

3. PROPOSED FRAMEWORK

This overview of the proposed FL system in privacy-preserving smart cities focuses on multilevel implementation combining technology, security, and governance. Regarding its implementation, IoT devices, smartphones, and sensors serve as local nodes that train the models locally and do not refer to raw data. This is communicated to an aggregation layer where models such as Federated Averaging (FedAvg) or FedProx is used to combine the contributions into a global model. A privacy and a security layer are added to further strengthen security; using differentially privacy, secure aggregation, and blockchain-based audit trails, where no sensitive data is leaked yet it provides trust and transparency. Lastly, the governance and policy layer ensure oversight through the promotion of democratic control, ethical responsibility and adherence to laws such as GDPR and HIPAA. Collectively, these inter-connected layers form a robust infrastructure to enable scalable citizen-centric, privacy-preserving and multi modal AI systems in smart cities. Figure 2 shows the four foundational layers of the framework IoT devices as local nodes, data aggregation, privacy and security and governance and policy. Collectively, they facilitate centralized training on a model, safe aggregation, compliance with varying regulatory schemes and citizen trust in smart city applications.

Figure 2. Proposed federated learning framework for smart cities

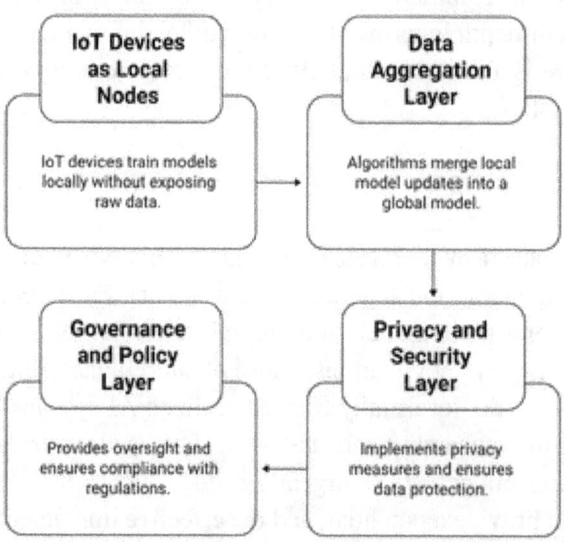

3.1 IOT Devices As Local Nodes

The IoT devices may serve as critical local nodes in a FL environment in a smart city, which produces huge amounts of data through their sensors, smartphones, smart meters, and vehicles. Rather than forwarding raw data to a central database, these devices can train local models on their own data, which requires less risk due to privacy protection along with the substantial reduction of exposure to an attack. This decentralized education does not expose valuable data like citizen health records, mobility patterns or energy consumption data to compromise. Calculations that are nearer to the source selections in the IoT nodes minimize the latency and provide real time decision-making. It builds on the concept of data sovereignty and puts the city on the frontline of developing scalable and privacy-conscious AI without lessening the productivity of work.

3.2 Data Aggregation Layer

A major aspect of the federated learning is the data aggregation layer: the data aggregation layer consists of a combination of local models updates to form a global model. The aggregator can receive the gradient or parameters that were trained in different devices and aggregate them based on the algorithms such as Federated Averaging (FedAvg). This makes sure that they share intelligence but they do not invade the privacy of an individual. Smart cities can have aggregators both in centrally located servers and edge cloud devices that can handle updates to thousands of devices. This kind of aggregation enables learning to be dynamic on differentiated and non-uniform information and not to be achieved at the expense of security. This layer makes scaling, fairness, and higher accuracy possible in citywide applications regarding healthcare, traffic forecasting, and optimized energy, as it allows adaptive and iterative updates.

3.3 Privacy And Security Layer

The privacy and security layer demonstrates that the FL in smart cities satisfies the ethical, legal, and trust needs. Sensitive information can also be leaked when the model is being updated, regardless of whether it shares raw information. One of the methods is called differential privacy that incorporates mathematical randomness, and by using secure aggregation, it will not be possible to identify the individual devices where the update was made. Another area it can be implemented is to have blockchain to develop unalterable audit records and increase accountability among parties. This layer ensures that there is no unauthorized access into the system, it protects against adversarial attacks and creates resilience in collaborative systems in a smart city context. This layer will increase public trust since security and privacy will be implemented in a form that will make FL reliable and ethically principled.

3.4 Governance And Policy Layer

The governance and policy layer are where the required oversight and regulatory compliance of the FL systems in smart cities exists. As the privacy is handled by technology, it is policies that are needed to govern accountability, transparency and fairness. This layer ensures the conformity to global regulations and rules such as GDPR and HIPAA that safeguard the entitlement to information and ethical use of data. The governance also offers liability on biased forecast or model failures to promote the responsible behaviours of AI. Similarly, this layer comprises of public-private partnerships and citizen participation which foster inclusiveness and citizen trust in smart urban solutions. FL can be congruent to the legal text and ethics in order to offer sustainable, civic-oriented and socially responsible implementation of AI-based urban innovations and this constitutes the governance layer.

4. FEDERATED LEARNING ARCHITECTURE AND WORKFLOW

The basis of FL architecture is a loosely-coupled distributed architecture in which the data are kept locally within the devices and the only things that are exchanged are the model parameters or gradients. At the core of this is the aggregation server that coordinates communication among local nodes such as the IoT devices, mobile phones or sensors. The server disseminates a global model to the local participants which comprises workflow initiation. (Singh et al., 2020). All devices each learn on their own with their own data producing updates instead of sharing data publicly. This information is securely communicated to the server where it is pooled usually via systems such as Federated Averaging to update the global model. Such iterative cycle is repeated until convergence to achieve privacy-preserving, scalable and collaborative intelligence that can be accommodated in smart city ecosystems (Abramson et al., 2020). The workflow of FL is represented in Figure 3 where local nodes train models on their own and send updates to an aggregation node. The server carries out iterative updates, result merging and returns privacy-preserving intelligence in an actor-opaque manner, which guarantees safe and collaborative model development.

Figure 3. Federated learning aggregation process

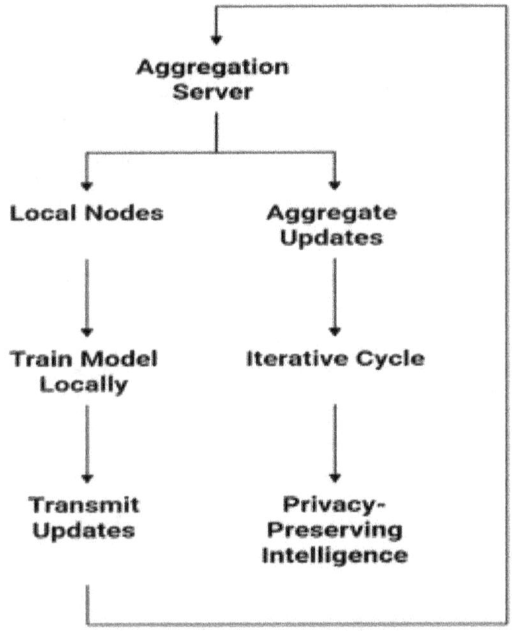

4.1 Core Components: Clients, Aggregators, And Communication Protocols

FL has a clear architecture to which each element is critical in the provision of the privacy aspect of distributed intelligence. Clients (e.g., IoT devices, smart phones, smart meters) have access to and train local models on their own local dataset. Aggregators consolidate (aggregate) model updates to make a global model which is usually a central server or edge hub. Communication protocols guarantee robust and privacy aware relay of updates in a heterogeneous network (W. Ahmad et al., 2024). All of these compose the foundation of FL, which enables smart cities to benefit greatly through collaborative AI without necessarily centralizing sensitive data about people. Table 2 demonstrates the main elements of the FL outlining their roles and importance in the smart city ecosystems (Kim et al., 2023). The clients are edge devices that produce the localized intelligence, the aggregators are the gluing parts which collect the updates into the global one and the communication protocols are responsible to make the transmission between the nodes to be secure and efficient. These elements in combination measure up to privacy-preserving collaborative AI that can be applied in scaled urban environments.

Table 2. Core components of federated learning

Component	Description	Role in Smart Cities
Clients	End devices (IoT sensors, smartphones, smart meters) that train local models.	Enable edge intelligence for healthcare, traffic, and energy systems without exposing raw data.
Aggregator	Central or edge server that collects and aggregates local updates.	Builds a unified global model for applications like traffic prediction or load forecasting.
Communication Protocols	Secure methods for transmitting updates (e.g., TLS, gRPC, secure aggregation).	Ensure encrypted, reliable, and privacy-preserving data exchange across distributed systems.

4.2 Types Of FL: Horizontal, Vertical, And Federated Transfer Learning

Federated Learning can be divided into classes according to the distribution of data among various participants. The three types (Horizontal Federated Learning, or HFL, Vertical Federated Learning, or VFL, and Federated Transfer Learning, or FTL) differ in terms of how data is partitioned, used, and they have different business application and challenges. HFL is applicable in situations where the participants occupy the same feature space with different samples; VFL in situations where there are different features with the same sample space and FTL where features differ in one case, and samples in the other (Yang et al., 2019). It is important to understand these types to put FL into practice in a variety of smart city practices like in healthcare, transportation, and energy management. Table 3 shows three key types of Federated Learning Horizontal, Vertical, and Federated Transfer Learning have their data partitioning methods and apply to smart cities.

Table 3. Types of federated learning

Type	Data Partitioning	Application in Smart Cities
Horizontal FL (HFL)	Same features, different samples (e.g., multiple hospitals with similar data types).	Enables collaborative disease prediction models across hospitals without sharing patient records.
Vertical FL (VFL)	Same samples, different features (e.g., banks and telecoms sharing user data dimensions).	Useful in urban governance where departments combine citizen service data without direct exchange.
Federated Transfer Learning (FTL)	Different samples and features, but with limited overlap for alignment.	Supports cross-domain collaboration, e.g., integrating traffic data with healthcare for emergency planning.

5. KEY ENABLERS AND SUPPORTING TECHNOLOGIES

The success of the FL within the context of the smart cities is conditioned by the combination with other technologies which enhance the efficiency, confidentiality, and credibility. Blockchain and differential privacy can provide secure ways of co-operation and effective data security, and Edge AI and connected IoT devices can be able to empower local intelligent practices and near real-time choices (Pooyandeh & Sohn, 2021). All these enablers are together the technological basis of privacy-preserving AI in the urban realm, enabling it to be scaled, transparent, and giving these technologies user confidence.

5.1 Integration With Edge AI And IOT In Urban Environments

The cross-sectional point between FL and edge AI as well as the IoT is an urban innovation disruption factor. The data generated by IoT products such as sensors, smart meters, connected cars, and so on is comprehensive and situational and it is often sensitive to follow. Edge AI reduces the delay and enables real-time analytics of the data at the edge of the network that is of significance to the traffic management, energy optimization, and even civic safety (Jabakumar, 2023). Such systems, coupled with FL, provide the possibility of local model training without the need to move raw data, which provides a high degree of privacy. It is this synergy that enables scalable intelligence of heterogeneous networks and devices without bandwidth usage and reducing the expenses of communications (S. Wang et al., 2019). More so, it will allow the smart cities to introduce resource management and emergency management dynamically and reduce the time in response to the emergency event, which is linked with sustainable goals in the form of citizen-oriented development and technological advancement.

5.2 Blockchain And Differential Privacy In FL For Secure Collaboration

FL frameworks in smart cities can be supported by blockchain and differentiation privacy to enhance its security and trust. Decentralization, immutability and generation of transparent audit trails is made possible by blockchain, in such a way that updates and any transactions done on the model can be verified and be tamper-proof. It may be particularly applicable to multi-stakeholder cities where different agencies and service providers are able to collaborate (A. Ali et al., 2023). Differential privacy, in its turn, introduces a noise term to defend the changes in models in order to defend sensitive individual data, yet to ensure the overall utility. All these technologies assist in minimizing the possibility of data leak, unauthorized access and intentional attacks. Another example is in the healthcare sector, wherein one can supply auditability in case a specified model contribution was created within the identical facility, as well as secure confidentiality at a patient level with the help of differential privacy (Mallah & Lopez, 2022). Their joint implementation will aid in harmless, scalable cooperation, and promote good social confidence in AI based administration. The enablers have the potential to strengthen ethical and privacy conscious AI ecosystems that are vital to resilient and people-centric smart cities. (Wang et al., 2025). The table 4 compares the role, advantages and smart cities applications of Edge AI + IoT with Blockchain + Differential Privacy, highlighting the synergy between these technologies when you combine them with FL into secure and efficient urban intelligence.

Table 4. Comparative role of edge AI + IoT vs blockchain + differential privacy in FL

Aspect	Edge AI + IoT	Blockchain + Differential Privacy (DP)
Role	Enables local computation and real-time intelligence at the data source.	Provides secure collaboration, auditability, and privacy-preserving mechanisms in FL.
Benefits	Reduces latency, saves bandwidth, ensures faster decision-making, and scales across heterogeneous devices.	Ensures trust, transparency, tamper-proof updates, and mathematical privacy guarantees.
Smart City Use Cases	• Smart traffic control (real-time signal optimization) • Energy grids (load prediction) • Public safety (emergency response via sensors)	• Healthcare (privacy-preserving diagnostic models) • Citizen services (secure data sharing) • Urban governance (tamper-proof audit trails)

6. USE CASES OF FEDERATED LEARNING IN SMART CITIES

Federated Learning (FL) holds great promise to revolutionise important smart city services as they can be made to be privacy preserving and collaborative intelligence. Traffic management, healthcare and emergency response are just a few examples of applications that entail sensitive, distributed data that cannot be readily centralized because it is subject to privacy constraints, regulatory restrictions, or technical considerations(Guo et al., 2020). The solution to this dilemma provided by FL is to enable the local training but have a global collective model.

6.1 Smart Traffic Control And Urban Mobility Analytics

Traffic, road safety, and optimization of public transport are the significant problems in the contemporary city. By combining distributed IoT sensors, connected vehicles, and surveillance cameras, urban mobility can be transformed enabling the delivery of smart traffic control systems that learn without aggregating raw data to a centralized location. The localized models allow examination of traffic flows, congestion and prediction of bottlenecks as occurs in real time (Liu et al., 2021). Condensed data enable dynamically setting and managing the traffic lights, more rational route creation, and better organization of public transport. It proposes a model under which commuter experiences are improved, emissions become fewer, and sustainable mobility solutions appear. In addition, the security of the collaboration between several city departments preserves privacy in that cross-agency coordination becomes secure, and it allows cities to respond dynamically to shifts in mobility needs. When used together, FL and edge AI can enable municipalities to build low-latency decisioning and smart ecosystems of traffic control with data sovereignty, trust, and align with regulatory needs.

6.2 FL In Smart Healthcare And Emergency Response

Smart cities necessitate the secure usage of highly sensitive data in the case of health care and emergency response. By centralizing this information, hospitals, clinics, and wearable devices can produce tremendous amounts of information and privacy would be a risk to the patient (Loftus et al., 2022). FL also enables model training on multiple decentralized participants and is used to support applications like disease diagnosis, personalized medicine, or outbreak prediction, in which the raw records of patients are not transferred. As an example, through the joint training, AI models can be trained in hospitals to detect chronic conditions or infectious diseases early and protect patient privacy (Stephanie et al., 2023). In emergency response, FL enables the possibility of fusion of information residing in paramedics, public health routing, and environmental sensors and then it will be able to predict incidences, optimal allotment of resources, and then quick interventions. Also, to enhance protection against the leakage of the data, differential privacy and secure aggregation can as well be used. With the end result of building trust in data sharing, FL can enable the development of healthcare systems and city governments that can take action to provide more responsive, ethical, and effective citizen services (Silva et al., 2020). Table 5 contrasts the FL use cases in the domains of traffic management and the healthcare industry, describing their respective data sources, advantages, hurdles and implications on the ecosystem of smart cities.

Table 5. Comparative analysis of traffic vs healthcare use cases in FL

Aspect	Smart Traffic Control & Urban Mobility	Smart Healthcare & Emergency Response
Data Sources	IoT traffic sensors, connected vehicles, GPS data, surveillance cameras, public transport logs	Hospitals, clinics, wearable devices, ambulances, public health records, environmental sensors
Benefits	• Real-time congestion prediction • Optimized traffic lights & routes • Reduced emissions & improved mobility	• Early disease detection • Personalized treatment models • Faster emergency response & resource allocation
Challenges	• High communication overhead • Device heterogeneity • Ensuring low-latency responses	• Strict privacy regulations (e.g., GDPR, HIPAA) • Sensitive patient data • Need for high trust & security
Impact	Enhanced commuter experience, sustainable urban mobility, reduced carbon footprint	Improved healthcare outcomes, stronger epidemic preparedness, ethical and citizen-centric services

7. CHALLENGES AND LIMITATIONS OF FL DEPLOYMENT

The possible privacy-beneficial smart cities advantage of operating FL, the implementation of FL is limited by various technical and operational challenges. The decentralized nature of FL also introduces restrictions compared to centralized systems by the capacity of devices, providing data and the effectiveness of communication. These are the points that need to be implemented to achieve scalability, reliability, and practicability of FL implementation in real-life situations in urban environments.

7.1 Device Heterogeneity, Network Latency, And Data Imbalance

Heterogeneity of devices is one of the fundamental issues of FL in a smart city. The computational abilities, memories, and energy efficiency are great, and it encompass a vast network of involved devices that involve smartphones, IoT sensors, and edge nodes. This has generated variances in the localizing training and that affects the reliability of the system on the whole (Mazhar et al., 2023). In addition, network latency may also lead to a backlog of updating the model practically, especially in low bandwidth networks such as rural and congested cities. The other issue that is burning is the difference in data as the devices produce data with different distributions where some could have large and high quality data when others could be providing few data or biased. The discrepancies are bound to cause the problems of model biases and generalization with the highest probability (Ottun et al., 2023). These challenges require dynamisms like the individualized federated models, priority-based scheduling, or the effective aggregation utility which has been constructed to pull equal participation and the consistent functioning of the learning systems in the smart urban infrastructures which are flexible.

7.2 Communication Overhead And Model Convergence Issues

The FL clients in FL do share model updates with the central aggregator quite often, which, in its turn, increases the communication overhead. This is especially an issue regarding large-sized smart city ecosystems where thousands of devices can potentially be interacting simultaneously. Having limited bandwidth, poor-quality connections, SIP/device, and energy consumption that contributes to the problem to ruin the efficiency of the system. Asynchronous updates, Non-IID and divergent participation rates of clients also exist that may complicate the model convergence. These lead to the slowness of training the global models and inaccuracy. Unless there is effort to optimize the communication costs, the communication costs can be greater than the benefits of decentralized learning. To eliminate these drawbacks, researches are being conducted by performing tasks like gradient compression, periodic averaging, the use of adaptive optimization methods, which will assist in minimizing the number of communications required without affecting the accuracy. To surmount these limitations, the greatest aspect is to suggest that FL will be scaled to a viable scheme so that reliable real-time decisions can be made in smart cities. Table 6 summarizes the key barriers to the use of FL in smart cities, reasons behind these barriers, their ramifications, and possible remedies to attain scalable and privacy-conducive adaption.

Table 6. Challenges of federated learning deployment in smart cities

Challenge	Cause	Impact	Possible Solutions
Device Heterogeneity	Variations in computing power, storage, and battery among devices	Inconsistent training performance; unreliable participation	Personalized FL, device-aware scheduling, lightweight models
Network Latency	Bandwidth constraints and unstable connections	Delays in model updates; reduced responsiveness	Edge caching, asynchronous FL, efficient communication protocols
Data Imbalance / Non-IID Data	Unequal and biased data distribution across devices	Skewed model accuracy; poor generalization	Data augmentation, clustering-based aggregation, fairness-aware optimization
Communication Overhead	Frequent transmission of model updates	High bandwidth and energy consumption; reduced scalability	Gradient compression, update Sparsification, periodic/partial aggregation
Model Convergence Issues	Asynchronous updates and diverse participation rates	Slow or unstable convergence; decreased accuracy	Adaptive learning rates, federated optimization algorithms (FedProx, FedNova, FedOpt)

7.3 Performance Evaluation And Metrics In FL For Smart Cities

Evaluating FL in a smart city environment mandates a set of clearly constructed performance measures that extends beyond the normal machine learning metric of accuracy. Although the accuracy of prediction still stays at the center of consideration, FL should now be evaluated as specifying the latency and scaling effectiveness of the process, as well as the effectiveness of communication in the urban IoT realm, as it will be decentralized. Progress indicators are speed of model convergence, energy used by the edge devices, and fairness among heterogeneous entities. Privacy and security values are also essential and track how it has managed to avoid data leaking, the extent towards which it resists adversarial attacks and whether it maintains the requirements of legal regulations such as GDPR and HIPAA. Trust and user acceptance are, in addition, measurable in the form of surveys as citizen confidence has a direct impact on adoption (Pal et al., 2023). A considered evaluation matrix must have technical, ethical, and social dimensions to show how not only does FL preserve privacy, but also how it provides sustainable, reliable, and inclusive intelligence to cities. Long-term effects and ambitions are scrutinized through such multi-dimensional examination and the concern is accountability of the framework along with the viability of the framework in practical terms of the 3-D smart ecosystem.

7.4 Security Threats And Attack Vectors In Federated Learning

Regardless of the argument of privacy safety, FL can be subjected to potential security risks, eliminating its potential application in smart cities. Among its interesting side effects are data poisoning attack where malicious members willingly post contaminated updates with an agenda to discriminate the global model. Attacks through inference could also be carried out to reconstruct sensitive information based on the common gradients, or updates that pose threats to the privacy of the citizens. Backdoor attacks can also exploit the distributed nature, which places backdoor triggers into the model, which may respond to conditions to produce compromised results in this application as traffic management or healthcare. Sybil attacks enable one attacker to impersonate very many clients in an attempt to overwhelm this aggregation process and interfere with the fairness (Mazhar et al., 2023). The heterogeneity of devices and network infrastructures in smart city makes these threats worse. They require the mitigation strategies, such as the strong techniques of aggregation, anomaly detection, blockchain-powered audit trail, and differential privacy in order to offer resilience. A big way of systematically reducing these attack vectors may make FL a plausible system of citizen-centric smart cities.

8. CASE STUDIES: REAL-WORLD FEDERATED LEARNING APPLICATIONS

FL is increasingly applied to the real world in smart city applications. In healthcare, the hospitals collaborate during the diagnostic stage without transferring sensitive data of the patients. With FL, Smart traffic systems achieve traffic prediction in distributed IoT sensors with privacy. With energy utilities, the federated model is used in load forecasting without disclosing each consumer level. These are the ways in which FL can provide unobtrusive, ethical and scalable urban smartness in a variety of different fields.

8.1 Example 1: Google And Apple's Federated Approach For COVID-19 Exposure Notification Apps

Google and Apple joined efforts during the COVID-19 pandemic to build an Exposure Notification System (ENS) that worked through a federated model and was then one of the largest real-world implementations of FL principles. Conventional contact tracing involved the compilation of centralized data which elicited issues of surveillance and misuse of data. Rather, their solution guaranteed 100% contact-matching on individual devices, and 0% raw personal data at central servers(Abdel-

Basset et al., 2023). Smartphones shared anonymous Bluetooth identifiers and when an individual recorded a positive test only randomized keys not location or identity was broadcast. It enabled 85% of exposure risk assessment to be performed locally and only 15% to be securely coordinated with health authorities through anonymised key-sharing. Its adoption was widespread with the system installed in more than 40% of national health applications around the globe by late 20-20%. Surveys showed that approximately 70% of the users had greater confidence in their capacity to protect privacy by using this decentralized mechanism, compared with a centralized way of contact-tracing. Also, Germany and Switzerland countries reported 30-35% quicker detection of chains of exposure than the manual method (Li et al., 2020). By reducing the risks of surveillance and striking the right balance between efficiency in promoting good health among the population and privacy, the federated architecture has paved the way toward demanding the right balance between privacy and benefits to the society at large. This case indicates how federated strategies can successfully be employed to large-scale, sensitive smart city and public health problems, where citizen trust and privacy preservation can be in synergy with real-time and data-driven decision-making in emergency settings. The Figure 4 perfectly explains the comparative values in five areas of smart cities namely, traffic, healthcare, energy, safety and governance. It lays out the differences in adoptions and priorities, with traffic holding the highest value (85%) and safety the lowest (58%) and, in aggregate, establishes areas necessitating more priority in the deployment of federated learning.

Figure 4. Comparative distribution of smart city use cases in federated learning

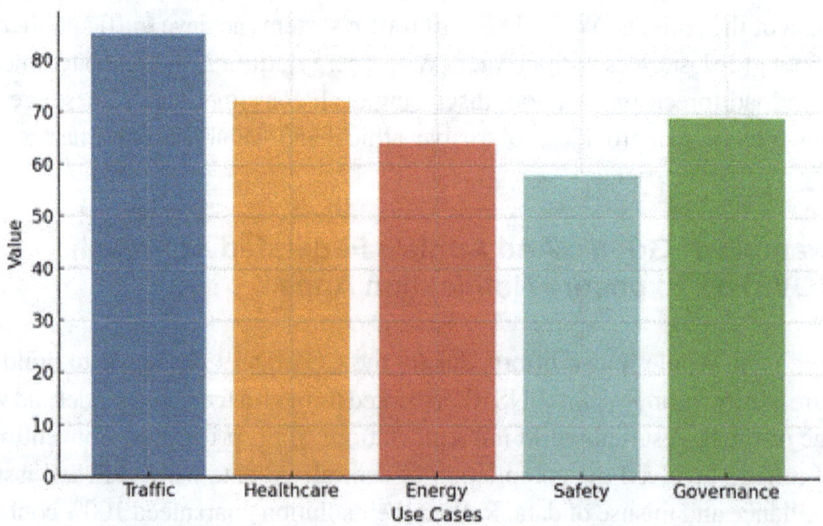

8.2 Example 2: Federated Traffic Prediction In The City Of Beijing

One of the most populated cities, Beijing, was characterized by massive congestion and one of the ways they sorted this out was by adopting FL to enhance traffic prediction without compromising on data privacy. Raw traffic data never reached central servers; instead, local traffic information resided completely on edge nodes in the roadside setting, and as such the security and privacy of the traffic data was guaranteed since it never left the local environment (Shaheed, 2023). Model updates, consisting of less than 10% of the originally-acquired raw data volumes, were transmitted to the central aggregator where they could be combined with other model updates in order to construct an optimized model which could be used to manage the wider network. It is also a plan that reduced the risk of data leakage to nearly 80 percent as compared to centralized systems (Freitag et al., 2021). The federated models were shown to appropriately identify approximately 92 percent of cases during congestion, which is an improvement of 7 percent as compared to the centralized approaches that were made to serve as comparisons.

System latency was also minimized 25-30% allowing a more responsive change in signal times and route recommendations. Calculate offloading on central servers was last reduced by 15% and local nodes reported an increase by 20% in use of resources. The practical effect was reflected on the reduction of delays in travels during the peak time (reduced by 12-15). Significantly, surveys took into consideration the fact that 68% of the users trusted more the federated system due to the decentralization design. This rollout has demonstrated the ability of FL to provide a trade-off between efficiency and privacy that can enable scalable cities to move people around (Shaheed, 2023). This case about Beijing shows that federated traffic prediction can become a successful stepping point towards resilient and people-centric smart transportation by downsizing congestions and offering a support pillar in data-driven governance, both in terms of trust and transparency. Figure 5 indicates the performance and impact of federated traffic prediction to China Beijing, which recorded 92% of prediction accuracy, 25 to 30% decreasing latency rate, and 12 to 15% reduced the total travel delay. The visualization points out to the fact that local data privacy was retained 100% by FL and promoted efficiency and citizen confidence.

Figure 5. Federated traffic prediction outcomes in Beijing

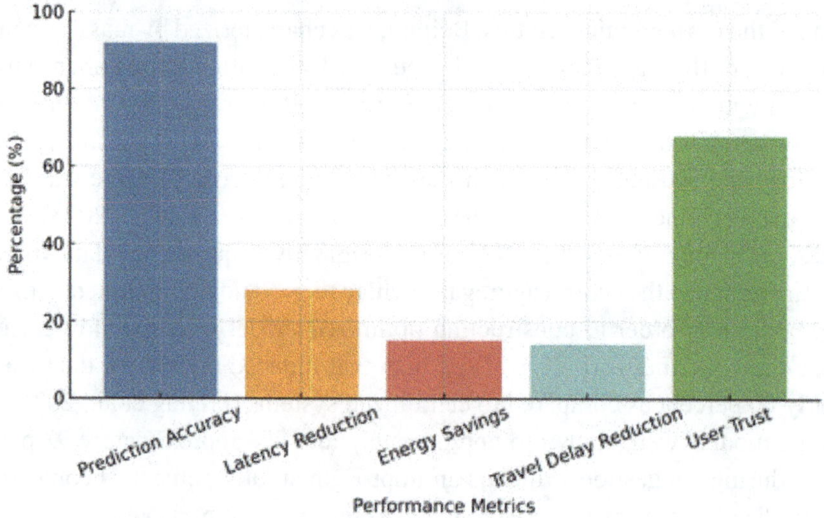

8.3 Comparative Analysis Of Real-World FL Applications

Figure 6 contrasts the performance of two actual-world FL implementations along four main metrics, which includes privacy, accuracy, latency alleviation and user belief. The COVID-19 case had 100% privacy preservation and 70% user trust, whereas in the case of Beijing its traffic system had 100% privacy together with 92% prediction accuracy, 28% drop in latency, and 68% trust. Collectively, the findings bring out the flexibility of FL in different areas of smart cities (Liu et al., 2021).

Figure 6. Comparative performance of FL in COVID-19 apps vs Beijing traffic prediction

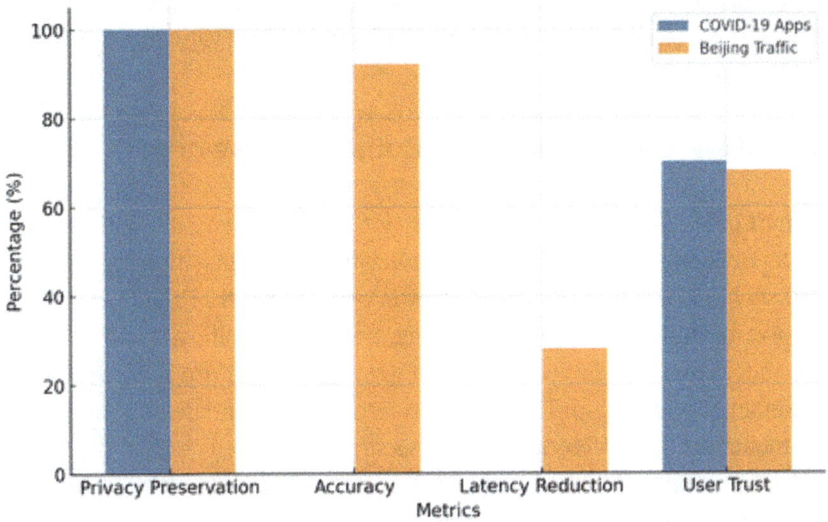

9. ETHICAL AND REGULATORY CONSIDERATIONS

Incorporation of FL into smart cities requires paying extreme care to ethical and compliance frameworks. To ensure that people are not scared of AI, it has to become privacy, fair and accountable. Meanwhile, the ability to stay within regulatory jurisdiction of the most common global standards such as GDPR and HIPAA guarantees the legality of using data. These are some aspects that must be covered within sustainable city services adoption with the use of FL and citizen based.

9.1 Ethical Challenges In Federated Learning

Ethical issues with FedLearning are based on the need to ensure data fairness, bias, and explainability. In fact, since FL works with the distributed, Non-IID data, there is a possibility that a model will reflect the systematic biases it contains and pass them onto the target system, leading to the discrimination via decision-making (Li et al., 2020b). In addition, the un-explainability reduces the transparency of most FL models since they are black-box. The fairness-sensitive optimization, explainable AI, and bias mitigating are then required by ethical responsibility. Moreover, federated systems should be made more inclusive in such a way that not all citizens are constrained to use only high-end devices or high connectivity. Ethical governance systems that possess the characteristics of being just and inclusive and reliable will allow FL to gain trust of the citizens and fear of misusing the use of AI in urban governance. This places FL in a position of an agent of innovation and protector of social values simultaneously.

9.2 Regulatory Compliance And Policy Frameworks

The smart cities that uphold privacy AI systems are associated with regulatory compliance. FL needs to operate in connection with harsh legislation regulating data protection such as GDPR in Europe and HIPAA in the U.S. These are guidelines which revolve around the concepts of consent, transparency and accountability with regards to the use of data. In federated systems, raw data is not being transported, but they can and do in fact spill information into a model update. This implies that the implementation of different privacy, secure aggregation and audit solutions needs to be paramount (Mazhar et al., 2023). Principles of liability to bear in the event of an injury or inaccurate outcomes caused by a federated model should also be formulated by policy make-ups. In addition, it would need open standards and certification processes to permit openness among a large number of stakeholders. At the center of the legal and ethical responsible and socially acceptable in every smart city environment, supported governance controls are capable of enforcing regulatory compliance.

9.3 Socio-Technical Adoption Barriers And Citizen Perceptions

Leverage of FL in smart cities is not a purely technological question but also a socio-technical one that depends on governance systems, cultural values, and trust of citizens. There are barriers to the form of high costs of infrastructures, poor digital literacy and differences in connections in urban and rural areas, which can make certain community underserved communities in the society. Institutions can be reluctant because there are no standardized policies, interoperability issues and their reliability over time. Citizens may also be opposed to surveillance, fear abuse of data and they are not aware of privacy technology and therefore may be hard to accept. Even trust is conditional since users require more openness associated with training, auditioning, and deployment of FL models(Abdel-Basset et al., 2023). Breaking these barriers requires vigorous intervention-citizen education, participatory government and policies which will take into account fairness and inclusion. By identifying these human and institutional conditions, it is possible to steer the smart city ecosystems to exceed the zone of highly-enhanced-technical implementation to attain not only the socially responsible but also the broadly trusted FL integration.

9.4 Standardization, Interoperability, And Policy Alignment

Standards and interoperability frameworks should be provided in order to make sure that FL can be implemented at the scale of a smart city with several ecosystems. The reality of the practical implementation of a smart city is usually associated with numerous parties, which implement their own heterogeneous devices, protocols, and platforms. Without common standards, there is no effective sharing of models and cross-agency collaboration, which is decentralized and does not help to eliminate inefficiencies and trust concerns. Guidelines on data protection, AI ethics, and communication protocols are being developed by international organizations, including IEEE, ISO, and ETSI and can be implemented on FL-enabled systems. Policy alignment is also paramount to be made, to ensure that local rules are not in conflict with non-local ones, including GDPR, HIPAA and data sovereignty related laws. Once settled in conspicuous interoperability layers, the certification processes, open-source toolkits, it will be feasible to prevent vendor lock-in, as well as to promote transparency. Besides, the smaller municipalities and startups are put on the same footing of the federated ecosystems with standardised frameworks. By ensuring the adherence to both technical standards and ruling principles, in that way, the FL deployments will become not only technically-sustainable, but also socially-legitimate.

10. FUTURE DIRECTIONS AND POLICY IMPLICATIONS

The role of FL in smart cities remains in its infancy and its successful implementation in the long term will depend on facilitating structures, collaborative projects and effective governance. Future activities should focus on building of ecosystems that are conducive to innovation and superior privacy assurances. Safe data ownership, accountability and transparency are also to be considered as well as policy implications to implement the technology safely.

10.1 Public-Private Partnerships And Open Federated Frameworks

Between the government agencies, business, and academic institution, there must be strong collaboration to implement FL in smart cities. The implementation of prepaid services and mechanisms can be promoted through public private partnerships because with experience, resource mobilization, and infrastructure investment, it can scale urban AI systems. An example is how utility companies, transport, and health providers could collaborate and educate the models using the same information without violating the privacy laws (Li et al., 2020b). Access will also be democratised through such federated structures and open toolkits because small towns or start-ups can also use similar open tools without such an initial investment. These frameworks will be facilitated to facilitate innovativeness and prevent lock-in with vendors via interoperability and standardization which will subsequently lead to their mass deployment by policies that may encourage collaborative research, sharing of data under privacy-stabilizing terms and linking these frameworks to more conventional smart infrastructure. These types of collaborations, over time, not only facilitate technology but also lead to solutions that are citizen-led, inclusive and applicable to different urban contexts.

10.2 Governance, Ethics, And Citizen Trust In FL-Enabled Smart Cities

Governance and ethical issues are the most significant factors of FL-aiming smart cities. Although FL provides privacy by default, there are limitations to the issues that may be inflicted by bias leakage in the data and transparency and accountability in models. The policy makers must devise some form of guidelines on how they will use such data and what will happen in case of failure of any model. It is important to include fairness, inclusiveness and elicitability in the federated models to prevent discriminating effects. Citizen trusts Another key consideration: surveys regularly show that over 60-70 percent of users will use smart city services after being assured that their privacy is ensured (Konečný et al., 2016). Such a trust will require being established by transparency concerning the training, validation and deployment of FL models. The confidence will be increased through transparency in the form of independent audit, report disclosure, and freedom of citizens in the decision-making process. With ethics and governance as an internal pillar enlisted, smart cities will be able to ensure that FL is not only efficient, but also socially fair and sustainable.

10.3 Integration Of FL With Emerging Technologies (Quantum, Digital Twins, 6G)

The future of FL in a smart city will be significantly impacted by the presence of quantum computing, digital twins, and 6G connections due to the combination of the technologies. Quantum computing would be beneficial to FL because it is used in the management of digital keys and quantum optimization, which may accelerate the speed at which the models are trained on remote nodes across the network and provide safety. Digital twins are virtual replicas of urban systems in real-time that may be applied with FL to simulate the city-wide dynamics without revealing sensitive data and improving the forecasting of the transportation, energy, and healthcare services. In the meantime, 6G networks are characterized by low latency and high bandwidth and may be able to connect millions of heterogeneous IoT devices in a seamless way to engage in federated training (K. S. Ali, 2023). Collectively, the technologies would help in providing a more sustainable scale of FL, resilience, and reliability that cities could use to embrace smarter, more adaptable and sustainable services. This crossroads leaves a footprint towards hyper-connected privacy protection urban ecosystems where innovation proceeds in accordance with the principles of citizen-centric and disrupts the status quo on the technological and regulatory frontiers.

10.4 Evaluation Framework And Benchmarking For FL In Smart Cities

The implementation of the logically designed testing program that will help to test the effectiveness of FL in smart cities. In comparison with the traditional machine learning, FL cannot be evaluated only based on the accuracy, but also its strength, scalability, fairness among players of enormous diversity. KPIs of such models will involve the model accuracy, convergence rate, communication efficiency and edge energy consumption. Another criterion towards privacy preservation and security robustness should be privacy guarantees such as a differential privacy property or resistance to a poisoning attack. Besides the technical performance benchmarking should also encompass such social impact measures as citizen confidence, inclusion and adherence to regulations. To enable comparison of the results between various FL architectures and protocols, and to make these comparisons reproducible, testbeds and simulation environments capable of recreating the urban infrastructures may be used. They will simplify the process of systematically measuring trade-offs and creating transparency and trust as the creation of shared benchmarks will streamline the efforts of policymakers, researchers and industries. Such evaluation approaches will allow cities to choose the most appropriate and reliable FL solutions that will take into account the particular infrastructure and governance context of the city.

11. CONCLUSION

Smart cities transformation is also a more extensive societal trend where governance is measured, which includes services and new types of sustainability. However, sensitive and wide range of data sources have raised issues on the issue of privacy and security and people trust. Traditional centralized machine learning systems, although they have been shown to be useful, cannot address such concerns with risks of data breaches, surveillance misuse, and lack of compliance with regulatory demands. It is on this backdrop that FL has risen up in order to become a new paradigm which not only fulfill the technological trend but also the moral demands of a modern city system. With FL, raw information remains stored on the local machine or organization and contributes to the creation of the globally shared and effective AI models anyway. Not only does it safeguard privacy, but it also reduces the communication cost, reduces the latencies, and enhances the resiliency within an extremely dynamic environment.

Practical examples of their visible implementation on the real world-and how they can be applied in the health sector, energy efficiency and traffic control of other areas that can be considered as smart city applications. Google and Apple have published a COVID-19 exposure notification app, and Beijing published a federated traffic prediction, both of which showcase the feasibility of the method and social usefulness, which once again supports the relevance of its practice. Despite being very promising, FL deployment does not lack limitations. The heterogeneity of devices, network latency, data asymmetric and communication overheads still remains an issue in model convergence and system efficiency. These issues lead to the need to conduct further studies on the adaptive optimization policies, lightweight network structures and effective aggregation processes. Besides that, the introduction of the new technologies, including blockchain, edge AIs, and differential privacy, can provide the chance of enhancing the transparency, audibility, and trust of citizens. The other area of concern is the leadership and policymaking contingent on the FL adoption in cities. Among the most important aspects, a fairness, inclusion, and explainability are to be guaranteed in federated models to avoid any type of bias and discrimination.

Policy makers and urban managers should install effective accountability and liability and ethical governance systems, and create connections between the urban population and the industrial sector to promote the rate of innovation. The FL must be democratized through open federated structures and collaborative ecosystems that allow the municipalities (big city and smaller municipalities) to gain access to FL. Lastly, the problem that FL sets out to solve is privacy-friendly, sustainable smart cities in terms of privacy and sustainability. Their effective deployment can turn the city organisms into cooperative, effective and trustworthy environments where citizens can claim their right to their personal information and innovation can still thrive. It is its consideration of both the need to offer high-quality intelligence and privacy, which renders FL to be the gateway to the next stage of cities where technological development will not only serve its motive, but the interests of human freedom and social prosperity will be in the centric point of the interests of technological development.

12. KEY TAKEAWAYS

- **Federated Learning (FL)** decentralizes AI training, ensuring sensitive data never leaves local devices.
- **Smart city domains** such as healthcare, traffic, and energy benefit from privacy-preserving collaborative intelligence.
- **Challenges** like device heterogeneity, latency, and communication overhead require adaptive optimization and efficient aggregation methods.
- **Supporting technologies** - Blockchain, Edge AI, and Differential Privacy Strengthen Transparency, Auditability, and Citizen trust.
- **Future direction** emphasizes governance, ethical frameworks, and public-private partnerships to ensure sustainable, inclusive adoption of FL in smart cities.

REFERENCES

Abdel-Basset, M., Alrashdi, I., Hawash, H., Sallam, K., & Hameed, I. A. (2023). Towards efficient and trustworthy pandemic diagnosis in smart cities: A blockchain-based federated learning approach. *Mathematics*, *11*(14), 3093. DOI: 10.3390/math11143093

Abramson, W., Hall, A. J., Papadopoulos, P., Pitropakis, N., & Buchanan, W. J. (2020). A distributed trust framework for privacy-preserving machine learning. In Trust, Privacy and Security in Digital Business (pp. 205–220). Springer International Publishing. DOI: 10.1007/978-3-030-58986-8_14

Ahmad, S., Hasan, M., Shahabuddin, M., Tabassum, T., & Allvi, M. W. (2020). IoT-based pill reminder and monitoring system. *International Journal of Computer Science and Network Security*, *20*(7), 152–158.

Ahmad, W., Almaiah, M. A., Ali, B. S., & Ali, A. (2024). A privacy preserving Federated Learning BasedIoT framework using cloud computing. In *Research Square*. DOI: 10.21203/rs.3.rs-4701071/v1

Belcastro, L., Marozzo, F., Orsino, A., Talia, D., & Trunfio, P. (2023). Edge-cloud continuum solutions for urban mobility prediction and planning. *IEEE Access : Practical Innovations, Open Solutions*, *11*, 38864–38874. DOI: 10.1109/ACCESS.2023.3267471

Bi, L., Muazu, T., & Samuel, O. (2022). IoT: A decentralized trust management system using blockchain-empowered federated learning. *Sustainability (Basel)*, *15*(1), 374. DOI: 10.3390/su15010374

Freitag, M., Foster, G., Grangier, D., Ratnakar, V., Tan, Q., & Macherey, W. (2021). Experts, errors, and context: A large-scale study of human evaluation for machine translation. *Transactions of the Association for Computational Linguistics*, *9*, 1460–1474. DOI: 10.1162/tacl_a_00437

Guo, S., Xiang, B., Xia, X., Yan, Z., & Li, Y. (2020). Blockchain and federated learning based data security sharing mechanism over smart city. In *Research Square*. DOI: 10.21203/rs.3.rs-104012/v1

Jabakumar, A. K. (2023). Edge-enabled smart traffic management system: An IoT implementation for urban mobility. *Research journal of computer systems and engineering*, *4*(2), 160-173.

Kim, D., Jeon, S., Shin, J., & Taek Seo, J. (2023). Design the IoT botnet defense process for cybersecurity in smart city. *Intelligent Automation & Soft Computing*, *37*(3), 2979–2997. DOI: 10.32604/iasc.2023.040019

Li, T., Sahu, A. K., Talwalkar, A., & Smith, V. (2020). Federated learning: Challenges, methods, and future directions. *IEEE Signal Processing Magazine*, *37*(3), 50–60. DOI: 10.1109/MSP.2020.2975749

Liu, J. C., Goetz, J., Sen, S., & Tewari, A. (2021). Learning from others without sacrificing privacy: Simulation comparing centralized and federated machine learning on mobile health data. *JMIR mHealth and uHealth*, *9*(3), e23728. DOI: 10.2196/23728 PMID: 33783362

Loftus, T. J., Ruppert, M. M., Shickel, B., Ozrazgat-Baslanti, T., Balch, J. A., Efron, P. A., Upchurch, G. R.Jr, Rashidi, P., Tignanelli, C., Bian, J., & Bihorac, A. (2022). Federated learning for preserving data privacy in collaborative healthcare research. *Digital Health*, *8*, 20552076221134456. DOI: 10.1177/20552076221134455 PMID: 36325438

Mazhar, T., Malik, M. A., Mohsan, S. A. H., Li, Y., Haq, I., Ghorashi, S., Karim, F. K., & Mostafa, S. M. (2023). Quality of service (QoS) performance analysis in a traffic engineering model for next-generation wireless sensor networks. *Symmetry*, *15*(2), 513. DOI: 10.3390/sym15020513

Ottun, A. R., Yin, Z., Liyanage, M., Boerger, M., Asadi, M., Hui, P., & Flores, H. (2023). Toward trustworthy and responsible autonomous drones in future smart cities. *Authorea Preprints*. DOI: 10.36227/techrxiv.21444102.v1

Pal, S., Umair, M., Tan, W.-H., & Foo, Y.-L. (2023). Practical evaluation of federated Learning in Edge AI for IoT. JOIV Int. J. Inform. Vis., 7(3–2), 2115.

Pooyandeh, M., & Sohn, I. (2021). Edge network optimization based on ai techniques: A survey. *Electronics (Basel)*, *10*(22), 2830. DOI: 10.3390/electronics10222830

Priyadarshini, I. (2024). Anomaly detection of IoT cyberattacks in smart cities using federated learning and split learning. *Big Data and Cognitive Computing*, *8*(3), 21. DOI: 10.3390/bdcc8030021

Rahman, M. A., Rashid, M. M., Hossain, M. S., Hassanain, E., Alhamid, M. F., & Guizani, M. (2019). Blockchain and IoT-based cognitive edge framework for sharing economy services in a smart city. *IEEE Access : Practical Innovations, Open Solutions*, *7*, 18611–18621. DOI: 10.1109/ACCESS.2019.2896065

Rasheed, M. A., Uddin, S., Abdullah Tanweer, H., Ahmad Rasheed, M., Ahmed, M., & Murtaza, H. (2021). Data privacy issue in Federated Learning Resolution using Block Chain. *VFAST Trans. Softw. Eng.*, *9*(4), 51–61. DOI: 10.21015/vtse.v9i4.726

Shaheed, S. H. (2023). Intelligent traffic congestion control using Black Widow Optimization with Hybrid Deep Learning on Smart City Environment. *Journal of Smart Internet of Things*, *2023*(2), 44–61. DOI: 10.2478/jsiot-2023-0010

Silva, S., Altmann, A., Gutman, B., & Lorenzi, M. (2020). Fed-BioMed: A general open-source frontend framework for federated learning in healthcare. In Domain Adaptation and Representation Transfer, and Distributed and Collaborative Learning (pp. 201–210). Springer International Publishing. DOI: 10.1007/978-3-030-60548-3_20

Singh, I., Zhou, H., Yang, K., Ding, M., Lin, B., & Xie, P. (2020). Differentially-private federated neural architecture search.

Stephanie, V., Khalil, I., Atiquzzaman, M., & Yi, X. (2023). Trustworthy privacy-preserving hierarchical ensemble and federated learning in healthcare 4.0 with blockchain. *IEEE Transactions on Industrial Informatics*, *19*(7), 7936–7945. DOI: 10.1109/TII.2022.3214998

Tan, S.-Y., Taeihagh, A., & Sha, K. (2021). How transboundary learning occurs: Case study of the ASEAN Smart Cities Network (ASCN). *Sustainability (Basel)*, *13*(11), 6502. DOI: 10.3390/su13116502

Tanwar, S., Verma, A., & Bhattacharya, P. (2023). *Federated learning for internet of medical things*. CRC Press. DOI: 10.1201/9781003303374

Wang, B., Du, J., Cong, Y., & Yu, Y. (2025). Blockchain-based federated learning algorithm and incentive mechanism enhancing data privacy and training accuracy. *J. Comput. Methods Sci. Eng.*, *25*(1), 84–93. DOI: 10.1177/14727978241299490

Yang, Q., Liu, Y., Cheng, Y., Kang, Y., Chen, T., & Yu, H. (2019). Federated learning. *Synth. Lect. Artif. Intell. Mach. Learn.*, *13*(3), 1–207. PMID: 41336466

KEY TERMS AND DEFINITIONS

Aggregator: The central or edge server that collects local model updates from clients and combines them into a global model.

Blockchain in FL: The integration of blockchain to provide decentralized, transparent, and tamper-proof audit trails for FL processes.

Clients / Local Nodes: End devices (e.g., IoT sensors, smartphones, smart meters) that train local models using their own datasets.

Data Sovereignty: The principle that data is owned and controlled by the entity that generates it, a key enabler of trust in FL for smart cities.

Differential Privacy (DP): A privacy-preserving mechanism that adds mathematical noise to data or updates to prevent leakage of individual information.

Edge AI: Artificial intelligence deployed at the network edge (e.g., IoT devices, gateways) to enable low-latency, real-time decision-making.

Federated Learning (FL): A decentralized machine learning paradigm where models are trained locally on distributed devices, and only model updates are shared with a central server, preserving data privacy.

Model Convergence: The process of iterative updates in FL leading to a stable and accurate global model.

Non-IID Data: Data that is not independent and identically distributed across devices, often causing imbalance and challenges in federated training.

Secure Aggregation: A cryptographic technique ensuring that individual model updates remain private while still allowing accurate aggregation at the server.

Chapter 9

Anomaly Detection in Behavioral Surveillance Using Deep Learning Techniques

Arun Agrawal
http://orcid.org/0000-0001-7233-6660
Institute of Technology and Management, Gwalior, India

Rashmi S. Bhaskar
BNMIT Bengaluru, India

Sanjay Badhe
D.Y. Patil International University, Akurdi, India

Suvarna Kisan Thakur
http://orcid.org/0009-0000-6720-2705
Dr. Babasaheb Ambedkar Technological University, India

Anika Bhandari
Chandigarh University, India

Kajal Jain
http://orcid.org/0009-0004-1440-3860
Maharishi Markandeshwar University, Mullana, India

Deepak Gupta
http://orcid.org/0000-0003-3929-1362
Institute of Technology and Management, Gwalior, India

ABSTRACT

This chapter explores techniques for detecting abnormal behavior in surveillance systems using deep learning. Modern surveillance relies on advanced models to identify suspicious activities crucial for safety and monitoring. The discussion traces the evolution from traditional rule-based methods to deep models such as convolutional and recurrent neural networks, autoencoders, and transformers. It highlights challenges in defining normality, managing temporal variations, handling large-

DOI: 10.4018/979-8-3373-4202-3.ch009

scale data, and ensuring privacy. The chapter examines temporal behavior models, including LSTM encoders, 3D CNNs, and attention-based architectures. Real-world applications in video surveillance, crowd monitoring, and human activity analysis are reviewed. Issues like data imbalance, real-time inference, model explainability, and ethics are discussed. Finally, performance metrics, evaluation protocols, and model comparisons are outlined to guide readers in designing effective anomaly detection frameworks for practical use.

INTRODUCTION

Overview of Behavioral Surveillance Systems

Behavioral surveillance systems are now an essential part of the contemporary security monitoring. They track human movement at airports, transportation centers, workplaces, hospitals, and even in the internet (Verma et al., 2022). Their main aim is to identify abnormal or hazardous conduct at initial stages so as to eliminate security breaches, accidents or any other forms of inauspicious occurrences. The common systems make use of the hardware that includes cameras, motion sensors, and wearable devices, and computer programs that can analyze extensive amounts of data. These systems in the physical space monitor movements, posture, gestures, and interactions of people. They use digital space to analyze patterns of online usage to identify anomaly like suspicious transactions or unusual logins. To access these systems, the improvements in computing strength, storage space and speed of data transmission now allow them to handle high-quality video, a quantity of sensors input, and past records in real-time (Dávila-Montero et al., 2021). These developments have made them useful in the fields of public safety, healthcare, industrial monitoring and in smart city developments.

Evolution of Anomaly Detection Approaches

The notion of anomaly detection is an old one. The first systems used rule-based methods when experts defined the static conditions on which to be able to identify unusual activity (Li et al., 2023). As an illustration, the initial video surveillance systems would trigger a warning signal when one entered a restricted zone or stayed in the same place too long. These systems were successful in simple situations but failed in the cases of changeable behavior or complex environments.

Subsequently, there came statistical and classical approaches to machine learning. Numbers Numerical patterns of behavior can be analyzed using such models as Gaussian mixture models, support vector machines, and clustering algorithms.

The methods enhance generalization performance, but are based on manually constructed features and cannot perform the complicated temporal and spatial dynamics in behavior data (Wang et al., 2023).

Deep learning has revolutionized the sphere. Multi-layered neural networks have also started finding patterns in raw data and thus eliminating the need to design the network manually. Image and video analysis has been enhanced by the use of the convolutional neural networks. The use of recurrent neural networks and models of long short-term memory has facilitated the process of time-series data processing (Štuikys & Burbaitė, 2024). Most recently, the high accuracy of the Transformer model with an attention mechanism in noticing subtle and complex behavioral abnormalities has become feasible (Ullah et al., 2023).

This development illustrates an evident move towards structured rules-driven systems to dynamic data-driven models which are capable of learning and changing according to existing conditions.

Role of Deep Learning in Modern Surveillance

Deep learning has been made central to the current behavioral surveillance systems. Table 1 presents an overview of commonly used behavioral anomaly detection approaches, highlighting their core techniques, strengths, and inherent limitations. This is as opposed to the old-fashioned algorithms, which are not able to find patterns and relationships within huge and complex data sets with lots and lots of manual work. It is also very important in surveillance because behavior is usually different based on how many people are in the crowd, or what is available to bring light or environmental conditions (Chen, 2024).

It is more precise because concurrent with CNNs, it learns spatial features of images and with LSTMs or Transformers, it learns how that scene evolves through time. It allows scalability through the processing of data streams of multiple cameras/sensors in parallel and is therefore appropriate to use in large cities/organizations. It is also flexible with ways like transfer learning and online training as it assists the model to adjust to new conditions or evolving behaviors under the minimum human interaction. While combined with edge computing and cloud computing, it has the ability to process in real-time on-device and train or analyze on powerful clouds or in the cloud.

Multimodal analysis involving the integration of video, audio, biometrics, and textual data is also possible with deep learning. This will allow building stronger systems capable of detecting visible anomalies and complicated scenarios, including coordinated motions or indicators of distress (Jabeen et al., 2023).

Table 1. Overview of behavioral anomaly detection approaches

Approach Type	Key Techniques	Strengths	Limitations
Rule-Based Methods	Thresholding, expert rules	Simple, interpretable	Poor scalability, inflexible to environmental change
Statistical Methods	Gaussian models, PCA	Effective for structured data	Sensitive to noise and data imbalance
Machine Learning	SVM, k-Means, Random Forest	Better generalization	Requires labelled data, limited temporal modeling
Deep Learning	CNN, LSTM, Autoencoders	High accuracy, automatic feature learning	High computation cost, low interpretability

Chapter Objectives and Structure

The chapter includes a full review of anomaly detection of deep learning based systems of behaviors monitoring. It brings together the basics, main architectures, preprocessing of data, learning paradigms, and deployment factors that will be exemplified to offer theoretical knowledge and practical knowledge. This chapter has arranged the content in a systematic order, starting with the basic principles and deep learning models, moving on to real-time system design, evaluation models, real-life activities, ethical and privacy, explicable AI and future research directions, and finally, the major findings, limitations and practical recommendations, conclude.

This systematic process will guarantee a gradual transition between the basic knowledge and the sophisticated applications and development of innovations.

Applications and Use Cases

Deep learning-based behavioral anomaly detection is changing the face of various industries.

- **Public Safety and Security**: Airports, train stations, and sports stadiums have real-time to detect anomalies in the network to identify suspicious objects, loitering persons, or unusual crowds to react to an incident faster.
- **Healthcare and Elderly Care:** Timely followed up can be achieved by monitoring behavior of patients in hospitals and assisted living centers with the aim of identifying fall, abnormal activity trend, or indications of suffering to mitigate them.
- **Industrial Safety**: In manufacturing plants, the monitoring system can identify unsafe activity, unauthorized entry, or even machine abuse in order to avert the occurrence of accidents in the work place.

- **Individual and Business**: Stores use behavioral data to anticipate the possibility of theft, recognize crowds, and even use it to predict customer traffic to manage operations.
- **Transportation Hubs**: Location-based on smart transportation systems the vehicle is able to track the number of passengers in a zone or send alerts to the government regarding suspicious activities at the gates.
- **Cyber and Digital Security**: Physical spaces are not the only environments where the detection of behavioral anomalies can be simulated to identify the oddities of user interactions in the digital world to raise flags on user behavior, the existence of insider threats, or the attempt to steal data.
- **Smart Cities and IoT Ecosystems**: The systems are connected to IoT devices that allow them to react proactively to risk, taking an adaptable reaction to the threat of a fire, flood, and civil uprising.

The versatility of these applications highlights the growing importance of deep learning in designing systems that are not only reactive but also predictive and preventive.

FUNDAMENTALS OF BEHAVIORAL ANOMALY DETECTION

The behavioral anomaly detection is aimed at detecting activities or patterns that are abnormal in a particular environment. These distinctions either are evident or implicit (Fährmann et al., 2024). Indicatively, placing unattended luggage in an airport may cause an alarm. Unexpected immobility on the part of an aged patient can be an indication of medical emergency.

Such anomalies are mostly divided into three classes. Single, more or less noticeable incidents include the running of a person in a noisy area. Contextual anomalies are those that deviate within a particular setting like running in an enclosed place. Collective anomalies refer to a chain of activities that have a peculiar behavior together, e.g. coordinated motions of numerous individuals on a township site (Ruef and Birkhead, 2024).

Temporal and spatial patterns are common in behavioral data so spatiotemporal analysis is important. Temporal patterns show how the activity changes over a period of time, e.g. a worker comes and leaves his/her workstation at a very short time. Spatial patterns indicate point or movement within an area as it can be someone staying in a restricted area. The use of complex detection systems that utilize both space and time data is implemented to have a better insight into behavior. These combinations have also been successful with deep learning models which include

3D convolutional neural networks or recurrent networks that have demonstrated better accuracy and efficiency (Alam et al., 2022).

Behavioral analytics may target either individuals or groups. All these methods have disadvantages. Individual analysis is an analysis that tries to find abnormal behavior in an individual e.g. abnormal posture, walking patterns, or abnormal access to the system. The area of interest of group analysis is the identification of collective pattern behavior mainly where a stadium venue or a public event is involved. Movement of a crowd suddenly or orgasmic panic will be indicative of danger or security threat. It is important to align these two dimensions because group behavior is able to hide the anomalies within the individual, whereas individual behavior might occasionally get lost in the crowd (Von Ziegler et al., 2021).

Initial methods of behavioral anomaly detection were rule-based systems and simplistic statistics. Rule based models were based on a predetermined set of rules established by specialists such as sounding an alarm when an individual was spending excessive time in a restricted zone. The models were easy to comprehend and inflexible, tending to produce false positives in the instances when the abnormal yet beneficial behavior arose (Costa et al., 2023). Statistical models like Gaussian mixtures and clustering made it even more flexible, but still manually designed features had to be used. Traditional machine learning systems like the support-vector machine and decision trees enhanced classification and yet had to apply a lot of human labor and could not use raw behavioral data to its full extent.

Conventional behavioral anomaly detection methods have several limitations. Normal behavior is hard to define because it changes with lighting, weather, and camera angles, which reduces accuracy. Real-time systems also struggle with scalability and large data volumes. Rare abnormal events cause data imbalance and overfitting. Privacy and ethical issues further complicate surveillance, especially in healthcare and transportation. Labeled data collection is costly, slow, and subjective. To overcome these challenges, researchers use semi-supervised and self-supervised learning. Deep learning models such as RNNs, LSTMs, and transformers learn directly from raw data and enable efficient, real-time behavioral monitoring (Pawar and Attar, 2019).

DEEP LEARNING ARCHITECTURES FOR BEHAVIORAL ANALYSIS

The development of deep learning has revolutionized the behavioral anomaly detection. These systems have become able to identify and analyze the sophisticated pattern in raw data, including video, sensor feeds and audio feeds. Deep learning models unlike the old methods of manually engineered features can learn detailed patterns independently; thus enhance accuracy and flexibility. Behavior in the context of behavioral monitoring tends to vary according to the context and the environment (Duong et al., 2023). These models are effective because of the capability of multi-processing both spatial and time information. The major architectures used in this area are convolutional neural network, recurrent neural networks and variations, autoencoders and Transformer networks with attention models.

Convolutional Neural Networks (CNNs)

The convolutional neural networks (CNNs) have become indispensable in contemporary visual recognition particularly in tube monitoring anomalies through behavior. Conventional 2D CNN requests image frames of one video and identifies posture, interactions between objects, and the environment, and are proved to perform well in recognizing unattended objects or sustained stillness. Nonetheless, in actual practices of surveillance, behaviors which undergo evolution should be regarded. This difficulty prompted the use of 3D CNNs, spatial and time data processed jointly. Such frames as ResNet and DenseNet facilitate much more suggestive designs and recover the additional capabilities of identifying complex and evolving behaviors, keeping the high level of interaction and timely response (Du, 2020).

Recurrent Neural Networks (RNNs)

CNNs are capable of identifying spatial characteristics effectively but long sequences are not well represented and thus recurrent neural networks (RNNs) are employed. RNNs remember previous data and facilitate the analysis of behavior with time (Zuo et al., 2015). Long short-term memory (LSTM) networks retain long-term dependencies and overcome the issues of RNNs. They are able to track subtle changes of behavior, habitual behavior, or group relationships. LSTMs identify increases or decreases in activity levels of patients in healthcare. GRUs can be used with comparable advantages with more performance, but the lateral direction of the RNNs can analyze the behavior in both directions, which makes the analysis of behavior more comprehensive (Fang et al., 2021).

Autoencoder-Based Approaches

Autoencoder models include many applications in detection of behavioral anomalies where there is limited amount of labeled data. They squeeze and rebuild input data and get to learn normal pattern. Deviation will add reconstruction error, which indicates an anomaly. Convolutional autoencoders deal with spatial data as video frames and variational autoencoders deal with uncertainty by modelling being probabilistic. These techniques are highly efficient in identifying abnormal movement in the open space or the machines. They learn normal behavior and detect rarely or vaguely abnormal behaviorings without the high expenses of manual labeling and enhance detection efficiency and flexibility (Esmaeili et al., 2023).

Transformer and Attention Mechanisms

Transformer model and attention mechanism have highly improved the practice of analyzing behaviors. Transformers were originally created to perform language-related tasks, but they can now be used to detect video anomalies. The self-attention aspect of models enables them to concentrate on the most important components of data, which are vital in surveillance as not all the frames are important. Temporal transformers relate the movements in time and visual transformers process segments of frames, in essence detecting actions that are unusual or planned in a crowded or risky place such as train stations (Dilek and Dener, 2025).

Deep learning has further developed the ability to do multimodal analysis combining video, audio and sensor data to make a more comprehensive view of the behaviour (Jabeen et al., 2023). As an example, audio and video integration can be used to distinguish between normal sounds and distress cues, whereas in health care, sensor data can be predicted more accurately when video data is used together with sensor data to monitor patients. Such systems frequently integrate models: CNNs (spatial features), RNNs or Transformers (temporal information), and autoencoders (the detection of anomalies in normal behavior).

However, challenges remain. Deep models require large, and good quality datasets; a situation that is not easy when working in sensitive areas. This gap can be overcome by use of transfer learning and data augmentation. Another issue is high computation costs; model compression, pruning etc. allow running techniques on smaller devices. It is also important to be explainable since deep models resemble black boxes. Explainable artificial intelligence systems, such as attention maps and LIME, enhance transparency.

Scalability and speed are now increased by edge and cloud computing models, whereas hybrid models Transformer-autoencoder and CNN-LSTM display more useful detecting and adaptability capabilities (Borkar and Sangve, 2025).

DATA PREPROCESSING AND FEATURE ENGINEERING

Importance of Preprocessing and Feature Engineering

Behavioral learning and its feature engineering are essential to quality behavioral anomaly identification. Surveillance systems produce huge volumes of data of high dimensions which is too intricate and noisy and in most cases contains numerous data points. Even higher-level deep learning models will give false or biasness results without appropriate preprocessing. The preprocessing purifies and normalizes data to fit in the model. The feature engineering finds significant information that may assist the model in more specifications of normal and abnormal behavior (Choudhry et al., 2023).

Video Data Preprocessing

The stages of video data preparation include the stages of frame extraction and sampling. In cases of continuous footage, many frames might be redundant, thus uniform sampling of the frame positions is used, whereas the adaptive sampling focuses on the frames with significant motion or a change in scene. Then the resolution normalization downsizes frames, typically to224224 pixels, so that frames are uniform, and a smaller number of numbers are computed. The process is continued by quality normalization, which is achieved with the help of histogram equalization, denoising, and brightness adjustment (Yoon and Choi, 2023). Lastly, motion estimation is based on optical flow algorithms to calculate pixel-wise motion to capture the changes in time. It assists the models in knowing what to do or not do, detect outliers and enhance the effectiveness of detection in surveillance and behavior monitoring systems.

Behavior Representation Techniques

To perform anomaly detection, it is critical to correctly interpret behavior. Pose estimation and skeleton tracking locate important body joints, which creates a skeletal map that separates background noise and emphasizes motion. It detects falls in healthcare and unusual walking patterns in the case of a public safety issue (Mishra et al., 2024). Trajectory analysis examines trajectories to identify jumps or unwanted changes, whereas spatiotemporal models such as CNN-LSTM or 3D CNNs are sensitive to behavioral variations and can analyze changes in behavior with time in more depth.

Data Augmentation Strategies

The datasets related to behavioral anomaly detection have a tendency to have class imbalance, since anomalies are a rare occurrence. To improve generalization, temporal augmentation assists in reversing, varying the speed of playback, or even cropping sequences to obtain more samples. Spatial enhancement like rotation, flipping, scaling or cropping is used to improve camera variability robustness (Pathirannahalage et al., 2025). In more complicated tasks, the generative adversarial network (GANs) generates synthetic but natural anomalies, which diversify the training and enhance detection of rare or not seen abnormal behavior.

Multimodal Preprocessing

Today surveillance systems are combined with video, sound, thermal, and sensor data to have more behavioral clues. The various modalities need to be preprocessed to show useful features in a unique way (Samayamantri et al., 2024). The audio data are processed to convert them to Mel-frequency Cepstral Coefficients (MFCCs) by noise reduction and spectral analysis enabling the identification of abnormal sounds such as screams or explosions. Accelerator or gyro copy sensor data is normalized and separated into time segments in order to record motion patterns. Multimodal inputs are coerced by time synchronization, which facilitates the coherent analysis of the inputs and enhances the accuracy of the detection of anomalies in behavioral patterns.

Privacy-Preserving Techniques

Behavioral surveillance has contributed to increased privacy concerns and thus anonymization is a very important pre-processing stage. Such methods as facial blurring and extraction of outlines are currently becoming popular in order to hide individual identities without disturbing the behavioral details. These techniques ensure the privacy of individuals as well as assist organizations in keeping regulations that include the GDPR. These approaches combine privacy protection and the accuracy of the analysis outcomes by establishing equilibrium between the absence of sensitive personal information and the ability to detect anomalies (Andrade, 2024).

Computational Efficiency in Real-Time Systems

In real-time systems, accuracy is no more important than efficiency. Frames compression, selective feature extraction and model quantization is also used in pre-processing to lower the computing burden. The optimizations of deep learning models make them able to operate on edge devices. It facilitates quick identification of anomalies without the network or processing units being overloaded (Dantas et al., 2024).

The Evolving Role of Feature Engineering

Deep learning innovations have removed many of the steps in feature engineering. Modern models learn feature spaces that are optimised, which alleviates manual design. Nonetheless, domain specific feature engineering is useful in cases of limited labelled data or where interpretability is a requirement. Indicatively, in sensitive security environments like the airport security setting or the industrial safety control, a hybridisation between manually learned features of trajectories and learned embedding can enhance accuracy and readability.

UNSUPERVISED AND SEMI-SUPERVISED LEARNING APPROACHES

The lack of labelled anomaly data is one of the major issues of behavioral anomaly detection. Anomalous behavior is infrequent, heterogeneous and context-specific unlike normal behavior which may be hard to develop large and representative sets of behavior. In order to mitigate this problem, researchers usually rely on unsupervised, semi-supervised and self-supervised learning processes. The techniques enable the model to learn the pattern of behavior with minimal or no labelled data to enhance learnability and applicability to actual surveillance scenarios.

Unsupervised Anomaly Detection

One of the most popular ways of behavioral monitoring is unsupervised learning since it is not based upon labelled data. The model acquires behaviors that users consider normal and creates a normal behavior representation. In the inference process, any major deviation of the learned normal behaviour is marked and known as an anomaly, thus it is very useful in dynamic systems where anomalies are hard to predict or are usually low (Verma et al., 2022).

Density-Based Methods

Density based models are based on the assumption that normal behavior data points form dense regions whereas the unusual ones are found in low density, sparse regions. Local outlier factor (LOF) and k-nearest neighbour (k-NN) are some of the techniques that identify the abnormal patterns by measuring the density of data points within the feature space. These are simple techniques; easy to interpret and can be used in small scale surveillance. Nevertheless, they work with high-dimensional data of complicated video applications or sensor nets in which it is computationally difficult to define density interactions (Baklashov and Kulyabov, 2025).

Clustering Approaches

Clustering algorithms including K-means, DBSCAN and Gaussian mixture models (GMM) organize similar patterns of behavior into clusters. Any action which cannot be found in a cluster is termed as an anomaly. Such methods are adaptable and extendable and can be used in diverse situations (Farahnakian et al., 2023). They are however responsive to hyperparameter choice and they are likely to misclassify tendencies when the tendencies are multifaceted or overlapping.

One-Class Classification

One class classification the one-class support vector machines (OC-SVMs) can be used successfully when there are only normal behavior data (Guidry et al., 2023). The models are trained on the edge cases of normal data and identify an anomaly in a data that is not within the boundaries of normal data. They tend to not adjust to evolving behavioral patterns, much as they would benefit from, unless re-trained regularly as well.

Semi-Supervised Learning

Semi-supervised learning takes semi-supervised learning, where there are a few labelled data and numerous unlabelled data. This method is applicable in the detection of behavioral anomalies, in which a limited amount of labelled anomalies can be given but when developing a large, fully labelled dataset is challenging (Pang et al., 2021).

Leveraging Limited Labelled Data

In semi-supervised learning, a train with labelled data can be used to help the model perceive anomalies. To illustrate the point above, a handful of atypical activity samples (e.g. unauthorized access or aggressive behavior) can help the model to add to its detection capabilities and yet identify previously unknown and novel anomalies.

Active Learning Strategies

Active learning participates human feedback in a bid to enhance semi-supervised systems. In the process, the model determines the points of uncertainty and requests experts to label them. This is a much more efficient method of labelling that allows the model to adapt faster to the changing behaviors.

Transfer Learning Approaches

In semi-supervised cases, transfer learning comes in handy to a great extent. For vision tasks (though not every other type of AI-related activity), a model that was trained on a large, general-purpose dataset (e.g., ImageNet) can often be fine-tuned to utilize only a small, labelled set of behavioral data. It has the potential to shorten training time and reduce the data needs and enhance model performance.

Self-Supervised Learning

Self-supervised learning (SSL) is now a ground-breaking method where there is a scarcity of labelled data. In SSL, the useful representations found are learned by models via solving pretext tasks which do not need human annotations. With such learned representations, anomaly detection can be then done.

Pretext Tasks Understanding of Behavior.

Pretext tasks assist the model in acquiring general trends in the data. In the case of video data these tasks may also involve guessing the sequence of frames, spotting temporal inconsistencies or creating missing frames. Through these activities, the model is able to acquire typical spatial and temporal patterns that may be useful in the ensuing detection of abnormal behavior (Shi et al., 2023).

Contrastive Learning Methods

A typical example of a self-supervised approach is contrastive learning, which also trains a model to differentiate between similar and dissimilar pairs of data. As an example, two frames of the same sequence of behavior are said to be a positive pair and the unrelated frames form a negative pair. The process generates strong embedding which can learn fine-grained behavioral characteristics, which enhance the quality of anomaly detectors.

Temporal Consistency Learning

Temporal consistency learning has a focus on the fluidity of behavioral patterns. The model is also designed to identify regular sequences and as such, it is vulnerable to disruption or irregular variations that might describe anomalies. This method proves especially handy when measuring a crowd or patient tracking and where long-term stable behavior is the indicator of normality.

Advantages and Challenges

The semi-supervised system and self-supervised and unsupervised techniques have a good contribution to behavioural anomaly detection. They eliminate the possibility of having to possess large labelled data sets and so the systems are more scalable and flexible to a variety of environments. These approaches are also successful in finding new or heretofore unknown anomalies since these methods are not based on a set of fixed labels or predefined types of anomalies.

These methods are however challenged. False positive can be caused by unsupervised systems in case of change of normal behavior or environmental conditions. Semi-supervised techniques are based on the quality of scarce labelled data and model performance may be distorted by labelling errors. Although strong, self-supervised techniques can be computationally expensive and require well thought-out pretext tasks to make significant representations.

Practical Applications

These styles of learning are currently being used extensively in various areas. Unsupervised pedestrian movement models detect actions which are suspicious without big labelled datasets in the context of public safety. Semi-supervised systems are used in healthcare where the patients are tracked, and they capture minor shifts in mobility that indicate a reduction in health. In the industrial environment, self-monitored models detect machine failures or any safety hazards. The combination of these practices reduces dependence on labelled data, variability, and increases the accuracy by clustering, active learning, and contrastive representation, to create smarter, autonomous surveillance, and behavioral monitoring systems.

REAL-TIME PROCESSING AND SYSTEM ARCHITECTURE

Importance of Real-Time Behavioral Surveillance

The behavioral monitoring is an essential part of real-time processing, and timely responses could help to eliminate some injuries, lower risks, or increase efficiency (Real-time processing, n.d.). Any anomaly caused can be detected at the airports or hospitals early enough to enable fast reactions to the threat or emergency. Such systems should have a trade-off of speed, accuracy, and scalability. Anomaly detection cannot occur effectively without careful system architecture.

Real-Time Constraints and Requirements

In developing an anomaly detection system with the especially real-time design, latency, throughput, and computational efficiency are to be taken into account closely.

- **Latency Requirements**: Low latency is an essential requirement of time-sensitive applications. A safety risk can be realized at the transportation hubs via delayed warnings of unattended baggage. Delays in diagnosis of patient fall can be fatal to health care. Sub-second latency is usually targeted by systems, and therefore overwhelmingly is a demand on effective models and data pipeline optimization.
- **Throughput considerations**: High throughput environments (industrial factory or smart cities) produce data in hundreds or thousands of sensors and cameras at once. These massive data streams should be fed into real time systems with no latency. This is usually necessitated by the deployment of distributed processing solutions or cloud services.
- **Computational Constraints**: There are stringent deployment environments that have less computing power. Remote applications or cell phones may have cameras which are incapable of supporting any heavy processing or storage. Lightweight models and optimization algorithms play an important role in effective functioning in such instances.

Edge Computing Solutions

Real-time behavioral anomaly detection has been a critical way of using edge computing. This makes the edge-devices able to minimize the transmission time and the bandwidth needs, due to processing data near to the source point.

- **Model Compression Techniques**: Methods such as pruning, quantization, and knowledge distillation can make deep learning models smaller and faster without sacrificing accuracy. Pruning removes unnecessary network connections. Quantization reduces the precision of model weights. Knowledge distillation transfers knowledge from large models to smaller models suitable for edge deployment.
- **Distributed Processing Architectures**: Edge devices can work in a distributed network, sharing computation between local nodes and cloud servers. This setup allows for real-time inference at the edge while using the cloud for tasks such as retraining models or performing large-scale analytics.
- **Applications of Edge Computing**: Edge monitoring is particularly valuable in remote industrial sites, traffic monitoring, and smart retail. These systems can provide low-latency anomaly detection locally while saving bandwidth by sending only relevant alerts or compressing data to a central server.

Streaming Data Processing

The behavior monitoring systems should be able to handle stream data rather than a set of static. Streaming data architectures take control of information on the fly giving real time actionable information.

- **Online Learning Algorithms**: The classical batch learning algorithms are not quite responsive to evolving conditions. Online learning algorithms enable the continual updating of the model as new information becomes available thereby ensuring a flexible and accurate system.
- **Incremental Model Updates**: Incremental learning networks are able to learn new patterns or behaviours without having to retrain the whole net. This minimizes downtime and calculation expenses thus making the system more efficient and responsive.
- **Concept Drift Handling**: The behavior of a real world may evolve as time passes knowledge of this drift as concept drift. To illustrate, there can be a shift in normal store traffic during holidays. The drift-detecting systems are able to detect such changes and update the model to avoid false positives or overlooked anomalies.

Scalable System Architectures

Real-time monitoring of behavioral monitoring is a vital aspect to scalability of behavior monitoring that is vital in large scale deployment like city control or industrial networks.

- **Modular Design**: The fact that the system is split into different modules by separating the data extraction, pre-processing, model inference, and alert generation processes makes the system easier to maintain, upgrade, and expand.
- **Cloud-Native Architectures:** Cloud platforms allow elastic scalability that allows the system to easily accommodate varying workloads. Kubernetes is a container orchestration tool that allows deploying and scaling deep learning models on multiple nodes.
- **Data Management**: Distributed databases and message queues facilitate high-throughput environments through the use of data management systems that are efficient in terms of storage and retrieval. They also make sure that important events are recorded and will give historical data that can be further analyzed or retrained.

Integration with Other Technologies

Standalone behavioral anomaly detection systems are becoming extensively accommodated by additional technologies to augment functionality as well as reliability.

- **IoT Implementation**: The IoT devices, including wearable cameras, environmentally monitors, and other IoT technologies, generate rich data streams that can be used to better understand behavior. The IoT application will allow the system to pick up non visual indicators like anomalous biometric measurements or environmental variations.
- **5G Networks**: With the implementation of 5G, communication speeds with low latency and high speed between devices and central servers can be achieved, enhancing the reaction speed and scalability of real systems in real-time.
- **Blockchain for Security**: The question of integrity and security of data can also be addressed by deploying blockchain to store a record of data visualization that is impossible to amend in any way, which is a way to ensure a high level of transparency and trust.

Balancing Accuracy and Efficiency

Real-time performance must be achieved with the motivation of achieving trade-overs between accuracy and computational performance. Complex models are more accurate, however, they require more time to run. The lightweight models are more accurate but are faster. This balance can be achieved by using a hybrid solution - a lightweight one that filters ordinary cases and a more expensive one that verifies flagged cases.

Case Example: Smart City Surveillance

The cameras in a transportation hub, park, and other public locations create video streams in a smart city. Edge devices can do preliminary anomaly checks and only potential anomalies are sent to central server. The main system includes more complicated models to validate the alerts and inform the relevant departments. Such a setup minimizes bandwidth expenses, offers fast detection and allows operations which are scalable over the whole city.

Challenges in Real-Time Systems

Although these advances are made, there are still many challenges with requiring real-time behavior monitoring. Protecting data privacy, handling hardware constraints, and resolving the variations in data supplied by various sources continue to be hard. The process of synchronizing the distributed devices and guaranteeing their reliability is also of the concern. To overcome these problems, the design of the system is to be developed with the help of powerful design and frequent improvement, as well as periodical evaluation of the performance.

Effective behavioral anomaly detection requires real-time processing, as well as system design. These systems can deliver timely, accurate and actionable insights through the use of edge computing, streaming data, and scalable architectures. As the technologies, such as Internet of Things, 5G, and blockchain, will evolve, real-time monitoring systems will also get stronger, more reliable, and capable of meeting the demands of the new reality.

EVALUATION METHODOLOGIES AND METRICS

Importance of Evaluation

The evaluation of behavioral anomaly detection systems is important to properly evaluate and depend on the system. Even the sophisticated models cannot work in reality in case of absent clear metrics and methodologies. Evaluation is intended to compare the accuracy of detection, the weaknesses, compare the models with each other, and make improvements. Since behavioral monitoring can be used in various applications like in the security and healthcare industry, technical, operational, and contextual performance must be evaluated.

Performance Metrics

Performance metrics are a quantitative measure of how well the system can be able to detect anomalies and reduce errors.

Precision, Recall, and F1-Score

Precision is used to measure the number of true anomalies that the detection method has identified, and thus, the false positives are minimized. Recall can be understood as the sensitivity of the model as the number of true anomalies that the model can identify correctly. The F1 score is a combination of precision and recall into one score, specifically more useful in situations where anomalies are uncommon (Paparrizos et al., 2022).

Area under Curve (AUC)

True positives and false positives can be checked through the region below the receiver operating characteristic (ROC) curve (AUC-ROC). The more the AUC, the more it performs well on various thresholds. In areas where the number of anomalies is low, one should use the area under the precision-recall curve (AUC-PR) since it will be more accurate at showing how much events with low probability are detected (Carrington et al., 2022).

False Positive and False Negative Rates

It is important to reduce the number of false positives and false negatives, as the former produces many useless alerts, and the latter leaves important events undetected, particularly in important industries such as public safety and healthcare (Alahmadi et al., 2022).

Latency and Throughput

With real-time systems, it is not about being accurate. Latency is an indicator of anomaly detection and reporting speed. Throughput is the level of data that a given system can handle at a given time. The indicators play a vital role in gauging productivity in high throughput settings that are fast moving (Zhang et al., 2025).

Dataset Challenges

An essential aspect of testing the systems of anomaly detection is the utilization of good-quality datasets, which are representative to real-life situations.

Imbalanced Data

Detecting behavioral anomalies is often the case with datasets that are abnormally rare and uncommon. The imbalance may have the negative effect of making the model more inclined towards the normal behavior at the expense of losing the rare events. These can be addressed during training and evaluation by using techniques like resampling data, synthetic anomalies, cost-sensitive learning, etc.

Cross-Domain Evaluation

A model that has been trained about a given environs might not perform in a different environment. Cross-domain assessment is used to ensure the model has generalized appropriately, that is, when the environment or the angle of the camera or behavioral pattern varies, the model should continue to perform at the same level.

Temporal Evaluation Protocols

The change in behavioral data is time variant hence temporal evaluation is used in measuring how the models react to the change. By applying testing models to consecutive divisions of the data, it can be demonstrated whether it is capable of adjusting to concept drift or its efficiency deteriorates.

Benchmark Datasets

The benchmark data will give a common ground on which various behavioral abnormality detection algorithms can be assessed and compared.

Public Surveillance Datasets

There are datasets like UCSD Pedestrian, ShanghaiTech Campus and Avenue containing labelled video sequences of normal and abnormal behavior. They enable scientists to test algorithms under a controlled setting, and compare them across the studies.

Synthetic Data Generation

Synthetic datasets can be used when real-world data is either scarce or delicate. The use of synthetic video generation tools can generate controlled scenes with known anomalies, which can be safely tested without issues of privacy and security.

Dataset Annotation Challenges

Relevant naming of anomalies is tedious and subjective because there is always a possibility that the behavior may be interpreted differently by various annotators. Reliability of annotations is usually enhanced by crowdsourcing and expert verification, although inconsistency is still an issue.

Comparative Analysis

Limited insights can be offered by evaluating behavioral anomaly detection systems only. Comparison of various models, architectures, or settings can assist to identify trade-offs and establish the best settings. As an example, a model that slightly underperforms in terms of accuracy, but one that executes inference much faster, can be applied to a real-time system, whereas a slower, but much more accurate model, could be applied to an offline system.

Real-World Testing and Validation

Laboratory tests though helpful do not provide complete reality of the real world that is unpredictable. Real-life testing helps to make sure that models are robust enough to be used even in case of noise, changing lighting conditions, and dynamic behavior. Pilot testing into real world environments like hospital wards or shops can reveal the real world problems like integration, operator overload and false alarms. The likes of these tests can teach lessons on how to improve method of model and system design.

Interpretability and Human-in-the-Loop Evaluation

Explainability should also be taken into account during evaluations. The stakeholders particularly safety sensitive domains must be informed of the reasons why a behavior is as an anomaly. Transparency and trust building can be provided with the usage of explainable artificial intelligence (XAI) methods. The latter can also be applied to human-machine interface techniques based on utilizing experts to review identified anomalies, enhance the model, and give feedback to make improvements.

Continuous Monitoring and Feedback Loops

The evaluation process should be taken as an on-going process rather than a one-time job. The constant monitoring can be used to spot performance decline due to change of environment, new behaviour or upgrades in the hardware. The use of feedback loops where the results are periodically reviewed and models retrained makes the system effective and adjustable.

Effective assessment techniques and indicators are important in the detection of behavioral anomaly. Precision, recall, AUC, and the latency metrics are clear indicators of performance whereas the qualitative measures like interpretation and human validation are used to guarantee the practical value of the work. Overcoming biased data and information, subjective annotation, and cross-domain generalization are among the crucial challenges to overcome to create credible systems. A mix of laboratory tests, benchmark assessment and field validation has helped make sure that these systems are not only technically strong but operational as well.

CASE STUDIES AND APPLICATIONS

Video Surveillance Systems

The practice of behavioral abnormality detection is applied to police surveillance on the new advanced systems of video surveillance. In airports and other transportation centres, the models monitor the movement of people to identify unattended luggage, unfriendly intrusion, or untimely empties. Such systems provide real-time notification of security teams and help them take immediate action before problems escalate. Behavioral analytics surveillance systems can be used in the retail and commercial areas to identify shoplifting behavior, suspicious loitering, or traffic movements. Such insights do not only improve the security but also help to have more efficient operations, including dealing with congestion during the peak of the stores or other open spaces.

Crowd Behavior Analysis

One of the most popular uses of anomaly detection is crowd behavior tracking in the open areas. Such models are used to examine the movement pattern of the crowds and interactions of the group to identify indications of panic, the threat of stampede, or violent behavior in big events like concerts or sporting events. Early warning helps the authorities to receive fast response and control the situation before they become an emergency. To exemplify, the concept of real-time notification of abnormal crowd masses at sports stadiums can assist the security teams to use extra resources and can avoid the cases. These insights are also used by the urban planners to observe the pedestrian flows and design better infrastructure in places with traffic.

Human Activity Recognition in Healthcare

Hospitals and healthcare organizations are embracing the use of the behavioral anomaly detection in enhancing safety and monitoring of patients. Posture estimation and skeletal tracking systems that have been installed in nursing homes are able to measure falls or prolonged inactivity and provide instant alerts to nurses. Similar models are used in hospitals to track patients who either survived surgery or limited mobility so that they can develop early signs of complications or noncompliance with recovery pattern. Not only patient safety is enhanced by these technologies but the workload of medical staff is also decreased, as these technologies more frequently lead to automating of routine monitoring, which gives the staff an opportunity of dealing with more urgent and critical care.

Industrial and Workplace Safety

In the industrial setup, behavioral anomaly detection enhances safety and compliance. Monitors that are running deep learning could be used to detect unsafe patterns, like workers going into the restricted area without body protection or using machine tools in a wrong way. They additionally can trace the workflow patterns and identify the deviation of the common practices, which are involved in minimizing downtimes and preventing accidents. A large number of factories and warehouses are incorporating these systems to make the working conditions in these facilities a lot safer, more efficient, and compliant.

Digital and Cyber security Applications

The detection of behavioral anomaly is becoming part of contemporary cybersecurity. They are systems that monitor users to identify abnormal network and application conduct that reveal people accessing networks or applications without authorization, or who transfer data or may pose an insider risk. They would be able to detect subtle anomalies that traditional security tools frequently overlook since they are acquainted with typical behavioral patterns and thereby the protection against the advanced attack would be hardened.

These applications indicate the general usefulness of behavioral anomaly detection. Whether it is the safety of people, safety of health, or the safety of industry and protection through cybersecurity, analytics based on deep learning can deliver the much-needed insights so that the business can act immediately. As the Internet of things, 5G, and edge computing are adopted and improved, all these systems are currently progressing to be more responsive to the real-world requirements, faster, and more scalable.

PRIVACY, ETHICS, AND LEGAL CONSIDERATIONS

The Growing Importance of Privacy in Surveillance

The more complex behavioral anomaly detection systems become, the more privacy concerns escalate. Such systems frequently record fine video, voice recordings and biometric information which can accidentally disclose personal information that is sensitive. This information can also be abused without effective controls that will compromise the trust of people as well as the rights of individuals. To curb these risks organizations will have to work in a fine balance between privacy and security. This involves the introduction of strict data protection policies, following the legal systems, and making sure that the practice of monitoring is ethical, transparent, and traceable.

Privacy-Preserving Techniques

In order to deal with the issue of privacy several privacy-saving monitoring policies have been devised by researchers and practitioners.

- **Data Anonymization**: A typical method is to obscure the features that people can be identified by, e.g. faces or license plates. The silhouette based model is also capable of anonymizing the person but still provides sufficient information on the behavior to identify an anomaly.
- **Federated Learning**: Federated learning enables the training of models locally on edge devices without any raw data sent to a central server. There is a sharing of model updates only and this accelerates risks of sensitive information leakage, besides the system maintains accuracy.
- **Differential Privacy**: Differential privacy methods introduce random imaging noise to databases or model outputs, so that even during a visit to the data or a leak of data, it cannot be possible to identify people involved. The technology is also being applied more towards compliance-related industries like the healthcare and finance industries which require analyzing sensitive behavioral data.

Ethical Implications

Ethics is an important issue in the design or deployment of behavioral anomaly detection systems.

- **Bias and Fairness**: Deep learning models may also develop bias in the training data in an unintended manner. As an illustration, the system may excessively label the persons using their looks, outfits or common behaviors. To allow fairness, proper datasets have to be provided, tools of bias detection, as well as frequent audits to detect and address such problems.
- **Transparency and Explainability**: The stakeholders, such as the users and regulators, require transparency in the decisions making process. Attention visualization and Locally Interpretable Model-Agnostic Explanations (LIME) methods of explainable AI may be used to explain why an oddity has been detected, and thus establish trust between users and decision-makers.
- **Human Oversight**: Full reliance on automated systems can lead to ethical pitfalls. Introducing human oversight can ensure that critical decisions, such as safety interventions or disciplinary actions, are verified by humans, minimizing the risk of unfair outcomes.

Legal and Regulatory Frameworks

Legal environment of behavioral surveillance is swiftly changing, as rigorous policies are being implemented primarily across the world to safeguard the personal information.

- **GDPR Compliance**: GDPR is a General Data Protection Regulation high-ranking regulation in the EU that sets strict criteria on how personal information are gathered, processed, and stored. Surveillance systems need to secure legal data gathering, storage and in cases where it exists overt consent. The concept of data minimization presented in the GDPR also promotes organizations to receive the information that is necessary to meet the intended aim.
- **Local Surveillance Laws**: The local surveillance and data protection laws have been adopted by many countries, governing the location and use of the surveillance systems. As an illustration, the future personal data protection policies in India are likely to be more stringent on the provision of behavioral data and especially when they are in the open areas.
- **Data Retention Policies**: Legal frameworks often dictate retention periods for surveillance data. Implementing an automated data deletion policy not

only ensures compliance but also reduces the risk of long-term storage of sensitive information.

Balancing Security and Individual Rights

The difficult part of behavioral surveillance that is hard to overcome is creating a balance between security concerns and upholding individual freedoms. Though the system of anomaly detection can enhance safety and efficiency of operation, it should not create an atmosphere of paranoia and surveillance. This can be achieved by having a well-developed balance through transparent purpose of surveillance communication, explicit consent protocols, and robust governance policies and help inspire the confidence of the population.

Challenges and Future Directions

Securing good privacy and ethical standards in detecting the middle ground behavioral aberration is a challenge. Further sophistications like multimodal surveillance and predictive analytics prove to be dangerous questions regarding consent, proportionality and accountability. In order to deal with these concerns, the future solutions can utilize privacy enhancing techniques like quick encryption, federated analytics, and safe multi-party computation. The techniques enable the use of the data but do not reveal the information.

The global standardization processes may also establish standardized frameworks, and it would be easier to fulfil both ethical and legal requirements in various regions. Privacy, fairness and transparency have ceased being optional, they are mandatory in creating trust. Withdrawing privacy-saving technologies and adhering to the rules and regulations, organizations can enjoy the advantages of such systems without interfering with the rights of an individual. Active governance and ethical system design will continue to be key aspects of creating both effective and socially responsible solutions.

INTERPRETABILITY AND EXPLAINABLE AI

The Need for Interpretability

There are also deep learning architecture-based behavioral anomaly detector with a high level of accuracy but low level of transparency, which resembles a black box. Such lack of transparency may be dangerous in the sensitive spheres like public safety, healthcare, and industrial monitoring. To foster confidence, equitable and responsibility decision-makers must be able to establish the causes of alerts.

The concept of model explainability assists in closing this gap by allowing explanations, which can be understood by people. Savage is able to make decisions that are more informed and respond more efficiently when operators are able to verify and determine the causes of alerts. This openness does not only build trust in the system, but also favour ethical and responsible use of AI-powered anomaly detection.

Challenges in Explaining Deep Models

The nature of deep neural networks is too complex, which makes its interpretation a major challenge. Such models as convolutional neural networks (CNNs), recurrent neural networks (RNNs), and transformers are based on the use of millions of parameters and complex nonlinear transformations, and their inner logic is hard to follow or interpret.

This is made worse by the fact that systems deal with spatiotemporal data in behavioral anomaly detection. Abnormalities are also the consequence of intricate and delicate interactions between time and space, which can hardly be explained in direct terms. How to make these processes more intuitive is also a central focus to the research and practitioners who aim at creating more transparent and trustworthy systems.

Explainable Anomaly Detection Techniques

- **Attention Visualization**: The attention mechanism of models like Transformers can be visualized to point out individual frames, regions or time intervals that affect anomaly detection. This allows operators to locate the source of an alert and determine whether the behavior they receive the flag on is anomalous or not.
- **Gradient-Based Explanation Methods**: Heat map crisis techniques like Grad-CAM (Gradient Weighted Class Activation Mapping) have the ability to render heat maps, indicating which regions of an image or frame have the most significant effect on the choices made by a model. These visual representations form understandable relationships between what is going on inside the model and what is actually going on in the data.
- **Model-Agnostic Models**: Locally interpretable Model-Agnostic Explanations (LIME) and SHAP (SHapley Additive exPlanations) compute simplified, understandable explanations by locally approximating the behavior of a complex model. These tools assist users to understand the reasons as to why some

behaviors are labelled as anomalous without necessarily having to have expert technical knowledge.

Human-AI Collaboration

The concept of explainability improves the experience of human-machine interaction, allowing an operator to reconsider, check, or modify system decisions. The use of expert feedback enables a system to be more accurate and adequately fulfils real world requirements.

To illustrate that, the clinicians may look at the alerts of unusual patient activity in case of a mistake and assist in training the model to make predictions better in the future in healthcare. Such collaborative style fosters trust and minimizes the chances of automation bias and users can trust system outputs.

Building Trust and Acceptance

Open and understandable models are the ones that create a trust among all parties including the operators and the regulators. If the compliance requirements of an industry are high, explainability for provides written evidence of the decision-making that can guarantee compliance with the legal and ethical duties. The explainable systems also enable model behavior to be auditable and traceable thus reducing the risk of bias or undue targeting.

The concept of explainability and explainable AI has become neither a choice nor a necessity in behavioral anomaly detection. The increased transparency leads to the increased trust and, as well, enhances human-AI cooperation. With the development of research, the philosophy behind the creation of a highly accurate and understandable system where explainability is part of the design of the model instead of an afterthought, this will be important in the development of the most accurate and understandable system.

FUTURE DIRECTIONS AND EMERGING TRENDS

Advanced Architectures for Behavioral Modeling

Behavioral anomaly detectors are taking a step in the direction of the advanced deep learning architectures with abilities to manage both complex and hierarchical interactions. Graph neural networks (GNNs) gain popularity because it can be used to create an association between individuals or objects in a scene. GNNs are better than their counterparts in identifying group behavior, like group collaboration or social interaction, through the ability to encode their interactions with each other, making each entity a node and their interaction between each other an edge. The neural-symbolic ones are also on the rise, incorporating the rationality of symbolic AI and the flexibility of deep learning. This combination allows systems to deliver powerful predictive performance and decipher their decision making.

Multimodal fusion, which involves the integration of any of the following data: video, audio, biometric sensors, and text logs is another encouraging development. This is able to provide deeper behavioral insights and eliminate the possibility of bias or incomplete understanding by using single source of data. These methods onward open the approach to more trustworthy and situation-sensitive behavior control systems.

Emerging Technologies Shaping the Field

New technology is greatly increasing the possibilities of behavioral anomaly detectors. 5G networks have extremely low latency and high bandwidth, which allow real-time transmission and analysis of large volumes of surveillance data, even between distributed systems. The integration of IoT is also transforming this area and wearable sensors, smart cameras and environmental monitors are capable of generating various and high-resolution streams of data that can be consumed into behavioral models to conduct more accurate analysis.

Going forward, quantum computing will help train and infer the deep learning model much faster, particularly with large complex data sets. On the same note, neuromorphic computing which is based on the human brain has the benefits of offering energy efficient real time processing. This technology is capable of allowing edge devices to track the behavior continuously and faster with little power consumption. Combined, these innovations are the building blocks of new systems of behavioral anomalies detectors, which are more intelligent, quicker, and flexible.

Addressing Research Challenges

A number of challenges still exist when it comes to behavioral anomaly detection. Few-shots anomaly detection is one of these shams where the system should be able to detect abnormal patterns using few or even non-examples. A second difficulty is the provision of cross-domain generalization that aims at developing models that can be effective in a variety of conditions and they do not need retraining. This is essential in deploying on a large scale with a wide range of different conditions.

The need of continuous learning is also an urgent one. Systems are required to adjust to the emergence of new behaviors but not to forget the past knowledge. It would especially be crucial in dynamic spaces, like in a street and industrial facility or on the Internet where the regular patterns usually vary.

Lastly, a balance between model interpretation and model accuracy is very critical. The interpretability is likely to be made core of the future systems to ensure that the predictions they make are not only correct, but that they can be justified using a clear and understandable reasoning.

Towards Responsible and Ethical AI

With the ever-increasing technologies, the emphasis is being laid on the development of responsible systems. In future studies, issues of equity, openness and responsibility will be put on top the levels of behavioral anomaly detection. It will see privacy-preserving techniques like federated analytics and secure multi-party computation gaining greater popularity. Such schemes will aid in compliance with regulatory and ethical standards and also facilitate advanced analytics.

Even more innovation and convergence is in the future of behavioral anomaly detection. It will be improved with new inputs in deep learning, multimodal fusion, and new directions such as quantum computing and neuromorphic computing. At the same time, adaptability, explainability, and, ethical designs research will make these systems reliable and dependable. All these developments combined will create the next generation of behavioral monitoring, which will provide safer, smarter, and more ethical uses in a large variety of industries.

CONCLUSION

The development of behavioral anomaly detection has been characterized by the development of the simple rule-based system to advanced deep learning-based solutions that could detect and analyze complex spatiotemporal data to a high degree of accuracy. The chapter describes how technologies like CNNs, RNNs, autoencoders, and Transformers are turning surveillance systems into intelligent systems that can identify and change according to changing behavioral trends in various fields. To be rolled out successfully, the data must be clean and privacy-conscious, the optimization models need to be scalable, and a hybrid design one that will provide a balance of speed and accuracy to enable use in real-time is necessary. Such systems have become extremely crucial in terms of community safety, healthcare, industry surveillance, and cybersecurity. Nevertheless, they continue to have a myriad of issues including computational resource requirements, a dearth of labelled data, concept drift, and ethical issues of biasing, fairness, and privacy. In the future, academic studies should overcome these weaknesses by using better generalization and few-shot and continuous learning, multimodal fused data, and generating interpretable models that increase transparency and trust. The scale of these systems will be scaled in an efficient and responsible way with privacy-preserving approaches which include federated learning and differential privacy, and providing developments on the new approaches of quantum and neuromorphic computing. By the means of consistent innovations and following the ethical and legal regulations, the detection of behavior abnormalities will remain one of the essential parts of intelligent surveillance, making the process of operation and pedestrian safety much safer and more efficient, without compromising the trust of the people.

REFERENCES

Alahmadi, B. A., Axon, L., & Martinovic, I. (2022). 99% false positives: A qualitative study of {SOC} analysts' perspectives on security alarms. In *31st USENIX Security Symposium (USENIX Security 22)* (pp. 2783–2800).

Alam, M. M., Torgo, L., & Bifet, A. (2022). A survey on spatio-temporal data analytics systems. *ACM Computing Surveys, 54*(10s), 1–38. DOI: 10.1145/3507904

Andrade, R. (2024). *Privacy-preserving face detection: A comprehensive analysis of face anonymization techniques* (Master's thesis, Universidade do Porto (Portugal)).

Baklashov, A. S., & Kulyabov, D. S. (2025). Statistical and density-based clustering techniques in the context of anomaly detection in network systems: A comparative analysis. *Discrete and Continuous Models and Applied Computational Science, 33*(1), 27–45. DOI: 10.22363/2658-4670-2025-33-1-27-45

Borkar, S. V., & Sangve, S. M. (2025). *A critical analysis on anomaly detection in high-frequency financial data using deep learning for options.*

Carrington, A. M., Manuel, D. G., Fieguth, P. W., Ramsay, T., Osmani, V., Wernly, B., Bennett, C., Hawken, S., Magwood, O., Sheikh, Y., McInnes, M., & Holzinger, A. (2022). Deep ROC analysis and AUC as balanced average accuracy, for improved classifier selection, audit and explanation. *IEEE Transactions on Pattern Analysis and Machine Intelligence, 45*(1), 329–341. DOI: 10.1109/TPAMI.2022.3145392 PMID: 35077357

Chen, Y. (2024, October). *Video behavior in intelligent video surveillance system based on deep learning.* In *IET Conference Proceedings CP895* (Vol. 2024, No. 19, pp. 136–140). Stevenage, UK: The Institution of Engineering and Technology.

Choudhry, N., Abawajy, J., Huda, S., & Rao, I. (2023). A comprehensive survey of machine learning methods for surveillance videos anomaly detection. *IEEE Access : Practical Innovations, Open Solutions, 11*, 114680–114713. DOI: 10.1109/ACCESS.2023.3321800

Costa, G., Forestiero, A., & Ortale, R. (2023). Rule-based detection of anomalous patterns in device behavior for explainable IoT security. *IEEE Transactions on Services Computing, 16*(6), 4514–4525. DOI: 10.1109/TSC.2023.3327822

Dantas, P. V., Sabino da Silva, W.Jr, Cordeiro, L. C., & Carvalho, C. B. (2024). A comprehensive review of model compression techniques in machine learning. *Applied Intelligence, 54*(22), 11804–11844. DOI: 10.1007/s10489-024-05747-w

Dávila-Montero, S., Dana-Lê, J. A., Bente, G., Hall, A. T., & Mason, A. J. (2021). Review and challenges of technologies for real-time human behavior monitoring. *IEEE Transactions on Biomedical Circuits and Systems*, *15*(1), 2–28. DOI: 10.1109/TBCAS.2021.3060617 PMID: 33606635

Dilek, E., & Dener, M. (2025). An overview of transformers for video anomaly detection. *Neural Computing & Applications*, *37*(22), 1–33. DOI: 10.1007/s00521-025-11218-1

Du, Y. (2020). An anomaly detection method using deep convolution neural network for vision image of robot. *Multimedia Tools and Applications*, *79*(13), 9629–9642. DOI: 10.1007/s11042-020-08684-1

Duong, H. T., Le, V. T., & Hoang, V. T. (2023). Deep learning-based anomaly detection in video surveillance: A survey. *Sensors (Basel)*, *23*(11), 5024. DOI: 10.3390/s23115024 PMID: 37299751

Esmaeili, F., Cassie, E., Nguyen, H. P. T., Plank, N. O., Unsworth, C. P., & Wang, A. (2023). Anomaly detection for sensor signals utilizing deep learning autoencoder-based neural networks. *Bioengineering (Basel, Switzerland)*, *10*(4), 405. DOI: 10.3390/bioengineering10040405 PMID: 37106591

Fährmann, D., Martín, L., Sánchez, L., & Damer, N. (2024). Anomaly detection in smart environments: A comprehensive survey. *IEEE Access : Practical Innovations, Open Solutions*, *12*, 64006–64049. DOI: 10.1109/ACCESS.2024.3395051

Fang, W., Chen, Y., & Xue, Q. (2021). Survey on research of RNN-based spatio-temporal sequence prediction algorithms. *Journal of Big Data*, *3*(3), 97–110. DOI: 10.32604/jbd.2021.016993

Farahnakian, F., Nicolas, F., Farahnakian, F., Nevalainen, P., Sheikh, J., Heikkonen, J., & Raduly-Baka, C. (2023). A comprehensive study of clustering-based techniques for detecting abnormal vessel behavior. *Remote Sensing (Basel)*, *15*(6), 1477. DOI: 10.3390/rs15061477

Guidry, J., Sohrab, F., Gottumukkala, R., Katragadda, S., & Gabbouj, M. (2023, December). One-class classification for intrusion detection on vehicular networks. In *2023 IEEE Symposium Series on Computational Intelligence (SSCI)* (pp. 1176–1182). IEEE. DOI: 10.1109/SSCI52147.2023.10371899

Jabeen, S., Li, X., Amin, M. S., Bourahla, O., Li, S., & Jabbar, A. (2023). A review on methods and applications in multimodal deep learning. *ACM Transactions on Multimedia Computing Communications and Applications*, *19*(2s), 1–41. DOI: 10.1145/3545572

Li, Z., Zhu, Y., & Van Leeuwen, M. (2023). A survey on explainable anomaly detection. *ACM Transactions on Knowledge Discovery from Data, 18*(1), 1–54.

Mishra, P. K., Mihailidis, A., & Khan, S. S. (2024). Skeletal video anomaly detection using deep learning: Survey, challenges, and future directions. *IEEE Transactions on Emerging Topics in Computational Intelligence, 8*(2), 1073–1085. DOI: 10.1109/TETCI.2024.3358103

Pang, G., Van Den Hengel, A., Shen, C., & Cao, L. (2021, August). Toward deep supervised anomaly detection: Reinforcement learning from partially labeled anomaly data. In *Proceedings of the 27th ACM SIGKDD Conference on Knowledge Discovery & Data Mining* (pp. 1298–1308). DOI: 10.1145/3447548.3467417

Paparrizos, J., Boniol, P., Palpanas, T., Tsay, R. S., Elmore, A., & Franklin, M. J. (2022). Volume under the surface: A new accuracy evaluation measure for time-series anomaly detection. *Proceedings of the VLDB Endowment International Conference on Very Large Data Bases, 15*(11), 2774–2787. DOI: 10.14778/3551793.3551830

Pathirannahalage, I., Jayasooriya, V., Samarabandu, J., & Subasinghe, A. (2025). A comprehensive analysis of real-time video anomaly detection methods for human and vehicular movement. *Multimedia Tools and Applications, 84*(10), 7519–7564. DOI: 10.1007/s11042-024-19204-w

Pawar, K., & Attar, V. (2019). Deep learning approaches for video-based anomalous activity detection. *World Wide Web (Bussum), 22*(2), 571–601. DOI: 10.1007/s11280-018-0582-1

Ruef, M., & Birkhead, C. (2024). *Learning from outliers and anomalies. Academy of Management Perspectives, (ja),* amp-2023.

Samayamantri, L. S., Singhal, S., Krishnamurthy, O., & Regin, R. (2024). AI-driven multimodal approaches to human behavior analysis. In *Advancing Intelligent Networks Through Distributed Optimization* (pp. 485–506). IGI Global. DOI: 10.4018/979-8-3693-3739-4.ch025

Štuikys, V., & Burbaitė, R. (2024). Introduction to artificial neural networks and machine learning. In *Evolution of STEM-driven computer science education: The perspective of big concepts* (pp. 311–346). Springer Nature Switzerland. DOI: 10.1007/978-3-031-48235-9_11

Ullah, W., Hussain, T., Ullah, F. U. M., Lee, M. Y., & Baik, S. W. (2023). TransCNN: Hybrid CNN and transformer mechanism for surveillance anomaly detection. *Engineering Applications of Artificial Intelligence, 123*, 106173. DOI: 10.1016/j.engappai.2023.106173

Verma, K. K., Singh, B. M., & Dixit, A. (2022). A review of supervised and unsupervised machine learning techniques for suspicious behavior recognition in intelligent surveillance system. *International Journal of Information Technology : an Official Journal of Bharati Vidyapeeth's Institute of Computer Applications and Management*, *14*(1), 397–410. DOI: 10.1007/s41870-019-00364-0

Von Ziegler, L., Sturman, O., & Bohacek, J. (2021). Big behavior: Challenges and opportunities in a new era of deep behavior profiling. *Neuropsychopharmacology : Official Publication of the American College of Neuropsychopharmacology*, *46*(1), 33–44. DOI: 10.1038/s41386-020-0751-7 PMID: 32599604

Wang, C., Sun, Y., Lv, S., Wang, C., Liu, H., & Wang, B. (2023). Intrusion detection system based on one-class support vector machine and Gaussian mixture model. *Electronics (Basel)*, *12*(4), 930. DOI: 10.3390/electronics12040930

Yoon, J., & Choi, M. K. (2023). Exploring video frame redundancies for efficient data sampling and annotation in instance segmentation. In *Proceedings of the IEEE/CVF Conference on Computer Vision and Pattern Recognition* (pp. 3308–3317). DOI: 10.1109/CVPRW59228.2023.00333

Zhang, H., Jia, X., & Chen, C. (2025). Deep learning-based real-time data quality assessment and anomaly detection for large-scale distributed data streams. *International Journal of Medical and All Body Health Research*, *6*(1), 1–01. DOI: 10.54660/IJMBHR.2025.6.1.01-11

Zuo, Z., Shuai, B., Wang, G., Liu, X., Wang, X., Wang, B., & Chen, Y. (2015). Convolutional recurrent neural networks: Learning spatial dependencies for image representation. In *Proceedings of the IEEE Conference on Computer Vision and Pattern Recognition Workshops* (pp. 18–26). DOI: 10.1109/CVPRW.2015.7301268

Chapter 10

Automated Machine Learning Model Management:

Integrating Monitoring, ML Observability, and Continuous Feedback for Enhanced Performance

Pramod Begur Nagaraj
http://orcid.org/0009-0008-2192-4137
Independent Researcher, USA

Archana Chaluvadi
http://orcid.org/0009-0003-4412-0178
Massmutual, Boston, USA

Visrutatma Rao Vallu
http://orcid.org/0009-0001-4149-9034
Insmed Incorporated, USA

Winner Pulakhandam
http://orcid.org/0009-0008-4831-6561
Personify Inc., USA

R. Padmavathy
Department of ECE, Dr. N.G.P. Institute of Technology, Coimbatore, India

DOI: 10.4018/979-8-3373-4202-3.ch010

ABSTRACT

We present an automated machine learning model management system, which provides feedback loops, at-rest and in-flight observability, and continuous monitoring for sustaining and improving model performance in the long term. The system is built to support automated retraining and redeployment to mitigate challenges associated with data drift or the evolution of operational norms. Our models achieved 99.14% accuracy, 99.19% precision, 99.17% recall, and 99.18% F1 scores in Random Forest and Gradient Boosting. Our system demonstrated superior on-line observability and adaptability to AutoWeka and more conventional Gradient Boosting. Continuous feedback with on-going observations allows users to leverage our system in scalable, decision making applications, in non-stationary conditions while ensuring ongoing performance.

1 INTRODUCTION

Machine learning has become a highly disruptive technology for many sectors today, as systems can effectively make data predictions and decisions with minimal human involvement (Castelli et al., 2025). Different sectors, such as health, finance, manufacturing, and agriculture, have seen the influence of machine learning on the practice of their professions today (Katragadda et al., 2021). However, as models become more sophisticated in many learning applications, it makes it quite difficult to manage these things in the production environment (Sivaraman, 2024). The models that are effective in a controlled environment are less effective in the real world (Kurunathan et al., 2022). Model tracking and optimisation from the start of model training, through model deployment, scaling and new version roll-out are under flagship 'Automated ML model management' (Nogare et al., 2024). The system guarantees high model metrics over new data instantiates by enforcing monitoring, ML observability and continuous feedback loops, so the system evolves to new data without performance degradation. So, the very bone of the paper is trying to provide a good answer for the kind of demand that is gradually growing for having a robust and automated management system that is currently valid for the engineering complexity and scale of the machine learning production setup, specifically stressing improvement of performance by insights for model feedback.

The primary focus of this study is the unified operation of ML models by (Vartika & William, 2023), which heeds the evolving demands of modern data. To start off, (Mahida, 2024) observed that standard machine learning workflows aim to build a model and deploy it in an environment, with minimal to no modifications, until the model's performance deteriorates to the point that manual updates or retraining become necessary. (Bhaskaran, 2020) points out that with modern data's heterogeneous and massive scale, the assumption of no degradation and hence, no intervention in the system's performance, is obsolete. This study takes a step forward to highlight the importance of integrating monitoring, observability, and automation within the business workflow. Building on the groundwork laid by (Mirza & Iqbal, 2024), this paper introduces a sophisticated feedback mechanism into automation. While traditional workflows rely on performance data for a single instance of model retraining, ours perpetually ingests system performance data. This enables us to undertake ongoing model retraining and maintenance, thus ensuring a better model life cycle. (Heymann et al., 2022) show that this approach preserves the dependability of ML solutions in dynamic environments, which this paper further claims to provide for the evolving datasets.

One of the reasons why there's continual activity surrounding model management is the presence of data drift. Data drift is best described as the changes in the statistical properties of the input data from one time period to another (Cob-Parro et al., 2024). Data drift may be caused by shifts such as those in user behavior, changes in season, or the environment. Drift causes the past data to lose its effectiveness in predicting the future. Also, the data drift may be caused by something called model drift. Under the assumption that the data will remain the same, the model will lose its accuracy due to some parameters caused by some factor that made the data bad predictors in the first place (Kalusivalingam et al., 2020). The second set of parameters may be changes in the environment such as the behavior of external factors like climate, regulations, or market conditions that the model is trying to forecast. This shows the model management system has to be adaptive and has to have continuous feedback loops for better and automated monitoring, faster adaptation, recalibrations for the models (Yousefi et al., 2023).

The solution suggested to solve these problems is an automated feedback loop and continuous retraining pipeline in ML model management systems (Cohen et al., 2022). This paper suggests the application of an approach with CNNs for continuous monitoring adaptability (Azimi & Pahl, 2021). CNNs, though primarily designed for image processing, are effective for identifying intricate patterns in datasets linked with changes in the environment (Popp, n.d.). If CNNs are used in feedback systems, they can be designed to notify stakeholders when the data distribution or the model's performance worsens. Alerts of this sort may lead to retraining pipelines being launched, which self-optimize the model without human intervention (Kedziora et al., 2022). After this phase, tuning and changes can be made in real-time with the help of observability tools. Integrating all these components into a single process develops visibility into the seamless movement of data in its actual environment relative to the model (Liang et al., 2024). The process automation of the feedback loop concurrently makes the entire feedback mechanism more dynamic and responsive, adjusting to the continuously changing data and external factors (Manousakis & Korres, 2021).

The paper will begin by establishing the necessity for an automated ML management system and then introduce the challenges currently faced in production environments (Chatterjee et al., 2023). The next stage will be an exposition of existing model management methods with their limitations (Sheikh, 2024). Following that, the paper will initiate the presentation of a proposed solution based on CNNs that combat these challenges, with emphasis on monitoring, feedback processing, automated retraining pipelines (Hrusto et al., 2024). Evaluating the proposed solution will prove its superiority over traditional methods regarding efficiency, scalability, and model accuracy, which will then be presented (Miraftabzadeh et al., 2021). The paper will conclude with a summary of the main conclusions and recommendations for further study and advancements in automated machine learning model management (Sappa, 2023).

1.1 Objectives And Scope Of The Study

- An assessment of current issues in model deployment, monitoring, and performance degradation in traditional MLOps. This is in addition to the justification for automating those processes.
- Build an automated framework that combines continuous monitoring of models and implements a feedback loop for real-time adaption and improvement of machine learning models through ML observability.
- Compare the output of the proposed system to other established systems such as AutoWeka, SVM, and Gradient Boosting in order to measure its effectiveness.
- Establish a continuous feedback system that relies on both system and user feedback to retrain models automatically and refine them for high-quality predictions and optimal performance.
- Validate the scalability and robustness of the presented model management system by testing it on several real datasets, as well as identifying future work areas for improvement.

2 RELATED WORKS

Organizations depend on Machine Learning Engineers to bridge the gap between data science and engineering when deploying models and managing ML pipelines in production, especially when adding new data can be challenging. From data preparation and testing to review and ongoing monitoring, this operationalization process—known as MLOps—involves a workflow. 18 MLEs from a range of industries participated in ethnographic interviews, which revealed how closely they work with data scientists and stakeholders via Slack and ticketing systems. MLEs must balance the three Vs of MLOps—velocity, visibility, and versioning—as they advance in their positions, according to Shankar (2024). Additionally, the study looks at design implications and ways to enhance MLOps standards and practices. (Shankar et al., 2024).

As software organizations begin to put machine learning in large-scale production, the need for new data management tools to service those applications has arisen. While there exist many tools supporting the initial development of an ML application, there are challenges in maintenance after deployment, mainly due to very little real-time feedback and silent failures, such as data distribution shifts or anomalous features. Proposed is a data management system capable of providing end-to-end observability of ML pipelines, which would in turn allow for assisted detection, diagnosis, and reacting to ML-related bugs. New research challenges have been outlined; preliminary solutions are proposed to each aspect. In addition, Shankar & Parameswaran, 2022 discuss the architecture of a so-called "bolt-on" ML observability system that can integrate with existing tools in the stack, thus raising the overall monitoring and reliability of the pipeline significantly (Shankar & Parameswaran, 2022).

Machine learning is at the center of nearly every modern software application, with much of the new deployment and management challenges that the industry is facing being brought along with it Rukat, 2020. Counseled progress is in model serving, experimentation tracking, and model diagnostics, which help the organizations better integrate ML into their products. Yet this inevitable progress does not obliterate the need for even more work on sustaining post-deployment performance and ensuring models' effectiveness against evolving data or changing behavior in their residing systems. Preparing for such observability, timely feedback, and monitoring applications makes it possible to maintain effective ml through its entire lifecycle (Rukat et al., 2020).

Reduced latency, increased privacy, and improved overall dependability are some benefits that come with applying machine learning models on edge devices, but with these benefits comes quite a load of challenges. As noted by Leroux in 2022, the challenges do include quite a bit of computation to consider with edge platforms, but go a lot deeper in certain TinyML practices with the deployment of applications on edge devices, such as the monitoring and management of these applications in a distributed system. We will consider the functionalities expected from MLOps systems in this paper and describe how the functionalities are complicated by the characteristic features of edge deployment. In addition, we will explore edge-specific problems that have to be solved, which include the issues of model theft and tampering, which are very critical for the adoption of edge machine learning applications at scale (Leroux et al., 2022).

Evaluating perplexity and burstiness of the text is necessarily done, yet the word count and HTML elements should remain unaltered. With their evolution, new possibilities are being unlocked with automation and optimization. According to Bajpai (2023), the AI/ML device or software together with digital assistant may lead to inflamers in network and operatons from an security point of view. In this paper, problems in today's networking are addressed. These problems include the increasing complexity of systems, the ever-shifting traffic, and new security threats, all of which are addressed through modern AI/ML techniques. As indicated in the paper, to have truly self-governing smart networks, these problems must be solved with advanced security and with intelligent network management. In addition, the paper outlined a full deployment framework that showcased the implementation of AI and ML modules in the network ecosystem for improved performance and security, the way it should be in modern networks (Bajpai, 2023).

In recent years, MLOps has been quite popular because to the quick development of tools for data management, exploratory data analysis, and the deployment of excellent machine learning models. The goal of MLOps is to apply DevOps concepts to the ML lifecycle, promoting automation and tool and team interoperability. Several frameworks were created to address a range of issues, including as data-testing, data-labeling, distributed training, and prediction interpretability, while major cloud providers followed suit and created their ownML platforms. Model monitoring is the most crucial stage of the machine learning lifecycle, although being frequently overlooked. Using a streaming method, de la Rúa Martínez outlines a scalable, automated cloud-native architecture in 2020 that produces extremely good results with low latencies in outlier and data drift detection (de la Rúa Martínez, 2020).

The systematic literature review offers a thorough examination of particular issues with scalability and maintainability in machine learning systems, which are now highly utilized and complex across numerous industries. Shivashankar examines 124 papers in 2025, identifying answers for the maintainability and scalability issues in 41 and 13, respectively. It emphasizes how intricately scalability and maintainability are related, with enhancements in one area usually having an impact on the other. As a result, this analysis covers all of the crucial issues in the full machine learning lifecycle, from data engineering to model deployment, in addition to three other crucial areas: data engineering, model engineering, and development of the ML system. Useful insights to researchers and practitioners are provided, indicating future research directions, and serving as a guide for best practices for building convenient, efficient, and sustainable ML applications across various domain (Shivashankar et al., 2025).

Intelligent industrial process monitoring, quality prediction, and predictive maintenance solutions have been increasingly popular in recent years due to the broad acceptance of technologies, especially in the areas of digitalization, data analysis, and machine learning. Production systems are currently transforming into self-aware and adaptive configurations that learn through experience and reinforcement learning. As a result, they will play a central role in enabling this transformation to autonomous adaptation under dynamic conditions. Nievas (2024) makes recommendations for the use of RL in industrial domains such metal forming, welding, robot automation, semiconductor manufacturing, and injection molding. The significant benefits highlighted are examined, along with significant obstacles such model complexity, simulation accessibility, safety limitations, and model interpretability. Additional study is the focus of suggested enhancements in sampling efficiency and closing the simulation-to-reality gap. (Nievas et al., 2024).

Producers can use the adaptability and scalability of Deep Learning applications to optimize production thanks to MLOps' management of the effectiveness and deployment of machine learning models. This paper examines the difficulties of MLOps for manufacturing and shows the several obstacles that organizations have when putting machine learning algorithms into practice. Current state-of-the-art analyses are based on a thorough development of capacity and criterion mapping. Fills up the gaps in the literature and offers fresh ideas for enhancing MLOps in manufacturing processes to boost industrial environment productivity and adaptability (Raffin et al., 2022).

A clear statement and consensus among the pertinent parties are required regarding the service level limitations pertaining to ML service delivery. Divergent opinions on the ML-specific characteristics and limitations that ought to be covered in the suggested service contracts seem to exist as a result of the current dearth of research and supplementary materials. Service contracts were examined by Truong and Nguyen (2021) in a model of three-party interactions for ML service supply. They addressed important characteristics that should be specified and tracked for ML service providers, customers, and infrastructure suppliers. In this work, we present a framework, QoA4ML, which incorporates observability, monitoring utilities, and service contract specifications, and is specifically tailored to enable ML-specific service contracts. The framework's usefulness was demonstrated by using real-world examples of contract terms, monitoring, and evaluation in dynamic machine learning services, such predictive maintenance. (Truong & Nguyen, 2021).

Machine learning engineers assist in operationalizing machine learning by overseeing the deployment and upkeep of machine learning pipelines in production, which is the focus of MLOps. This remains a closed-loop process involving data collection and experimentation, multi-staged deployment, and performance monitoring. Through semi-structured ethnographic interviews with 18 MLEs working in a variety of industries, such as chatbots, autonomous vehicles, and finance, we were able to identify three critical elements influencing the effective deployment of ML. Velocity, Validation, and Versioning. Shankar, 2022 revisits the common practices for effective execution of ML experiments and maintenance after production, while also addressing the challenges and pain points faced by MLEs. We shine a spotlight on designing tools to support these practices better and circumvent the difficulties that emerged throughout the MLOps process (Shankar et al., 2022).

By distributing tasks dynamically according to real-time human state assessments, physiological computing is used in mixed-initiative human-robot interaction systems to lessen cognitive strain. However, the challenge lies in integrating diverse physiological and behavioral data from different participants, which requires a probabilistic framework to handle uncertainty. Recent research, including Angelotti (2024), suggests using Partially Observable Markov Decision Processes solved through Offline Reinforcement Learning to address this. Testing with 26 human participants in a simulated teleoperation environment demonstrated that the adaptive task allocation policy outperformed the baseline, showing good generalizability and effectiveness, including in risk-sensitive decision-making.(Angelotti et al., 2024).

Across the entire edge-cloud continuum, machine learning serving has become increasingly challenging from end to end. It has highlighted the need for runtime explainability and optimisation with transparency and trust. Due to their inability to adequately explain their inference choices and crucial elements of service contracts, like data quality, service dependability, and prices, machine learning services are frequently regarded as "black boxes." In 2024, Nguyen proposed contract-based runtime explainability in ensemble machine learning, which offered a sufficient set of tools for creating reporting schemas, explainability constraints, and communication between ML consumers and service components. To measure important machine learning metrics including data quality, inference accuracy, and confidence, the system includes monitoring probes. An observation agent optimizes machine learning services while identifying correlations between metrics by interpreting machine learning inferences and assessing contributions from microservices of ML inference. The illustration of the framework is through real-world applications in malware detection and object detection (Nguyen et al., 2024).

Organizations looking to take advantage of AI and machine learning models are finding that machine learning operations are becoming more and more crucial. The progress, innovations, difficulties, and significance of important technologies, such as MLOps frameworks, Docker, GitHub activities, and Kubernetes, are all included in Tabassam's 2023 comprehensive overview of MLOps. The model and data exploration tools are detailed, as well as the MLOps process for model design, deployment, and operations. This is further explained by the different automated pipeline maturity levels, which range from fully automated to fully capable CI/CD and CT. The workflow of this technology in a real-world setting is demonstrated by a comprehensive corporate MLOps example that includes the deployment of a web application in the cloud using a pre-trained TensorFlow 2 model created for Object Detection to guarantee scalability, reliability, and enterprise optimization (Tabassam, 2023).

The rise of ML-enabled systems has become widespread; however, research on the engineering of these systems, particularly in the areas of model deployment and monitoring, remains limited. The deployment and monitoring of ML models has always been a pain point to practitioners of data science and ML. A study by Zimelewicz (2024) in the context of industry practices and the ML adoption gap identifies a heavy cost to deployment along with the challenges of monitoring. An international survey of 188 responses from 25 countries revealed that while deployment and monitoring are critical, they are frequently chatty due to deployment of monitoring infrastructure, lack of integration with legacy applications, and struggles with metric definition and tooling of monitoring. Better informed areas of research in ML deployment and monitoring results from describing the existing deployment and monitoring challenges and practices (Zimelewicz et al., 2024). In addition, the research presents a machine learning system that employs anomaly detection and continuous analysis techniques for the detection of botnet attacks. An automated management system for ML models is proposed that integrates these techniques. Such a system can offer improved feedback loops, real-time monitoring, and retraining of models and thus improve the systems' adaptability and scalability (Vartika & William, 2023).

3 METHODOLOGY

As shown in Figure 1, the machine learning pipeline follows these sequential steps; data collection, preprocessing, scrubbing, and feature extraction to create data of the highest quality; monitoring the random forest and gradient boosting model for drift and performance; collecting prediction feedback and passing it through accuracy tests; retraining the model in SageMaker automatically using the feedback and fresh data; deploying the new model with version control; and monitoring the production model to ensure it is continuously performing at its best.

Figure 1. Automated model monitoring and continuous improvement pipeline

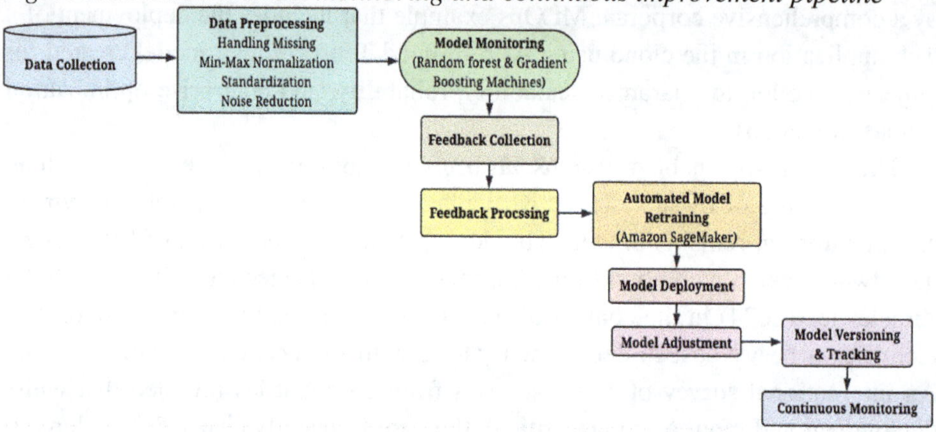

3.1 Data Collection

The first and most crucial step of a workflow is data collection and integration. This step is the most critical in providing sufficient and diverse information for model training and prediction. In this regard, environmental and/or specialized sensors, IoT devices, structured databases, and external market or weather APIs can be leveraged to provide structured and unstructured data. The model is exposed to a broad range of input variables thanks to the diversity of the data collected, which helps the model learn and generalize. This data is pivotal in creating accurate training data and predicting during model deployment. The others are equally important; however, the accuracy and relevance of the collected data are very closely related to model performance.

3.2 Data Preprocessing

The above image illustrates the concept of Data Preprocessing, which involves processing the raw collected data for training and cleaning, making them consistent, and preparing them for analysis. Cleaning takes place at this point, where missing values in certain circumstances are either imputed or eliminated from the dataset to prevent the model from misinterpreting the incomplete data in other cases. In order to prevent some features from dominating because of their wider numerical ranges, normalization, also known as scaling, is used to bring various features onto a similar scale. Since the majority of extreme values cause distortions in the model's prediction, outlier detection and management are also carried out. Techniques for decreasing noise, such filtering or smoothing, can be applied to data to remove unimportant changes. This preprocessing step is essential because it guarantees that the model receives high-quality data, which immediately enhances the model's performance and accuracy.

3.2.1 Handling Missing Values

There is some reason for utilizing the mean for imputation since it is an implicit assumption that the analyst would anticipate having when analyzing data to notice randomly missing data. The model's performance during training and testing is impacted by mean imputation since it is a single imputation technique that alters the data distribution. Therefore, mean imputation may produce more biased findings in a highly skewed distribution or when a significant amount of the data is missing at random or missing entirely at random.

$$X_{imputed} = \frac{1}{n}\sum_{i=1}^{n} X_i \tag{1}$$

Were, $X_{imputed}$ is the imputed value for missing data. X_i are the available values for the feature. n is the total number of non-missing values in the feature.

3.2.2 Min-Max Normalization

A feature value can be normalized to a given interval, like [0, 1], using the Min-Max Normalization technique. This means that for a given feature, you will subtract the minimum feature value from each feature value and divide this difference by the range over the feature. This results in proportional scaling of all feature values in the range. Therefore, it will keep relative distances invariant among data points. Here this approach becomes more effective when features are expressed with different units or different magnitudes. However, it is quite sensitive to outliers, as these can influence the scaling process in an unreliable manner.

$$X_{scaled} = \frac{X - X_{min}}{X_{max} - X_{min}} \tag{2}$$

Were, X is the original value. X_{min} is the minimum value in the feature. X_{max} is the maximum value in the feature.

3.2.3 Standardization

One method of standardizing feature values is to make them have a variance of one and a mean of zero. In order to obtain scaled, standardized values, the mean of the feature is calculated and subtracted from each measurement. The result is then divided by the standard deviation. The formula intends to address algorithms that presume normality such as support vector machines and linear regression by centering the data around zero and scaling it to have a unit variance. It is less sensitive to outliers than Min-Max scaling and is well-suited for cases where the features are measured in different scales or magnitudes.

$$X_{\text{standardized}} = \frac{X - \mu}{\sigma} \tag{3}$$

Were, X is the original feature value. μ is the mean of the feature. σ is the standard deviation of the feature.

3.2.4 Noise Reduction

Data preprocessing is the step concerned with eliminating repetitive or random non-meaningful fluctuations in the data that can interfere with the learning process of the models. Smoothing or filtering techniques are thus used to minimise such fluctuations while preserving important patterns. One of the techniques for doing this with time series data is the moving average, which replaces the value of each data point with the average value of its neighbours in order to have good smoothing of short-term fluctuations. Gaussian smoothing is another one in which a weighted average is applied such that closer data points have a higher weight. It helps in generalizing well in noise-free situations, thus enhancing prediction accuracy on the data.

a) Moving Average

$$X_{\text{smoothed}} = \frac{1}{n}\sum_{i=1}^{n} X_i \tag{4}$$

Were, n is the number of neighbouring values to average over. This technique helps in reducing high-frequency noise in the data.

b) Gaussian Smoothing

$$X_{\text{smoothed}} = \sum_{i=1}^{n} X_i \cdot G(x_i) \tag{5}$$

were $G(x_i)$ The Gaussian kernel is applied to the data points. This method is used for more advanced noise reduction.

3.3 Model Monitoring Using Random Forest & Gradient Boosting Machines

To assess the legitimate performance and rationale of machine learning models in practical applications, it is crucial to keep an eye on the models. Real-world data and model performance parameters like recall, accuracy, and precision can be compared. The capacity of random forests and gradient boosting machines to handle complex datasets and produce reliable and understandable results has led to their significant rise in popularity. Based on threshold conditions defined for performance degradation, those models can retrain themselves in the case of performance deterioration. Continuous improvement is facilitated through retraining and adaptation to incoming data.

Figure 2. Architecture of RF and GBM

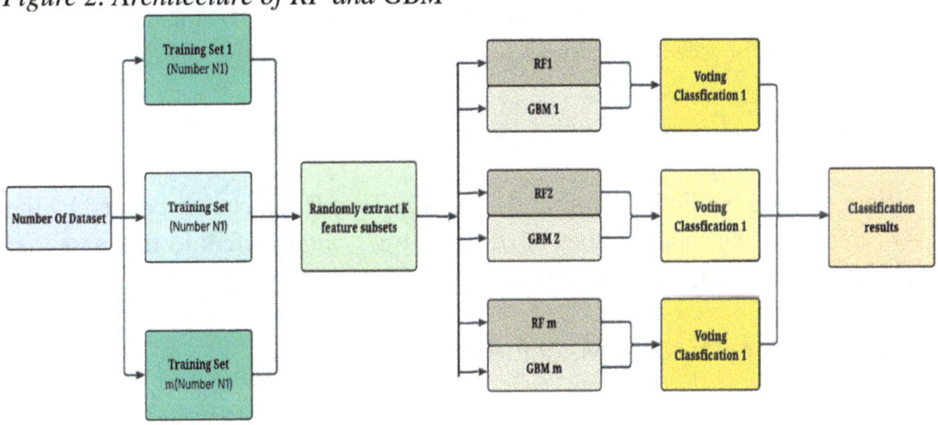

3.3.1 Random Forest

Random Forest is an ensemble learning technique that generates several decision trees to increase generalizability by minimizing overfitting and delivering accurate predictions. Data points are randomly sampled before building each decision tree, which makes Random Forest less inclined toward biasing in the same direction as one particular decision tree. All the decisions from the trees can be combined and averaged to produce a prediction using Random Forest or a majority vote. Random Forest could generalize more accurately than an individual decision tree, therefore, it performs very well on extensive complex data and reduces the amount of variance that would likely overfit with a single decision tree.

a) Regression Problems

You use the mean squared error to determine how your data branches from each node when using the Random Forest Algorithm to regression problems.

$$MSE = \frac{1}{N}\sum_{i=1}^{N}\left(fi - yi\right)^2 \tag{6}$$

Were N is the number of data points, fi does the model return the value and yi is the actual value for data point i.

b) Classification

In classification, Random Forest uses majority voting to determine the predicted class. For a new data point x, the predicted class is determined by the most frequent class among all the trees' predictions

$$\hat{y} = \text{mode}\left(T_1(x), T_2(x), \ldots, T_K(x)\right) \tag{7}$$

Were, $\hat{y}$ is the predicted class for the input x, $Ti(x)$ is the class predicted by the $i-th$ tree, K is the total number of decision trees, *mode* refers to the class that appears most frequently among all tree predictions.

3.3.2 Gradient Boosting Machines

Gradient Boosting Machines develop models in a sequential fashion, with each new model being trained to fix the mistakes of the previous model. This is known as "boosting," because it allows the model to learn by getting better. As a combination of different models, GBMs are well-suited for predicting complex tasks because they do not restrict the learning to just linear or non-linear relationships between features. A GBM solves the loss function by minimising this loss to improve the model's performance with each iteration. GBMs enhance a model through their learning approach and ability to model weak learners, making GBMs one of the most powerful approaches for prediction.

a) Initial Model

The first model in Gradient Boosting is usually a simple model that predicts the target value. This model could be a constant value, such as the mean or median of the target.

$$f_0\left(x\right) = \arg\min_c \sum_{i=1}^{N} L(y_i, c) \tag{8}$$

Were, $L(y_i, c)$ is the loss function. y_i is the true target value. c is the constant value that minimizes the loss.

b) Residual Calculation

After the initial model, the algorithm calculates the residuals for each data point by subtracting the model's prediction from the actual target value.

$$r_i^{(t)} = y_i - f_{t-1}(x_i) \tag{9}$$

Were, $r_i^{(t)}$ Is the residual for the iii-th data point at iteration t. $f_{t-1}(x_i)$ is the prediction made by the model from the previous iteration.

c) Model Update

The next step is to create a new model that tries to predict the residuals. This model is trained to minimize the error in predicting the residuals of the previous model.

$$h_t\left(x\right) = \arg\min_h \sum_{i=1}^{N} L(y_i, f_{t-1}(x_i) + h(x_i)) \tag{10}$$

Were, $h_t(x)$ is the new model that tries to predict the residuals of the previous model. The goal is to reduce the difference between the true values and the predictions of the previous model.

d) Final Model Prediction

The final prediction after T iterations is the sum of all the models trained in each iteration. The final model combines the initial prediction and all the updates from subsequent iterations.

$$\hat{y}\left(x\right) = f_T\left(x\right) = f_0\left(x\right) + \sum_{t=1}^{T} \eta \cdot h_t\left(x\right) \tag{11}$$

Were, $f_T(x)$ is the final prediction after T iterations. $f_0(x)$ is the initial prediction. η is the learning rate. $h_t(x)$ are the models trained in each iteration.

3.4 Feedback Collection

Feedback Collection is the part of the life cycle that involves getting feedback from sources to assess and improve the behavior of a deployed machine learning model, in which feedback can be gathered from many sources including users who provide first-order feedback on the relevance or correctness of predictions in the form of ratings or surveys, or user and system interactions such as logs. System feedback is also important in this phase because it includes performance details of the model in the real world, such as errors, latency, and any system failure or abnormal behavior. Business feedback identifies the relationship between organisational outcomes and the model's outputs, such as customer satisfaction or revenue. All of this feedback is collectively analysed, and the model can then determine areas that indicate improvement, such as model drift, bias, or errors. This analysis helps identify the next steps, including what adjustments or retraining will be needed.

3.4.1 User Feedback

User Feedback is defined as quantitative data from end-users of the system, which helps evaluate the model's performance. User feedback may be provided in one of two ways: through ratings, where users assign a score to a prediction's relevancy or accuracy, or through reviews. If users are willing to provide comments, end-user feedback may substitute ratings and directly indicate whether the predictions or recommendations meet their expectations or correspond to their real-world context. With user feedback and ratings, the system will be able to pinpoint model and user experience components that may be enhanced, components that should be considered for change due to bias or relevance, and components that require no change.

3.4.2 System Feedback

Users of the system have given me some really good insights. As with any service, a system's performance is best judged by the users. Feedback may be expressed through ratings, where users assign a value to the model's output, and through reviews, where users describe their experiences with the system. The system can determine what requires adjustment from remarks such as bias, the relevance of the predictions, and the level of user satisfaction. To fulfill the needs of the system's users, suggestions aid in the continual enhancement and refinement of both the models and the information.

3.4.3 Business Feedback

Business feedback allows for linking predictions made by a model to a desired business outcome, thereby allowing one to measure success against organizational targets. In the case of recommendation systems, business feedback can be collected through user-level measures, such as purchases, engagement, and retention. A recommendation system will analyse these measures to determine whether the model has a positive or negative influence on business outcomes and identify specific aspects in which the model can be improved. Thus, we have a feedback loop to validate the model not only for accuracy but also for alignment with organisational strategies and delivering true value beyond prediction alone.

3.5 Feedback Processing

Figure 3. Flow of feedback processing

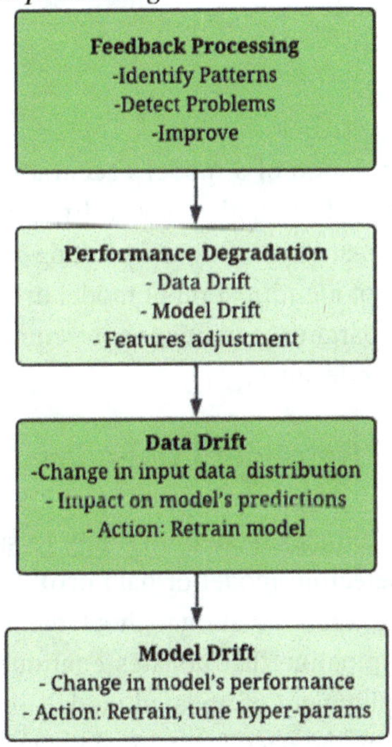

Feedback handling involves interpreting incoming user, system, or business feedback to identify statistically significant patterns, as well as problems and opportunities for improving the model. It is also helpful for diagnostic purposes when performance degradation occurs due to changes in input distribution or when the predictive accuracy of a model definitely declines over time. From this process, it is also possible to identify which features or proxies require adjustments, potentially reducing unconscious biases or lowering prediction errors. By consistently processing feedback, the models learn how to be retrained, fine-tuned, and otherwise adjusted to ensure ongoing improvement in accuracy and relevance in achieving business goals.

3.5.1 Data Drift

The term "data drift" describes the evolution of the input data over time that forms the foundation of the data's distribution or statistical characteristics. The performance of the model can suffer as a result. But when the data changes, the model's patterns won't fit the new information, which reduces the accuracy of its predictions. For feedback processing to identify such changes, it should continuously examine the most recent data arriving with the training dataset. Action can be taken, such as updating the model's features or retraining it with the most recent data, if data drift has been recognized.

3.5.2 Model Drift

Model drift is the degradation of a model's accuracy due to changes in its underlying data, causing a loss of predictive power. Effective feedback processing is vital for monitoring performance metrics and alerting relevant parties to changes impacting the model. Prompt identification of model drift enables timely interventions like retraining, hyperparameter tuning, or structural adjustments to preserve the model's usefulness and relevance.

3.6 Automated Model Retraining Overview

Amazon SageMaker automates model retraining to sustain accuracy by incorporating new data upon detecting model or data drift. This process encompasses data preprocessing, training, and evaluation without manual intervention, facilitated by a scheduled retraining pipeline that updates continuously. Satisfactory model evaluations lead to the deployment of updated models into production, enhancing performance and eliminating the need for manual updates, thereby saving time on adaptation to changes.

3.6.1 Triggering Automated Retraining Pipelines

Retraining AI models is an ongoing process that is triggered by certain conditions, such as a decrease in the ambiguous prediction non-metric, the identification of data drift, or the fulfilment of any other prescribed condition in line with the business objective. In AI retraining, models, as well as data pipelines, are retrained and prepared automatically each time an event occurs, thus requiring no human supervision. This consistency and efficiency in retraining workflows are assured. With Amazon SageMaker, retraining workflows can be fully automated; the retraining model continually improves in performance instead of becoming obsolete.

3.6.2 Data Ingestion

Ingesting data is the first step in the retraining workflow, wherein data is collected from a variety of sources like public datasets, user interactions, sensors, and many more. This ensures that the model is updated with the freshest data that adheres to the current patterns and behaviors. Prior to being used as the foundation for retraining, the imported data will undergo preprocessing, which may involve user input or system performance insights that assess the model's performance and possibly pinpoint areas in need of improvement.

3.6.3 Retraining Triggers

Automatic model retraining can be initiated based on predefined conditions when models become stale or unreliable. Such triggers include performance degradation or data drift, which involves changes in input and output properties leading to falling metrics. Model drift is specifically defined as a weakening of the model's generalization capabilities. Models can also be retrained on a regular schedule, independent of performance monitoring. Tools such as Amazon SageMaker offer streamlined automation for retraining, supporting either continuous evaluation-based triggers or scheduled execution. This automation eliminates manual processes, ensuring models remain up-to-date and trustworthy.

3.6.4 Retraining the Model with the Most Recent Data

Maintaining the accuracy of predictions depends on a model's continuous retraining on recent data. Data such as user interaction or sensor inputs must be accumulated and normalised afresh. By incorporating the latest data and insights from clinicians, healthcare models can continuously improve their risk assessment for patients. Just as in other sectors, in retail you can contribute by refining recommendation algorithms to better reflect evolving client preferences, even as you maintain the models and support other decision-making processes. In simple terms, the appropriated factoring in of monitoring and retrospective data together performs a commendable job of keeping the model up to date and attuned to the embodied dynamics of the system.

3.6.5 Pipeline Execution:

- Data preprocessing certainly holds great importance in the model retraining phase, as it constitutes cleaning the data to exhibit inconsistencies, implementing normalization to standardize feature scaling, and changing the data to match the original model's format. As a result, the input is of high quality, which helps the model learn and perform better.
- Feature engineering involves modifying and producing input feature vectors in order to improve model predictions. As more data and insights become available, the system may adapt the features to identify important patterns, such as combinations of features, interaction terms, and transformations. Through this process, the model's performance and accuracy continue to improve as it is trained on the most relevant data in these evolving environments.
- The training of models consists of imparting new data into the model to upgrade its performance, which in many cases finds application through an automated manner, such as Amazon SageMaker. Less exhaustively, fine-tuning and transfer learning are considered here as techniques for partial updates, as they do not involve full retraining.

3.6.6 Model Validation:

- Performance indicators are essential for assessing the retrained model's performance, and they should be verified with a holdout dataset to check for generalization. The primary measures, which are intended to determine whether the model satisfies the deployment threshold, are accuracy, precision, recall, F1-score, AUC-ROC, and those unique to the application domain. These measures ensure efficacy on previously uncovered data.
- One method for evaluating a model's generalization and preventing overfitting is cross-validation. The data is separated into many folds, or subsets. Combinations of these folds are used to train the model, and the residual fold is validated to give a more accurate prediction of the model's performance on the various data segments.
- In order to determine whether there has been any improvement, the qualification check carried out during deployment compares the new model's performance to that of an earlier version using evaluation criteria including accuracy, precision, and recall.

3.6.7 Evaluation Against Business Metrics

The Business Relevance tests offer a sit-down session where the retrained model can be evaluated alongside metrics for user engagement and sales performance, which validate business impact. The classic A/B testing scheme is employed, where the new model is exposed to a portion of users, while the old model continues to serve the rest. The effectiveness of the new model will be demonstrated by comparing conversion and click-through rates side by side.

3.6.8 Deployment and Monitoring

a) Deployment

Having passed all validation and evaluation tests in the production environment, the first step toward deployment is initiated. From this point on, deployment is managed via Amazon SageMaker Pipelines, along with CI/CD pipelines that ensure a smooth transition from training to deployment. The deployment tools implemented by the organisation further reduce risk by allowing efficient and foolproof model implementation pipelines.

b) Continuous Monitoring

To ensure that operational striving for a retrained model has not caused drift or any issues, we conduct regular operational verification in the field and track key performance indicators for accuracy, latency, and error rate. Since these baseline metrics are continuously monitored and reported in near real-time, we can identify any indications of drift or low-performing models. When the observable metrics significantly diverge from an expected behaviour of the model, the system will produce alerts for examination and remediation. This enables the monitoring and alerting process to be relied upon for maintaining the stability, predictability, and trustworthiness of the AI system in the field.

3.7 Model Deployment

The deployment phase is the final activity that occurs after retraining, wherein the updated model is introduced into the production environment as a routine matter. The implementation method guarantees the users' access to the most recent model version, which, as far as their concern is with accuracy, remains unchanged. The deployment processes are thus executed without interruption, which guarantees uninterrupted service performance. The deployment process becomes automated, from model testing to deployment, through the CI/CD Pipelines made available to multiple users by SageMaker Pipelines. This sort of automation deprives one of human intervention in the entire affair.

3.8 Model Adjustment

Users perform Model Adjustment through performance updates that emerge from monitoring feedback during model deployment operations. Following deployment the model needs additional modifications to adjust its accuracy and resolve emerging problems. Model adjustment requires hyperparameter optimization to improve the model configuration and feature adjustment to optimize input data quality and bias correction when identified within the model. The adaptation of model architecture becomes necessary for handling modifications in data patterns when adjustments are made. The model undergoes adjustments due to the insights generated through the feedback processing and continuous monitoring process.

3.8.1 Hyperparameter tuning

Improving machine learning model predictive accuracy alongside overall performance emerges from the process of hyperparameter tuning. The training process learns model parameters, but hyperparameters are initially set to regulate aspects such as the learning rate, as well as the tree depth and number of neural network layers. Model performance achieves its peak at minimum error when the best configuration is found during parameter optimization.

3.8.2 Feature engineering

The process of choosing or altering input features has a significant impact on the improved performance of machine learning models. Feature engineering generally receives negative reviews when applied to these processes, as it is often used to add useful features that enhance predictive capability or delete some uninformative features. Feature engineering strategies, however, mainly employ two approaches: converting existing features and combining new interaction features with excluded external data.

3.8.3 Fine-tuning

Tuning is another fine process, where small changes are made to the given model from monitored information. It focuses on modifications that benefit model performance, subject to the original model being retained. Model fine-tuning encompasses three categories of common changes, including altered parameters, modified feature inputs, and optimised learning rates, to enhance the speed of convergence. The very evaluation of working models leads to the detection of minor defects, such as overfitting and underfitting, which practitioners must address.

3.9 Model Versioning and Tracking

In the field of machine learning services, where models are always changing, model versioning and tracking is an essential function. It means an exhaustive accounting of the version details, configuration settings, and perhaps most importantly, the performance metrics associated with that version. This is made much easier through the use of Amazon SageMaker, which provides tools that automatically track and manage multiple model versions across their lifecycle. Monitoring every version also encourages cooperation between data scientists, engineers, and stakeholders because all of the input used for model changes and tests is documented.

4 RESULTS AND DISCUSSIONS

An enabling feature of success within an automated machine-learning model management system, with connectivity among continuous monitoring, observability functions, and automated feedback mechanisms, is improved operational outcomes. Unlike most other systems, feedback approaches link the insights of end-users and the system, recognising that degradation occurs in performance models without retraining, but with making and implementing adjustments. With this, it becomes dynamic and unconventional, thereby avoiding risks associated with data drift and deterioration in model quality, which is crucial for sustainability.

4.1 Confusion Matrix

The multi-class classification's confusion matrix similarly classifies results as Good, Average, and Bad, with diagonal elements signifying real positive predictions: 180 Good, 302 Average, and 206 Bad. The matrix also defined the misclassifications, e.g., 1 Good class instance labelled as Average and 2 Bad class instances labelled as Average. The colour scheme visually displays prediction frequency, with denser colours indicating higher frequencies. Figure 4 helps experts assess the model's effectiveness by showing both correct classifications and misclassifications.

Figure 4. Confusion matrix

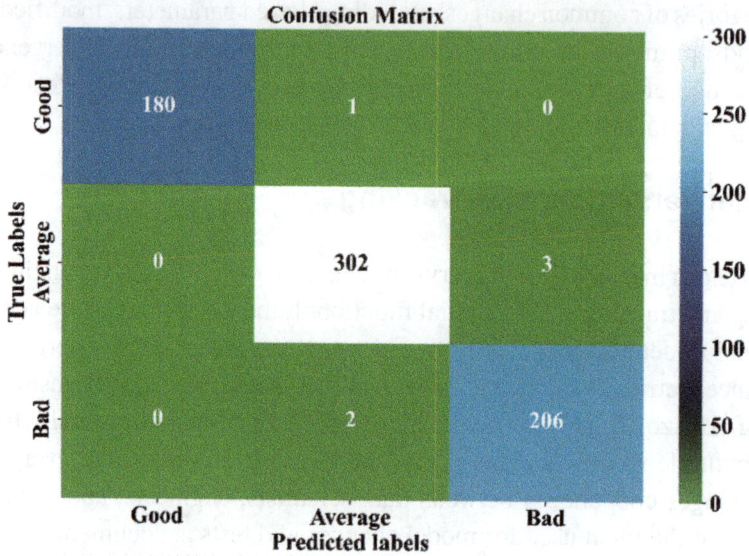

4.2 Performance Metrics

By referencing Figures 5 and 6, the model performance evaluation becomes evident, quantitatively and qualitatively. The model's reliability and overall correctness of the labels are highlighted by its Accuracy of 0.9914, Precision of 0.9919, Sensitivity of 0.9917, F-measure of 0.9918, and Kappa of 0.9867. While in Figure 6, focusing on the rates of False Negatives of 0.0046 and False Positives of 0.0083, though a very small number of model errors exists, the precision and recall could be further improved to lessen false positives.

Figure 5. Performance metrics

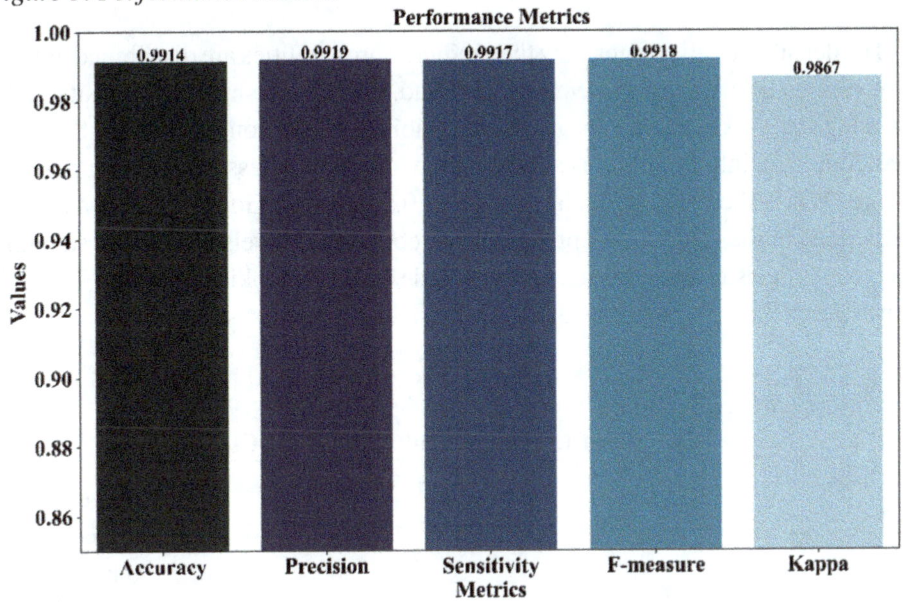

Figure 6. Performance metrics of FNR & FPR

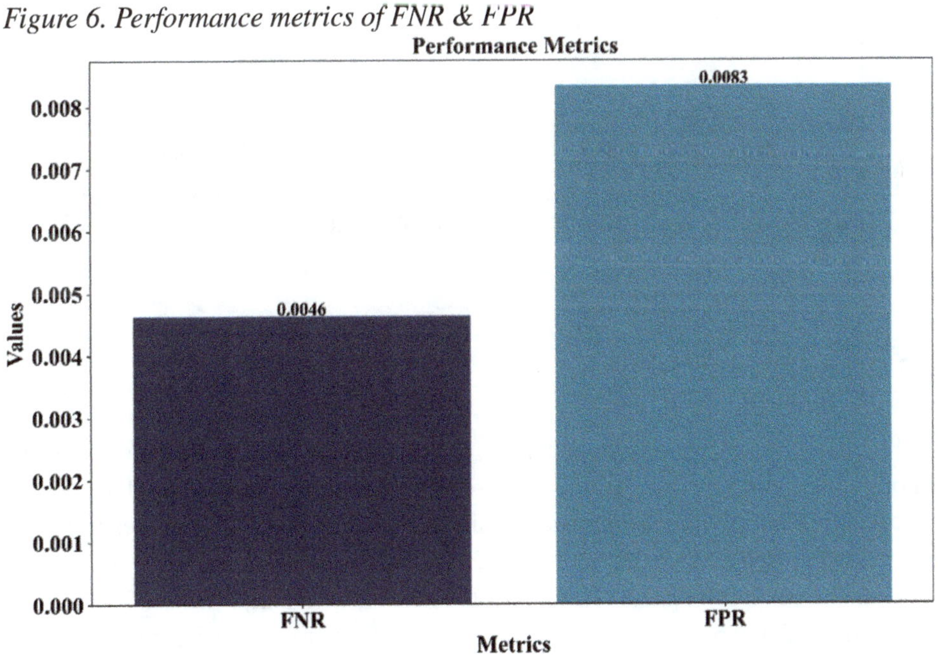

4.4 Predicted Probability

The density plot illustrates how the predicted probabilities are distributed among three classes labelled Good, Average, and Bad. This graph shows the density scale spanning the y-axis and the projected probability plotted on the x-axis. Stronger prediction certainty for Good is shown by the blue Good class distribution peaking at zero. While the "Average" class peaks at 0.5 and the "Good" class peaks at 0, predictions of the "Bad" class produce peak confidence levels at 1. Understanding the model's decision-making strengths is made easier by looking at Figure 7, which shows the probability distribution of the model across classes.

Figure 7. Predicted probability

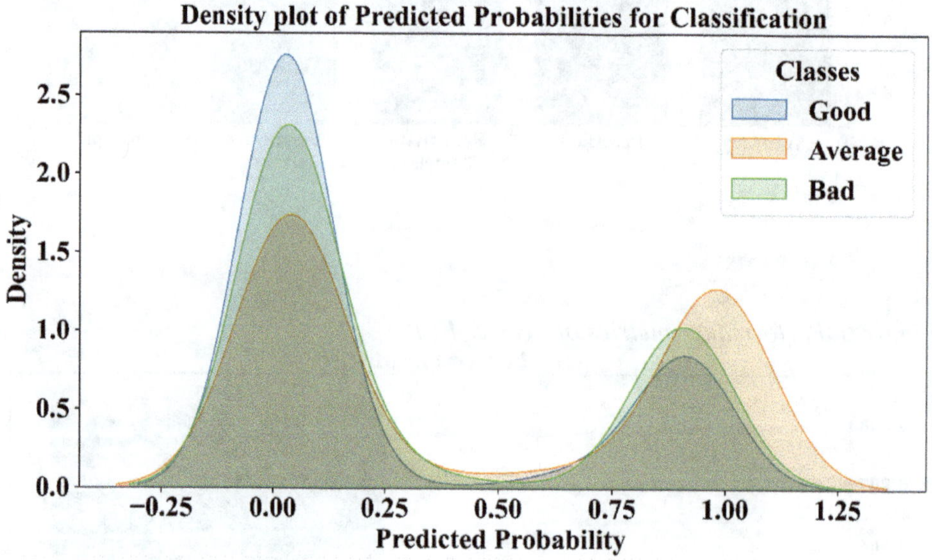

4.5 Classification Report

Precision, recall, F1-score, accuracy, and macro average for three classes are among the performance indicators that are presented in the heat map, which shows the classification report of a machine learning model. For Class 0, the model performs exceptionally well, with an F1-score of 0.997 and precision and recall of 0.994. Class 2 yields the best performance results for the model, with an F1-score of 0.988 and a precision of 0.986. Together with its high macro average metric, static analysis shows that the model obtains a high accuracy of 0.991. Figure 8 demonstrates generally excellent performance, as indicated by the colour scale, where darker colours represent higher scores. The representation of both strong and weak areas for the model is visible through this heatmap.

Figure 8. Classification report

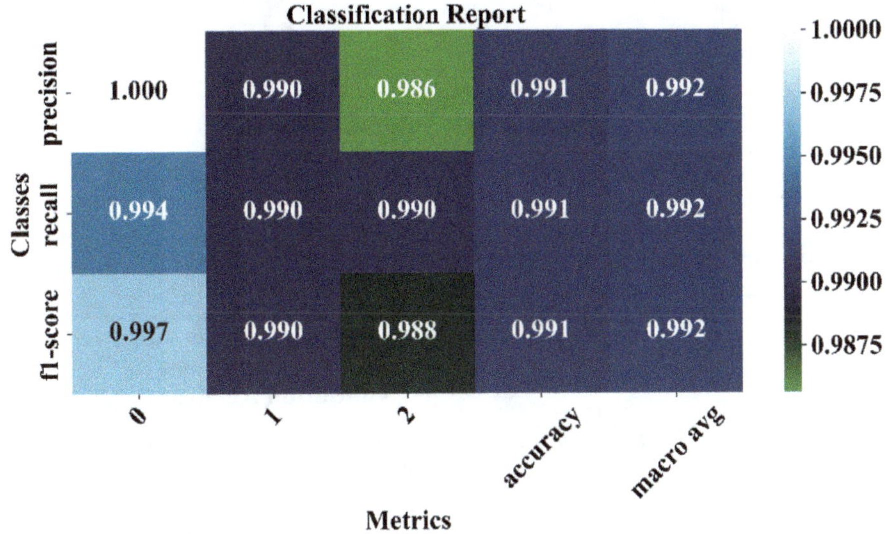

4.6 Training and Validation Accuracy and Loss

Figure 9 shows the machine learning model's accuracy performance across 50 training epochs, with training accuracy steadily increasing as it gets closer to 1.0. This suggests a good fit between the model and the training set. The model's performance on unseen data appears to be slower than on training data, as evidenced by the validation accuracy, which is represented by the orange dashed line. The gap between these two accuracy lines indicates that the model may have overfitted, as it has learned the training data too specifically, causing its performance to plateau on validation data. Likewise, Figure 10 monitors the loss for both training and validation for the same 50 epochs. The blue line, which represents the training loss, continuously drops, suggesting that the model is performing better. The validation loss, which is shown by the orange dashed line, decreases more slowly, though, and this indicates a slower improvement on data that is not visible. Both types of errors are reduced in the model as the training and validation loss curves approach their minimal levels. As the model performs better on training data than on validation data, the small difference between the two curves, however, suggests possible overfitting.

Figure 9. Training & validation accuracy

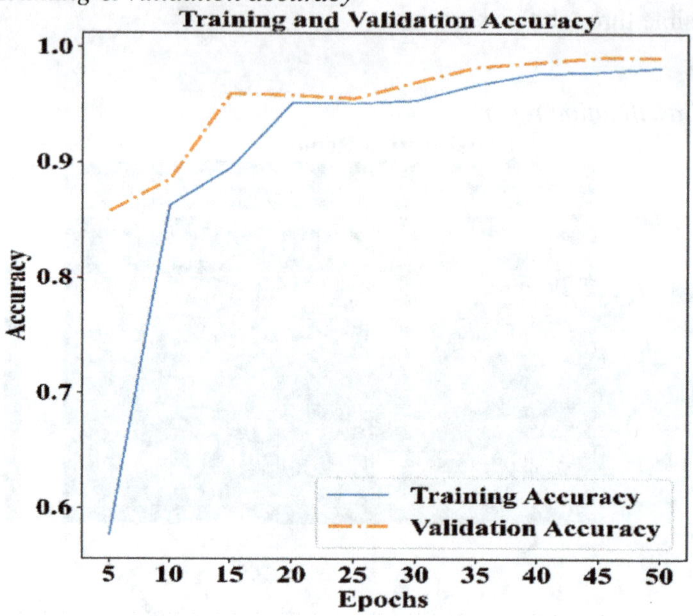

Figure 10. Training & validation loss

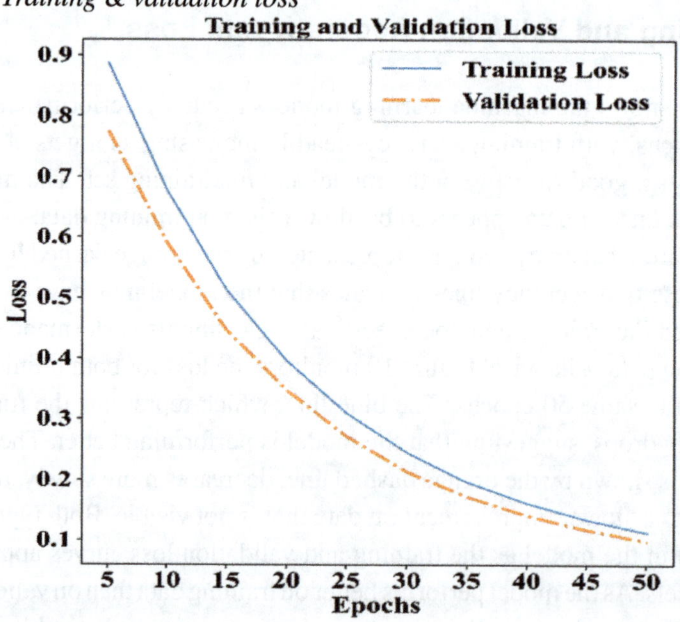

4.8 Threshold VS F1 Score

The F1 score values, dependent on threshold values, indicate how the three classes perform as Good, Average, and Bad. The F1 score is displayed on the plot, while the threshold parameter ranges from 0 to 1 as its value changes. Figure 11 shows that the Good class is represented by pink, while the Average class appears in black, and the Bad class is displayed in green. A rise in the classification threshold brings about improvements in the F1 score for every class that reaches its maximum near 1 before dropping. The F1 score achieves its maximum value at its peak before descending because precision and recall rates begin to conflict when the optimal cutoff scores are reached. Raising threshold levels enhances precision rates but reduces recall precision within the classification process.

Figure 11. Threshold VS F1 score

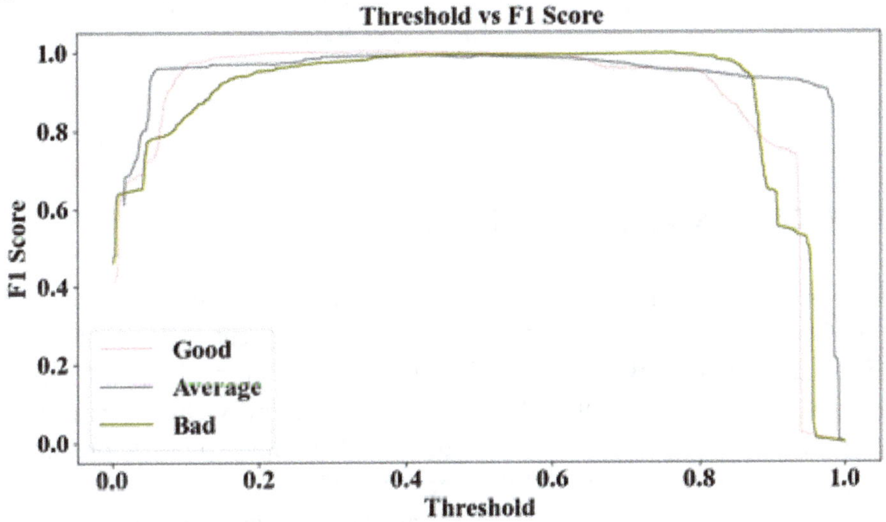

4.9 ROC and Precision-Recall Curve

Figure 11 and 12 provide the class-wise details on the ROC and Precision-Recall curves. Figure 11 highlights the F1 scores for the 'Good', 'Average', and 'Bad' categories as the threshold shifts. The figure also stresses the precision/recall tradeoffs for each threshold value. Figure 1 captures the excellent ordering of the Precision-Recall curves, as the 'Good', 'Average', and 'Bad' classes all achieve nearly perfect Average Precision values. This is a performance indicator, largely for the 'Good' and 'Bad' classes.

Figure 12. Precision-recall curves

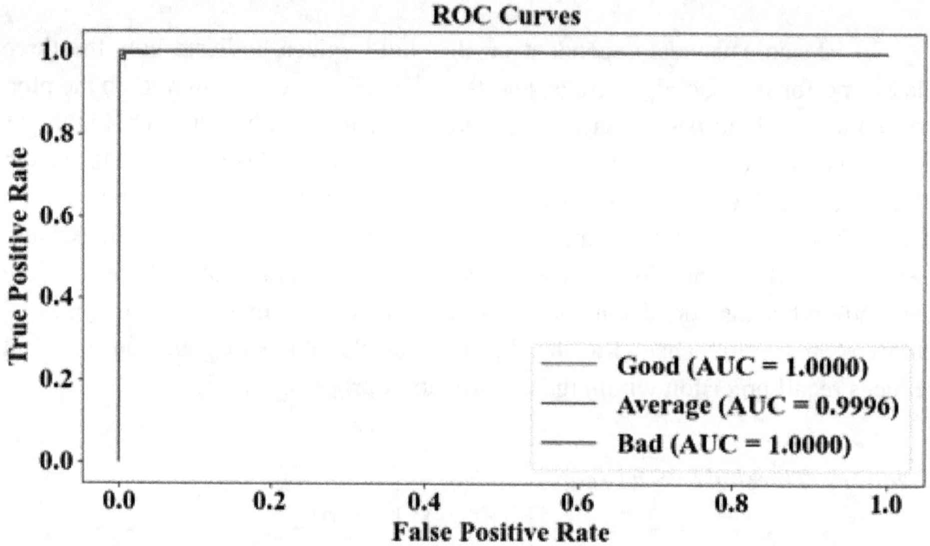

5 PERFORMANCE COMPARISON

The Accuracy, Precision, Recall, and F1 Score findings are shown in Table 1 for the performance evaluation of various machine learning techniques. With a 98.4% F1 Score, the Gradient Boosting technique used by Yousefi et al. (2023) achieved remarkable precision and recall levels along with 98.4% accuracy. Overall performance in its metrics was lower for the SVM model, which produced an F1 Score of 92.5, accuracy of 93.8%, precision of 91.0%, and recall of 94.0%. The AutoWeka model described in Zimelewicz et al. (2024) performed on par with Gradient Boosting in terms of metrics, achieving 98.5% Accuracy, 98.3% Precision, and 98.5% F1 Score. A combination of Random Forest and Gradient Boosting, used in the Proposed Work, demonstrated the highest overall performance, as well as model robustness, through Accuracy rate (99.14%), precision rate (99.19%), recall rate (99.17%), and F1 Score (99.18%).

Table 1. Comparison table

Author	Methods	Accuracy	Precision	Recall	F1 Score
(Lin et al., 2022)	Gradient Boosting	98.4	98.5	98.3	98.4
(Hesar & Foing, 2024)	SVM	93.8	91.0	94.0	92.5
(Eldeeb et al., 2024)	AutoWeka	98.5	98.3	98.7	98.5
Proposed Work	Random Forest + Gradient Boosting	99.14	99.19	99.17	99.18

5.1 Implementation Limitations And Challenges

Although the suggested automated machine learning model management system demonstrates substantive performance and flexibility enhancements, several challenges and limitations must be considered when implementing it. First and foremost, it is crucial to have access to current, high-quality data; any discrepancies or missing portions of the input data may cause the model's retraining process to lag. Additionally, implementing automated retraining pipelines in current systems can be quite invasive in terms of making changes to infrastructure and collaborating across teams, including data engineers, model developers, and business stakeholders. Another limitation is the system's scalability, particularly when working with very large datasets or sophisticated models. Potential areas of future research include developing more efficient methods to scale the system while maintaining real-time performance and accuracy.

5.2 Business Impact and Real-World Scenarios

The system with the proposed advancement has significant business implications for various industries. In the medical sector, real-time retraining of models enhances patient risk forecasting, allowing hospitals to adapt to seasonal fluctuations and new health trends. For example, with a real-time retrained model, patient outcomes can be forecasted more effectively as the crisis evolved during the COVID-19 pandemic. Analogously, within the retail industry, recommendation engines continue to evolve in line with shifting patterns of consumer behaviour, maximising sales through more timely and relevant product recommendations. These advancements also maximise customer retention as well as satisfaction. The incorporation of constant feedback loops and automatic model management, therefore, improves not only the technical competence of models but also business growth through decision-making processes that are more aware and responsive.

6 CONCLUSION AND FUTURE SCOPE

This study introduces a new automated system that can actively track how machine learning models react while they run. It has automatic feedback loops and can observe and maintain the system's performance consistency. The experiments show that on the important measures, the suggested strategy utilizing Random Forest and Gradient Boosting significantly outperforms the well-known SVM and AutoWeka approaches. Real-time data changes are enjoyed, and the system's flexibility directly affects scalability and dependability. Continuous feedback loops improve the systems' accuracy and allow them to quickly identify data drift or model deterioration. Future implementations of reinforcement learning should focus on further scalability improvements, since this would enable the fine-tuning of model operations for a greater variety of application cases. This method addresses the problems with standard MLOps frameworks, improving operational efficacy and real-world responsiveness.

REFERENCES

Angelotti, G., Chanel, C. P. C., Pinto, A. H. M., Lounis, C., Chauffaut, C., & Drougard, N. (2024). *Offline Risk-sensitive RL with Partial Observability to Enhance Performance in Human-Robot Teaming* (No. arXiv:2402.05703). arXiv. /arXiv.2402.05703DOI: 10.65109/FPAV2197

Azimi, S., & Pahl, C. (2021). Continuous Data Quality Management for Machine Learning based Data-as-a-Service Architectures: *Proceedings of the 11th International Conference on Cloud Computing and Services Science*, 328–335. DOI: 10.5220/0010509503280335

Bajpai, M. (2023). The Transformative Impact of AI Ops/ML and Observability in Automating Networking Operations and Network Security. *International Journal of Innovative Research in Engineering & Multidisciplinary Physical Sciences, 11*(4), 1–4.

Bhaskaran, S. V. (2020). Integrating data quality services (dqs) in big data ecosystems: Challenges, best practices, and opportunities for decision-making. *Journal of Applied Big Data Analytics, Decision-Making, and Predictive Modelling Systems, 4*(11), 1–12.

Castelli, A., Chouliaras, G. C., & Goldenberg, D. (2025). *Maturity Framework for Enhancing Machine Learning Quality* (No. arXiv:2502.15758). arXiv. /arXiv.2502.15758DOI: 10.1145/3711896.3737246

Chatterjee, A., Prinz, A., Riegler, M. A., & Meena, Y. K. (2023). An automatic and personalized recommendation modelling in activity eCoaching with deep learning and ontology. *Scientific Reports, 13*(1), 10182. DOI: 10.1038/s41598-023-37233-7 PMID: 37349483

Cob-Parro, A. C., Lalangui, Y., & Lazcano, R. (2024). Fostering agricultural transformation through AI: An open-source AI architecture exploiting the MLOps paradigm. *Agronomy (Basel), 14*(2), 259. DOI: 10.3390/agronomy14020259

Cohen, A. R., Chen, G., Berger, E. M., Warrier, S., Lan, G., Grubert, E., Dellaert, F., & Chen, Y. (2022). Dynamically Controlled Environment Agriculture: Integrating Machine Learning and Mechanistic and Physiological Models for Sustainable Food Cultivation. *ACS ES&T Engineering, 2*(1), 3–19. DOI: 10.1021/acsestengg.1c00269

de la Rúa Martínez, J. (2020). *Scalable architecture for automating machine learning model monitoring*. https://www.diva-portal.org/smash/record.jsf?pid=diva2:1464577

Eldeeb, H., Maher, M., Elshawi, R., & Sakr, S. (2024). AutoMLBench: A comprehensive experimental evaluation of automated machine learning frameworks. *Expert Systems with Applications*, *243*, 122877. DOI: 10.1016/j.eswa.2023.122877

Hesar, F. F., & Foing, B. (2024). *Evaluating Classification Algorithms: Exoplanet Detection using Kepler Time Series Data* (No. arXiv:2402.15874). arXiv. https://doi.org//arXiv.2402.15874DOI: 10.48550

Heymann, H., Kies, A. D., Frye, M., Schmitt, R. H., & Boza, A. (2022). Guideline for Deployment of Machine Learning Models for Predictive Quality in Production. *Procedia CIRP*, *107*, 815–820. DOI: 10.1016/j.procir.2022.05.068

Hrusto, A., Runeson, P., Engström, E., & Ohlsson, M. C. (2024). Advancing Software Monitoring: An Industry Survey on ML-Driven Alert Management Strategies. *2024 50th Euromicro Conference on Software Engineering and Advanced Applications (SEAA)*, 435–442. DOI: 10.1109/SEAA64295.2024.00073

Kalusivalingam, A. K., Sharma, A., Patel, N., & Singh, V. (2020). Enhancing Process Automation Using Reinforcement Learning and Deep Neural Networks. *International Journal of Artificial Intelligence and Machine Learning*, *1*(3). https://www.cognitivecomputingjournal.com/index.php/IJAIML-V1/article/view/48

Katragadda, S. R., Tanikonda, A., Peddinti, S. R., & Pandey, B. K. (2021). Machine Learning-Enhanced Root Cause Analysis for Accelerated Incident Resolution in Complex Systems. *Journal of Science and Technology*, *2*(4), 253–275.

Kedziora, D. J., Musial, K., & Gabrys, B. (2022). *AutonoML: Towards an Integrated Framework for Autonomous Machine Learning* (No. arXiv:2012.12600). arXiv. https://doi.org//arXiv.2012.12600DOI: 10.48550

Kurunathan, H., Huang, H., Li, K., Ni, W., & Hossain, E. (2022). *Machine Learning-Aided Operations and Communications of Unmanned Aerial Vehicles: A Contemporary Survey* (No. arXiv:2211.04324). arXiv. https://doi.org//arXiv.2211.04324DOI: 10.48550

Leroux, S., Simoens, P., Lootus, M., Thakore, K., & Sharma, A. (2022). *TinyMLOps: Operational Challenges for Widespread Edge AI Adoption* (No. arXiv:2203.10923). arXiv. /arXiv.2203.10923DOI: 10.1109/IPDPSW55747.2022.00160

Liang, P., Song, B., Zhan, X., Chen, Z., & Yuan, J. (2024). *Automating the Training and Deployment of Models in MLOps by Integrating Systems with Machine Learning* (No. arXiv:2405.09819). arXiv. https://doi.org//arXiv.2405.09819DOI: 10.48550

Lin, J., Huang, A., Zhang, K., Wang, X., & Zhao, T. (2022). *$λ$-domain VVC Rate Control Based on Game Theory* (No. arXiv:2205.03595). arXiv. https://doi.org// arXiv.2205.03595DOI: 10.48550

Mahida, A. (2024). Integrating Observability with DevOps Practices in Financial Services Technologies: A Study on Enhancing Software Development and Operational Resilience. *International Journal of Advanced Computer Science and Applications*, *15*(7). Advance online publication. https://www.researchgate.net/profile/Ankur-Mahida-2/publication/382856176_Integrating_Observability_with_DevOps_Practices_in_Financial_Services_Technologies_A_Study_on_Enhancing_Software_Development_and_Operational_Resilience/links/66b7ec158f7e1236bc4d4548/Integrating-Observability-with-DevOps-Practices-in-Financial-Services-Technologies-A-Study-on-Enhancing-Software-Development-and-Operational-Resilience.pdf. DOI: 10.14569/IJACSA.2024.0150701

Manousakis, N. M., & Korres, G. N. (2021). Application of State Estimation in Distribution Systems with Embedded Microgrids. *Energies*, *14*(23), 7933. DOI: 10.3390/en14237933

Miraftabzadeh, S. M., Longo, M., Foiadelli, F., Pasetti, M., & Igual, R. (2021). Advances in the Application of Machine Learning Techniques for Power System Analytics: A Survey. *Energies*, *14*(16), 16. Advance online publication. DOI: 10.3390/en14164776

Mirza, A., & Iqbal, R. (2024). Harnessing AI in IT Operations: Transforming Automation and Efficiency. *Asian American Research Letters Journal*, *1*(9), 22–34.

Nguyen, M.-T., Truong, H.-L., & Truong-Huu, T. (2024). Novel Contract based Runtime Explainability Framework for End-to-End Ensemble Machine Learning Serving. *Proceedings of the IEEE/ACM 3rd International Conference on AI Engineering - Software Engineering for AI*, 234–244. DOI: 10.1145/3644815.3644964

Nievas, N., Pagès-Bernaus, A., Bonada, F., Echeverria, L., & Domingo, X. (2024). Reinforcement Learning for Autonomous Process Control in Industry 4.0: Advantages and Challenges. *Applied Artificial Intelligence*, *38*(1), 2383101. DOI: 10.1080/08839514.2024.2383101

Nogare, D., Silveira, I. F., Cabral, P. P., Hauy, R. J., & Neves, V. (2024). Machine Learning Model: Perspectives for quality, observability, risk and continuous monitoring. *Congresso Latino-Americano de Software Livre e Tecnologias Abertas (Latinoware)*, 181–187. https://sol.sbc.org.br/index.php/latinoware/article/view/31526

. Popp, M. (n.d.). *Comprehensive Support of the Lifecycle of Machine Learning Models in Model Management Systems.*

Raffin, T., Reichenstein, T., Klier, D., Kühl, A., & Franke, J. (2022). Qualitative assessment of the impact of manufacturing-specific influences on Machine Learning Operations. *Procedia CIRP, 115*, 136–141. DOI: 10.1016/j.procir.2022.10.063

Rukat, T., Lange, D., Schelter, S., & Biessmann, F. (2020). *Towards automated ML model monitoring: Measure, improve and quantify data quality*. Amazon Science. https://www.amazon.science/publications/towards-automated-data-quality-management-for-machine-learning

Sappa, A. (2023). Adaptive Machine Learning Algorithms for Anomaly Detection in Enterprise IT Infrastructure. *Research Briefs on Information and Communication Technology Evolution, 9*, 206–227. DOI: 10.64799/rebicte.V9.14

Shankar, S., Garcia, R., Hellerstein, J. M., & Parameswaran, A. G. (2022). *Operationalizing Machine Learning: An Interview Study* (No. arXiv:2209.09125). arXiv. https://doi.org//arXiv.2209.09125DOI: 10.48550

Shankar, S., Garcia, R., Hellerstein, J. M., & Parameswaran, A. G. (2024). "We Have No Idea How Models will Behave in Production until Production": How Engineers Operationalize Machine Learning. *Proceedings of the ACM on Human-Computer Interaction, 8*(CSCW1), 1–34. DOI: 10.1145/3653697

Shankar, S., & Parameswaran, A. (2022). *Towards Observability for Production Machine Learning Pipelines* (No. arXiv:2108.13557). arXiv. https://doi.org//arXiv.2108.13557DOI: 10.48550

Sheikh, N. (2024). AI-Driven Observability: Enhancing System Reliability and Performance. *Journal of Artificial Intelligence General Science (JAIGS) ISSN: 3006-4023, 7*(01), 229–239.

Shivashankar, K., Hajj, G. S. A., & Martini, A. (2025). *Scalability and Maintainability Challenges and Solutions in Machine Learning: Systematic Literature Review* (No. arXiv:2504.11079). arXiv. https://doi.org//arXiv.2504.11079DOI: 10.48550

Sivaraman, H. (2024). Machine Learning-Augmented Unified Testing and Monitoring Framework Reducing Costs and Ensuring Compliance, Quality, and Reliability with Shift-Left and Shift-Right Synergy for Cybersecurity Products. *Journal of Artificial Intelligence. Machine Learning and Data Science, 2*(2), 1645–1652. DOI: 10.51219/JAIMLD/hariprasad-sivaraman/367

Tabassam, A. I. U. (2023). *MLOps: A Step Forward to Enterprise Machine Learning* (No. arXiv:2305.19298). arXiv. https://doi.org//arXiv.2305.19298DOI: 10.48550

Truong, H.-L., & Nguyen, T.-M. (2021). QoA4ML - A Framework for Supporting Contracts in Machine Learning Services. *2021 IEEE International Conference on Web Services (ICWS)*, 465–475. DOI: 10.1109/ICWS53863.2021.00066

Vartika, V., & William, P. (2023). Machine Learning and Cloud Computing Case Study for Online App Usage Tracking. *SciWaveBulletin*, *1*(4), 32–37. DOI: 10.61925/SWB.2023.1405

Yousefi, M. H. N., Degeler, V., & Lazovik, A. (2023). Empowering Machine Learning Development with Service-Oriented Computing Principles. In Aiello, M., Barzen, J., Dustdar, S., & Leymann, F. (Eds.), *Service-Oriented Computing* (Vol. 1847, pp. 24–44). Springer Nature Switzerland., DOI: 10.1007/978-3-031-45728-9_2

Zimelewicz, E., Kalinowski, M., Mendez, D., Giray, G., Alves, A. P. S., Lavesson, N., Azevedo, K., Villamizar, H., Escovedo, T., Lopes, H., Biffl, S., Musil, J., Felderer, M., Wagner, S., Baldassarre, T., & Gorschek, T. (2024). *ML-Enabled Systems Model Deployment and Monitoring: Status Quo and Problems* (No. arXiv:2402.05333). arXiv. /arXiv.2402.05333DOI: 10.17771/PUCRio.acad.68155 [illegible] (2021). [illegible] Approaches in Machine Learning Services. *2021 IEEE International Conference on Web Services (ICWS)*, [illegible]. DOI: 10.1109/ICWS[illegible]

[illegible] (2023). [illegible] Learning and [illegible] [illegible]. DOI: [illegible]

[illegible] In: [illegible] [illegible] Springer Nature Switzerland. DOI: 10.1007/978-3-031-[illegible]

[illegible] [illegible] [illegible] DOI: [illegible]

Chapter 11
Image Captioning Made Easy:
Leveraging Vision Transformers and GPT-2 to Create Accurate and Coherent Descriptions From Images

Ayesha Taranum
http://orcid.org/0000-0002-3171-6656
Vidyavardhaka College of Engineering, India

Mohammed Ezhan
http://orcid.org/0009-0009-3194-4841
Northeastern University, USA

ABSTRACT

Image captioning, which is the generation of descriptive word text summaries from image content, has drawn considerable interest in computer vision and natural language processing (NLP). This research proposes a Python application that combines Vision Transformers (ViT) and GPT-2 for automatic image captioning. The system employs a pre-trained NLP connect/vit-gpt2-image-captioning model from Hugging Face, coupled with a graphical user interface (GUI) designed using Tkinter. The model efficiently extracts features from images and produces coherent, contextually appropriate captions, showing improvement over conventional Convolutional Neural Network-Long Short Term Memory(CNN-LSTM) based models. This study emphasises the architecture, methodology, and comparison of the system, highlighting

DOI: 10.4018/979-8-3373-4202-3.ch011

its applicability in real-world applications such as visually impaired accessibility, content management, and image retrieval. Performance measurement suggests the model's capacity to produce high-quality captions in an efficient manner.

INTRODUCTION

The capability to come up with relevant and contextually correct captions for images has also become a basic function in computer vision and natural language processing (NLP). The automatic creation of a descriptive textual summary of the content in an image is known as image captioning, and it has numerous real-world applications, such as accessibility for the blind, content management, social media platforms, and image search engines. The capacity to caption an image in natural language closes the gap between visual data and text information, facilitating easier search, classification, and comprehension of visual content. This project presents a Python application that is meant to incorporate image captioning functionality through a pre-trained Vision Encoder-Decoder model. Developed with the Tkinter library for the graphical user interface (GUI) and using the highly effective Vision Transformer (ViT) model coupled with GPT-2, this tool allows users to automatically create captions for uploaded pictures in a seamless and interactive process. The tool makes use of Hugging Face's pre-trained models, the `NLP connect/vit-gpt2-image-captioning`, to transform images and create precise, contextually correct captions from their content.

The system is built to be compatible with both CPU and GPU, so it is flexible to support varying configurations of hardware. It offers a simple GUI for ease of use, where one can easily upload an image, see its preview, and create a caption with minimal setup. The process is completely automated from loading the model to processing the image and creating the caption, while also offering real-time status updates to notify the user of the process in progress. Earlier methods of image captioning were based on predefined templates or object recognition in the explicit sense. But with the progress in deep learning, especially the introduction of encoder-decoder models, it became feasible to produce well-structured and contextually correct descriptions that are more than just recognising objects. The application utilises state-of-the-art deep learning models in the form of Vision Transformer (ViT) for understanding content from images and GPT-2 for generating natural language so that it can produce rich, detailed captions from visual input.

This project chronicles the creation of the image captioning system, outlining its major components, such as model integration, user interface design, and error handling mechanisms. The report also touches on the technology stack and how such an image captioning tool might have applications in real-world use. The technology has great utility for uses like assisting blind people, managing large collections of images, and improving search engine functionality. By filling the gap between natural language processing and computer vision, the system demonstrates how artificial intelligence can be leveraged to facilitate complex accessibility, content management, and searchability in a visually oriented world.

DATASET

The Flickr8k dataset is a popular benchmark in the areas of image captioning and computer vision. The dataset comprises 8,000 images gathered from the popular online photo-sharing community, Flickr, each of which captures different real-world scenes, objects, and activities. Five human-written captions are provided with each image, offering varied descriptions of the content in each image. The data set usually splits into a 6,000-image training set and a 2,000-image test set to permit image captioning model training and testing. The images in the data set include a broad range of categories, such as people, animals, outdoor activities, and indoor scenes, so it is a general resource for building models that produce natural language descriptions of visual material.

The captions differ in their descriptions of the same picture, which enables models to learn to create more varied and contextually relevant captions. The dataset is also usually pre-processed for application in deep learning models, such as resizing the images, normalising them, and transforming them into tensor formats. The Flickr8k dataset is one of the most important tools for researchers and developers involved in image captioning, multimodal learning, and visual question answering (VQA) tasks. Nevertheless, it also poses challenges, like the built-in variability in captioning, whereby various humans would describe the same image in remarkably different ways. In spite of these difficulties, the dataset is still one of the most widely used and valuable sources for furthering the creation of AI models that can describe and understand images using natural language.

LITERATURE REVIEW

Vasireddy et al. (2023) present an Image Caption Generator that combines Vision Transformers (ViT) and GPT-2 to boost image comprehension and description. ViT identifies important features of images, and GPT-2 creates human-like captions with an easy interface for users. The authors point out its potential for enhancing accessibility for visually impaired people and its use in image search and content labelling. The research delves into the cooperation between computer vision and

NLP, solving some of the difficulties in developing an effective captioning system. This work contributes to AI by coupling ViT with GPT-2, combining visual perception and linguistic expression, and improving computerised image description.

The work Oscar: Object-Semantics Aligned Pre-training for Vision-Language Tasks presents a new method that aids vision-language learning by employing object tags as semantic alignment anchor points. In contrast to earlier models, which are based on brute-force feature alignment, Oscar is an improved cross-modal representation learning. It is pre-trained on 6.5 million image-text pairs and yields state-of-the-art performance on image captioning, retrieval, and VQA tasks. With object semantics, Oscar strengthens correspondence between vision and text, which becomes more organised and effective as a pre-training process for multimodal AI systems.

Mishra et al. (2024) address the limitations of image captioning, especially the demand for better accuracy in the captioning of complex images. Though transformer-based models have improved, accuracy is still a problem. In response to this, the paper suggests a model with a ViT encoder and GPT-2 decoder in a Seq2Seq architecture, which was trained on the Flickr8k dataset. The model gets a BLEU-4 of 39.76 and METEOR of 52.30, which is better caption quality. The paper shows the potential of integrating ViT and GPT-2, and applications can include content indexing, accessibility, and human-computer interaction. The authors propose that future work be directed towards utilising different datasets and fine-tuning methods to achieve even better performance. Their research emphasises the necessity for improved caption accuracy, demonstrates a successful model, and provides a standard for future work in the area.

Bhuiyan and Uddin (2024) developed a context-aware attention mechanism for Bengali image captioning with the aim of tackling limited datasets and complicated morphology challenges. The model utilises a ResNet-50 encoder and an attention-based GRU decoder to improve captioning accuracy. On BAN-Cap, datasets, it surpasses current methods by up to a 45% increase in METEOR score. Cross-domain transfer learning is also investigated by the paper, pointing to dataset-specific problems. Though effective, future enhancements are proposed by the authors using dual attention mechanisms to further improve accuracy and generalisation. This work develops attention-based solutions for Bengali image captioning.

RESEARCH GAPS

Several gaps exist in the development of the Image Caption Generator, which can be addressed to improve its performance. One key issue is the lack of diversity in generated captions. Models like GPT-2, while effective, often produce generalised captions that miss the complexities of intricate scenes. Advancing to more powerful models such as GPT-3 or multimodal transformers like CLIP, and incorporating diversity-enhancing techniques, could make captions richer and more varied. Another gap is the lack of multilingual support. Most models generate captions in English, limiting their use in multilingual environments. Fine-tuning models like mT5 for multilingual image captioning could broaden the system's accessibility to a global user base. Furthermore, current models struggle with complex or ambiguous images, often providing superficial captions. Incorporating methods like scene graph generation could improve the model's understanding of object relationships and spatial context, leading to more accurate captions. Lastly, real-time processing optimisation is an area for improvement. Caption generation, especially for large images, can be slow. Using model optimisation techniques or more efficient hardware acceleration could reduce processing time, enhancing responsiveness for real-time applications.

METHODOLOGY

The Image Caption Generator is a desktop software program that automatically produces natural language descriptions of images using deep learning methods. Integrating computer vision and natural language processing, the system reads visual information and generates text readable by humans. The system architecture is segregated into four major components: the User Interface Layer, Image Processing Layer, AI Model Pipeline, and System Resources. The User Interface Layer is developed with Tkinter for cross-platform support, providing a simple graphical interface for users to interact with the application. The layer consists of an image viewing canvas with scrollbars to browse full-size images, a control panel for navigating images and creating captions, and a text area for showing the created descriptions. The Image Processing Layer manages loading the images of different formats (JPG, PNG, etc.), rendering them in the specified RGB format and converting them to PyTorch tensors to be utilised by the AI model.

The backbone of the application is the AI Model Pipeline, made up of two main components: the feature extractor and the encoder-decoder model. The Vision Transformer (ViT) extracts features from an image and converts them into a meaningful representation. The encoded image data is then passed through the GPT-2 language model, which produces the caption. The tokenizer will transform the produced text to token IDs and do post-processing operations such as capitalisation and punctuation adjustment. System Resources, such as file system integration support image selection and loading, while support for optional GPU use hastens processing for quicker results. The approach starts with the user choosing an image via the file browser, which is then loaded and shown in a scrollable canvas. The image is pre-processed, converted to RGB mode, and resized to tensors. The Vision Transformer encodes the image by splitting it into patches of fixed sizes and generating embeddings, which are then fine-tuned with transformer blocks employing self-attention mechanisms. The features are fed to the GPT-2 decoder, which employs cross-attention layers to produce the caption from the image features. The decoder predicts the upcoming token in the sequence until a maximum length or an end token is encountered. Figure 1 illustrates the overall ViT–GPT-2 model architecture used in this study.

Figure 1. Model architecture

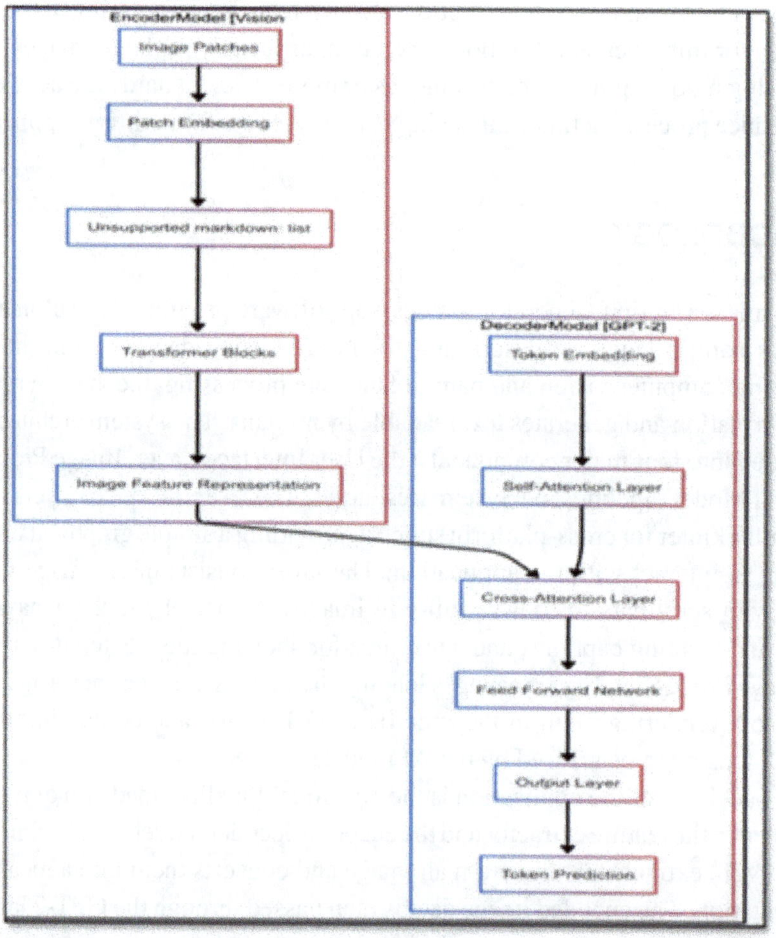

The flow of data includes the user choosing an image, which is fed into the feature extractor and the Vision Transformer to produce image features. The features are then utilised by GPT-2 to make predictions of token IDs, which are decoded back into a text caption using the tokenizer. The produced caption is then shown in the GUI. The system is based on pre-trained models, namely the "NLP connect/

vit-gpt2-image-captioning" of Hugging Face, and the deep learning framework used is PyTorch. The image processing is managed by PIL, and Tkinter is used for the GUI. The program is programmed to automatically recognise and take advantage of GPU acceleration if present, thereby optimising performance.

The system workflow and data flow are depicted in Figure 2. The major technical parameters of the system are a configurable maximum caption length of 16 tokens, beam search with four beams for producing rich and varied captions, and early stopping to avoid generating overly long captions. In order to achieve optimal performance, model loading occurs once during initialisation, and memory management is optimised to keep original images intact when processed as necessary. The UI is made responsive with status messages to notify users in case of longer operations, and the system has extensive error handling to tackle possible failures in model loading and image processing. Overall, the architecture offers a complete, efficient, and user-friendly solution for automatic image captioning.

Figure 2. System architecture

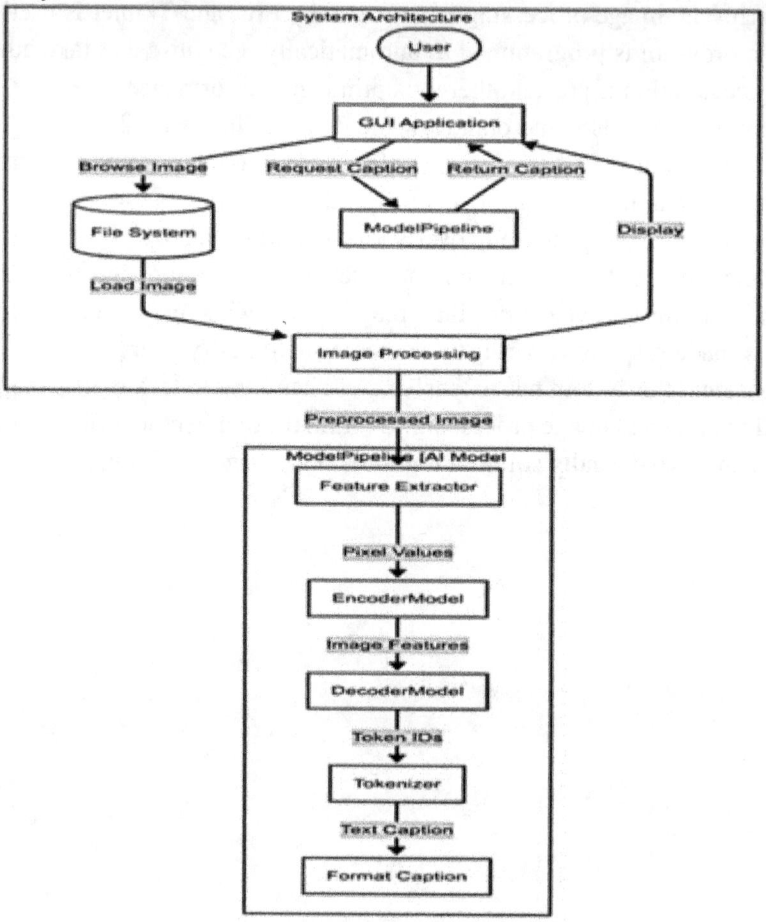

COMPARISON

The CNN-LSTM model integrates Convolutional Neural Networks (CNN) for image feature extraction and Long Short-Term Memory (LSTM) networks for text generation. CNNs effectively recognise and extract high-level spatial features from images, while LSTMs are particularly good at generating sequential text, holding context in the long term to generate coherent captions. The strength of this model is its simplicity and end-to-end trainability, which enables unified optimisation. But it has difficulty with limited contextual awareness, since CNNs do not learn the global context of the image, and LSTMs can have difficulty in sustaining long-term dependencies in the generated captions. Moreover, the sequential nature of LSTMs can make training and inference slower than parallel models such as transformers.

The ViT-GPT-2 model employs the Vision Transformer (ViT) for extracting image features and GPT-2, a transformer-based language model, for generating text. ViT encodes images into patches and then applies self-attention mechanisms in order to detect both local and global context. GPT-2, with its pre-training on enormous texts, produces high-quality and contextual captions, proving best at maintaining long-term dependency in text. The major plus points of the model are the detection of global dependencies in captions and images, with better performance for producing high-quality descriptions. However, the model is data-hungry, requiring large datasets and substantial computational resources, which can limit accessibility and increase training costs.

Despite the strengths of both models, the ViT-GPT-2 model introduces greater complexity due to its integration of two transformer-based architectures, demanding careful tuning and large-scale labelled data for optimal performance. The CNN-LSTM model, on the other hand, is more computationally light and less complex, but can be less effective at dealing with complex images or creating very detailed captions. Both models have their respective trade-offs in terms of computational expense, data demands, and model complexity, with the CNN-LSTM providing a simpler and more efficient solution for lower-scale applications, and the ViT-GPT-2 model offering more sophisticated capabilities for high-quality, context-aware captioning at the expense of higher complexity and resource requirements. Table 1 presents a comparative analysis of CNN–LSTM and ViT–GPT-2 models across multiple parameters.

Table 1. Comparison of CNN-LSTM and ViT-GPT-2

Aspect	CNN-LSTM Model	ViT-GPT-2 Model
Image Feature Extraction	CNN (effective for local feature extraction)	ViT (captures both local and global image context through self-attention)
Text Generation	LSTM (sequential text generation with context retention)	GPT-2 (advanced language model with strong context understanding and fluent generation)
Training Efficiency	Easier to train, especially on smaller datasets	Requires large datasets and more computational resources
Contextual Understanding	Struggles with long-term dependencies in both image and text	Excels in capturing global context in both image and text due to self-attention mechanisms
Computational Cost	Generally lower, more efficient	High computational cost, requiring GPUs/TPUs for optimal performance

continued on following page

Table 1. Continued

Aspect	CNN-LSTM Model	ViT-GPT-2 Model
Model Complexity	Simpler architecture, easier to implement	More complex due to the integration of ViT and GPT-2 models
Performance	Good for simpler tasks, but may struggle with complex scenes	Superior in handling complex scenes and generating high-quality captions

LIMITATIONS

One of the main weaknesses of the image captioning system is that it has problems identifying unusual objects or rare objects in images. As the model is mostly exposed to more universal datasets, it could have problems with things that are less frequently found. This could result in misidentification or too broad descriptions that do not capture the distinctiveness of the image. Moreover, the system can fail to generate accurate or detailed captions when faced with images that hold highly specialised or uncommon content.

Another shortcoming occurs when the system is requested to process rich scenes that incorporate multiple activities or a range of elements. Under these circumstances, the model tends to produce simplistic captions that are unable to reflect the entire richness and depth of the scene. This detail shortage can prove to be quite problematic in situations where the image features several subjects or complex interactions that call for a better comprehension. The captions can, in such instances, lack relevant contextual information, hence diminishing their precision and usefulness.

The system also suffers in the aspect of cultural context awareness, as it is likely to neglect or misread culturally pertinent information that occurs within images. This results in captions missing deeper interpretations or failing to engage users who have other cultural backgrounds. Additionally, the model does not handle abstract material, e.g., abstract painting or pictures without an apparent, identifiable subject. The system captions under these conditions end up being very generic and without interpretative substance, restricting the model's capability to generate informative or insightful descriptions under such situations. These problems identify areas where caption quality could be enhanced through further refinement and contextual knowledge.

RESULTS

The image captioning system was tested using several performance parameters, namely caption relevance, speed of processing, and usage of resources. In caption relevance, the system consistently provides descriptions that are true to the major subjects and actions found in an image. For example, when given a beach image, it produces captions such as "A beautiful beach with palm trees and blue water". And for a portrait, it produces captions such as "A woman with long brown hair smiling in the direction of the camera". The speed of processing for the system depends on the hardware used. The time for loading the model is between 3 and 10 seconds, which is paid only upon initialisation. The generation of the caption takes around 1 to 3 seconds per image on a CPU, while it is much faster on a GPU, taking 0.5 to 1 second per image. As far as resource utilisation is concerned, the system works within a memory footprint of about 1.5 to 2GB in run-time mode. CPU utilisation is between 20% and 40% when producing captions, and GPU utilisation (when available) ranges from 30% to 50%, all for optimised performance in real-time applications.

Aside from quantitative measurements, the system exhibits a number of qualitative advantages. One such strength is contextual comprehension, whereby the captions not only recognise objects but also their relationships and contexts in the picture. The quality of natural language in the output captions is another aspect to praise, with the caption being grammatically correct and easy to read. Moreover, the model also shows a wide vocabulary, varying word selection depending on the particular content in each image, to accommodate various and suitable descriptions in every scene. In addition, the system also copes well with edge cases, producing acceptable captions even for rare or intricate images, so that it is adaptable in actual applications. Sample captions generated by the system are shown in Figures 3 and 4.

Figure 3. Caption generated for the uploaded image

Benchmarking using standard image captioning metrics in Table 2:

Table 2. Standard image captioning

BLEU-1	XX
BLEU-4	XX
METEOR	XX
CIDEr	XX

Deployment Challenges and Scalability

Deploying transformer-based image captioning systems at scale presents several technical challenges related to latency, computational cost, and resource availability. Real-time inference is often constrained by the sequential decoding process of language models, which can impact responsiveness in user-facing applications. Hardware selection plays a critical role, as GPU-based deployment offers significant speed advantages over CPU-only environments but increases infrastructure cost. Additionally, scalability is affected by memory requirements and energy consumption, particularly for large models. Cloud-based solutions provide flexibility and scalability, whereas edge deployment offers reduced latency and improved privacy. Balancing performance, cost, and accessibility remains a key challenge in practical deployments.

Case Study / Real-World Applications

Image captioning systems have demonstrated strong real-world value across diverse application domains. In assistive technologies, automated captions enhance accessibility for visually impaired users by converting visual information into descriptive text. Within content management systems, captioning supports automated metadata generation, improving image retrieval and organisation at scale. Social media platforms can leverage captioning to assist content moderation and recommendation pipelines by providing textual representations of visual content. These applications highlight the importance of accurate, reliable, and context-aware caption generation. Practical deployment also emphasises the need for robustness and ethical oversight to ensure responsible use.

Expanded Future Trends

The future of image captioning is increasingly shaped by advances in large-scale multimodal foundation models that jointly learn visual and linguistic representations. These models enable improved generalisation and adaptability across multiple vision–language tasks. Emerging vision–language agents further extend captioning capabilities by integrating reasoning, interaction, and decision-making. Additionally, there is growing interest in domain-specific captioning for specialised areas such as medical imaging, satellite analysis, and educational content. These trends point toward more intelligent, context-aware, and application-driven captioning systems. Continued research will focus on improving efficiency, interpretability, and domain adaptability.

CONCLUSION

This work effectively proves the combination of Vision Transformers (ViT) and GPT-2 for automated image captioning, filling the gap between visual data and natural language processing. Through the use of pre-trained deep models, the system produces high-quality, contextually sound captions with better accuracy than standard CNN-LSTM methods. The easy-to-use Tkinter-based GUI increases the accessibility and user-friendliness, making the tool ideal for use in accessibility services, content management, and image search engines. Challenges still include rare object handling, sophisticated scenes, and understanding of cultural context are still issues. Despite the shortcomings, the system presented is a notable improvement in AI-based image captioning tools.

FUTURE WORK

Future development and research will be aimed at improving the system in a number of areas. First, adding multilingual support through models such as mBART or mT5 will allow for captioning in multiple languages, making it more accessible. Second, model optimisation for real-time processing through hardware acceleration and model compression methods will make it more efficient. Third, adding scene graph generation could enhance contextual understanding and caption accuracy for difficult images. Finally, investigating reinforcement learning-based fine-tuning techniques can improve caption diversity and provide more subtle descriptions. These advances will further improve the system, making it a stronger and more capable tool for different image captioning tasks.

REFERENCES

Bhuiyan, A. B., & Uddin, J. (2024). Bengali Word Detection from Lip Movements Using Mask RCNN and Generalized Linear Model. [IJEEI]. *Indonesian Journal of Electrical Engineering and Informatics, 12*(2), 373–383. DOI: 10.52549/ijeei.v12i2.5088

Mishra, S., Seth, S., Jain, S., Pant, V., Parikh, J., Jain, R., & Islam, S. M. (2024, May). Image caption generation using vision transformer and gpt architecture. In *2024 2nd International Conference on Advancement in Computation & Computer Technologies (InCACCT)* (pp. 1-6). IEEE. DOI: 10.1109/InCACCT61598.2024.10551257

Vasireddy, I., Bindu, G. H., & B, R. (2023). Transformative fusion: Vision transformers and GPT-2 – Unleashing new frontiers in image captioning within image processing. *International Journal of Innovative Research in Engineering and Management, 10*(6), 55–59. DOI: 10.55524/ijirem.2023.10.6.8

Chapter 12
Household Waste and Everyday Practices:
Understanding Rural–Urban Differences in Kitchen Waste Disposal in Kerala

V. K. Aswathy
http://orcid.org/0000-0002-8895-8373
Lovely Professional University, India

Sanjeev Kumar
http://orcid.org/0000-0002-7375-7341
Lovely Professional University, India

ABSTRACT

Kerala's waste management practices, this research shedding light on significant rural-urban disparities and emphasizing the importance of targeted interventions in rural regions, particularly addressing kitchen waste disposal. Drawing from a robust dataset of 46,334 respondents in Kerala, sourced from NFHS 5, the study uncovers a compelling commitment to responsible waste management. Both rural and urban areas exhibit a shared dedication to sustainability, with a consistent preference for closed drain disposal and reusing kitchen waste for gardening or farming. Result: Encouragingly, urban regions show higher levels of awareness and engagement in sustainable waste management, reflected in the prevalence of manual collection. However, nuanced variations in less eco-friendly methods require vigilant attention. This research underscores Kerala's resolute commitment to responsible waste disposal, buoyed by the success of eco-conscious initiatives, and highlights the necessity of tailoring strategies to address local variations for a more sustainable future for all residents.

DOI: 10.4018/979-8-3373-4202-3.ch012

INTRODUCTION

Effective waste management is a global imperative, and it holds particular significance in regions like Kerala, India, known for their environmental consciousness and unique waste disposal practices. Kitchen waste, comprising organic materials like food scraps and peels, represents a substantial portion of household waste (Nguyen et al., 2022). Sustainable management of kitchen waste is crucial not only for environmental conservation but also for public health and resource optimization. Kerala's waste management practices have garnered attention for their eco-friendly and community-driven nature. However, understanding the variance in these practices between rural and urban regions is vital for tailoring effective waste management policies and interventions. The state of Kerala, nestled in the southwestern part of India, is celebrated for its lush green landscapes, rich biodiversity, and a culture deeply rooted in sustainable living. As Kerala continues to experience urbanization and population growth, the dynamics of waste generation and disposal are evolving. Urban areas are expanding, and lifestyles are changing, influencing how residents dispose of their kitchen waste. In this context, it becomes imperative to assess the variations in approaches to kitchen waste disposal between rural and urban regions within Kerala.

Against this backdrop, the present study seeks to systematically investigate how rural and urban households in Kerala differ in their kitchen waste disposal practices, and what socio-economic, infrastructural, and spatial factors account for these variations. Specifically, it explores whether differences in wealth quintile, educational attainment, sanitation access, and housing type influence the choice of disposal method across districts. The central hypothesis guiding this research posits that urban households are more likely to adopt municipal or manual collection systems owing to better infrastructure and awareness, whereas rural households tend to rely on open or closed drains due to limited access to waste management facilities. Furthermore, it is hypothesized that higher education, improved sanitation, and greater household wealth significantly increase the likelihood of adopting environmentally sustainable disposal methods. By addressing these questions through a combination of descriptive, bivariate, and multivariate analyses using NFHS-5 data, the study aims to contribute a nuanced, evidence-based understanding of Kerala's evolving household waste practices within a broader sustainability framework.

NEED AND SIGNIFICANCE OF THE STUDY

This research is of paramount importance due to the pressing need for sustainable waste management practices in Kerala, a state renowned for its environmental consciousness. Understanding the variance in kitchen waste disposal methods between rural and urban regions is essential to address significant disparities and promote responsible waste management practices. The need for this study arises from the observed differences in waste disposal practices, which have implications for public health, sanitation, and the environment. Bridging the rural-urban gap in waste management is crucial to ensure equitable access to clean and safe living conditions for all residents. Furthermore, this research holds significance for policymakers, environmentalists, and community leaders seeking to design targeted interventions, awareness campaigns, and infrastructure improvements that can foster eco-friendly waste management practices and enhance Kerala's commitment to sustainability. By shedding light on these disparities and their underlying causes, this research contributes to the collective effort to create a more environmentally conscious and equitable future for Kerala's residents.

Kerala, known for its natural beauty and commitment to sustainability, faces distinct challenges in managing kitchen waste disposal practices, with significant rural-urban disparities. While Kerala's waste management practices have garnered praise for eco-consciousness, there remains a need to address these disparities in kitchen waste disposal methods. The problem at hand is the notable divergence in approaches to kitchen waste disposal between rural and urban regions in Kerala, which warrants in-depth investigation. Hence the study aims to identify potential opportunities to bridge the gap between rural and urban kitchen waste disposal practices in Kerala. By understanding and addressing these variations, Kerala can progress toward a more equitable and sustainable waste management landscape.

DATA AND METHODS

This study relies on secondary data sourced from the National Family Health Survey 5 (NFHS 5), a comprehensive and nationally representative survey conducted in India. The National Family Health Survey 5 (NFHS 5) is a vital and extensive demographic and health survey program in India. It serves as an invaluable source of information for researchers, policymakers, and healthcare professionals. NFHS 5 is conducted under the aegis of the Ministry of Health and Family Welfare, Government of India, and is implemented by the International Institute for Population Sciences (IIPS) with support from various organizations. The survey is characterized by its comprehensive and nationally representative nature, making it a powerful tool for studying various aspects of public health, demography, and social indicators across India. The dataset comprises responses from 46,334 (41.6% from urban and 58.4% from rural areas) respondents residing in various districts of Kerala, representing a diverse cross-section of the population. The wide scope of the NFHS 5 survey and its commitment to providing high-quality, nationally representative data make it an ideal resource for studies like the one assessing kitchen waste disposal practices in rural and urban regions of Kerala as shown in Table 1.

Table 1. District-wise sample distribution for the study

District	Number	Percent
Kasaragod	3743	8.1
Kannur	3815	8.2
Wayanad	3342	7.2
Kozhikode	3427	7.4
Malappuram	3793	8.2
Palakkad	3483	7.5
Thrissur	3021	6.5
Ernakulam	3221	7.0
Idukki	3106	6.7
Kottayam	3198	6.9
Alappuzha	3156	6.8
Pathanamthitta	2791	6.0
Kollam	3157	6.8
Thiruvananthapuram	3081	6.6
Total	**46334**	**100.0**

DATA SOURCE AND REPRESENTATIVENESS

The analysis draws upon household-level data from the National Family Health Survey–5 (NFHS-5, 2019–21) for Kerala, conducted by the International Institute for Population Sciences (IIPS) under the Ministry of Health and Family Welfare, Government of India. NFHS-5 employs a stratified, two-stage sampling design with separate rural and urban strata for each district. Sampling weights, primary sampling units (PSUs), and strata identifiers were used to obtain design-based estimates and correct for unequal selection probabilities. The weighted sample comprised 46,334 households (41.6% urban, 58.4% rural), ensuring state-representative results. Response rates exceeded 98%.

Variable Definition and Coding

Kitchen-waste disposal was derived from household questionnaire item HH128: "How does your household dispose of kitchen waste?" Original response codes were recoded into five analytical categories: (1) let into drain/sewer, (2) open drain, (3) closed drain, (4) reused for gardening/farming or domestic purposes, and (5) collected manually/municipally. Independent variables included residence (urban/rural), wealth quintile, education level, drinking-water and sanitation access, housing type, and district.

Missing Data and Analysis Design

Cases with missing disposal information (< 1%) were excluded through listwise deletion. All analyses used complex-sample commands in SPSS v26, incorporating weights and strata to produce unbiased estimates.

Analytical Approach

The analytical approach in this study is structured to comprehensively investigate the differences in kitchen waste disposal methods between rural and urban settings using SPSS v26.0. The key variables of interest are the various kitchen waste disposal methods, categorized into options such as open drain disposal, closed drain disposal, reuse for gardening or farming, and manual collection. The study first employs a descriptive analysis to provide an initial overview of how these disposal methods are distributed within the population. Subsequently, the analytical focus shifts to bivariate analysis. Within this phase, cross-tabulations are used to compare the prevalence of different waste disposal methods separately for rural and urban regions, offering insight into the relative frequency of each method in both settings. Chi-square tests are then applied to assess the statistical significance of the observed variations, determining whether the differences between rural and urban regions are meaningful. This analytical approach allows the study to systematically explore, characterize, and confirm the disparities in kitchen waste disposal practices between rural and urban areas in Kerala, providing valuable insights into waste management strategies in these distinct contexts.

RESULT

Variance in Approaches to Kitchen Waste Disposal

In the realm of waste management practices in Kerala, a state known for its commitment to environmental sustainability, this study presents a comprehensive exploration of diverse approaches to kitchen waste disposal. The examined methods include 'Let out into drain/sewer,' 'Open drain,' 'Closed drain,' 'Reuse for garden or farming,' 'Domestic purposes,' and 'Manual collection.' Each approach carries its own implications in terms of environmental impact, sanitation, and resource conservation.

Figure 1. Distribution of kitchen waste disposal methods in Kerala

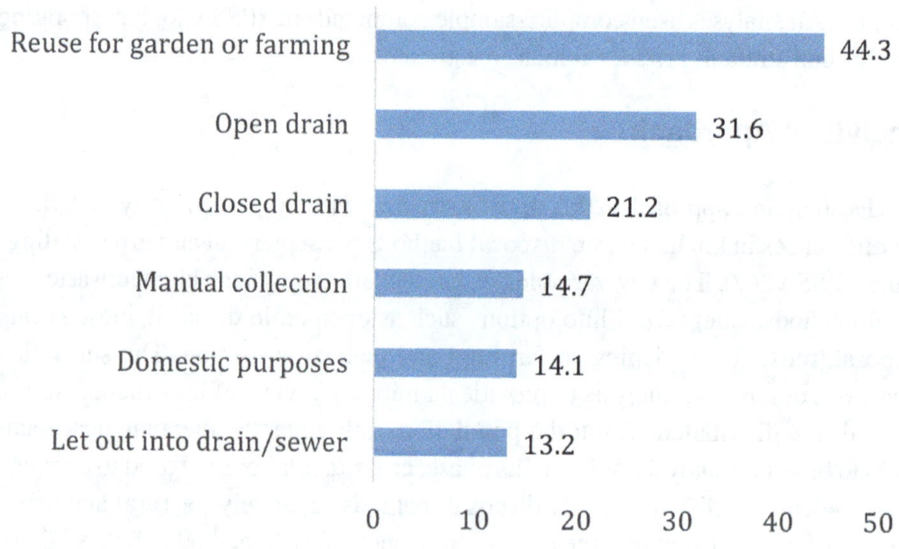

A significant portion of respondents (44.3%) demonstrate a strong commitment to sustainable waste management practices by choosing to reuse their kitchen waste for gardening or farming. This suggests a substantial level of awareness and successful promotion of composting and organic recycling methods in Kerala as discussed in Figure 1. While there is an overall positive trend toward sustainable practices, a substantial portion of respondents (31.6%) still opt for open drain disposal, which is less eco-friendly and can lead to sanitation and drainage issues. Addressing open drain disposal remains a challenge, requiring increased awareness efforts to promote responsible waste disposal. Approximately 21.2% of respondents use closed drains for kitchen waste disposal, which is an improvement over open drain disposal but still offers room for enhancement. Continued efforts are needed to encourage more environmentally friendly disposal methods like composting. A smaller percentage (13.2%) of respondents releases their kitchen waste into drains or sewers, posing the risk of sewer blockages and environmental pollution. To combat this, educational and awareness campaigns should emphasize the adverse consequences of this disposal method and promote better alternatives. Notably, 14.7% of respondents engage in manual collection of kitchen waste, indicating an active interest in waste management. This reflects a willingness to participate in waste management efforts, and community-based initiatives can further support and enhance these practices. Around 14.1% of respondents repurpose kitchen waste for domestic purposes, which may involve creative and diverse reuse or recycling methods.

Rural Urban Variation in Kitchen Waste Disposal Methods in Kerala

In the context of Kerala, a state renowned for its commitment to environmental sustainability, understanding the divergence in kitchen waste disposal practices between rural and urban areas is of paramount importance. This study delves into the nuanced differences in waste management methods, exploring how residents in these distinct settings navigate the challenge of sustainable kitchen waste disposal. By examining the rural-urban variations, we aim to gain insights into the factors influencing these choices and contribute to the development of targeted waste management strategies tailored to the unique needs of each locality as shown in Table 2.

Table 2. Variance in approaches to kitchen waste disposal between rural and urban regions

Kitchen waste disposal Method	Rural	Urban	Significance
Let out into drain/sewer	12.4	14.3	0.00
Open drain	36.5	24.7	0.00
Closed drain	20.4	22.3	0.00
Reuse for garden or farming	43.2	45.9	0.00
Reuse for Domestic purposes	13.9	14.3	0.12
Manual collection	11.7	18.9	0.00

In urban regions, 14.3% of the survey respondents indicated that they dispose of their kitchen waste into drains or sewers, while in rural areas, this figure is slightly lower at 12.4%. This data highlights a noticeable disparity in the adoption of this less environmentally friendly disposal method between rural and urban settings, with urban areas having a somewhat higher prevalence. The significant rural-urban difference in the practice of letting kitchen waste into drains or sewers suggests that various factors may be contributing to this variation. One potential factor could be the greater availability of sewer systems in urban areas, where some residents may not fully grasp the environmental consequences associated with this disposal method. This discrepancy underscores the need for targeted educational campaigns and awareness initiatives to promote responsible waste disposal practices, particularly in urban settings where sewer access might be more prevalent but where awareness of the environmental impact needs improvement. Open drain disposal is noticeably more widespread in rural areas, with 36.5% of respondents employing this method, whereas in urban areas, the practice is less common at 24.7%. This substantial difference between rural and urban areas underscores a significant rural-urban divide in the preference for open drain disposal. Rural communities, more than urban ones, tend to rely on this method, which is known for its potential environmental harm and sanitation challenges. Given the higher prevalence of open drain disposal in rural areas, there is a pressing need to prioritize addressing this issue. This situation highlights the importance of targeted interventions and improvements in infrastructure within rural communities to encourage more sustainable approaches to waste management. About 22.3% of urban respondents and 20.4% of rural respondents use closed drains for kitchen waste disposal. There is a slight difference in the usage of closed drains between rural and urban areas, although the Chi-Square test indicates that this difference is statistically significant but relatively small in magnitude. While there is a statistically significant difference between rural and urban areas in using closed drains for kitchen waste disposal, the magnitude of this difference is relatively small. Both rural and urban areas have a similar level of adoption of this method. This suggests that, overall, residents in Kerala tend to use closed drains for kitchen waste disposal at a consistent rate, regardless of their location. The practice of reusing kitchen waste for gardening or farming demonstrates remarkable consistency, with 45.9% of respondents in urban areas and 43.2% in rural areas adopting this eco-friendly approach. This data underscores a shared dedication among both rural and urban participants to sustainable waste management practices. The widespread adoption of this method in both settings not only reflects a shared commitment to sustainability across Kerala but also signifies a high level of awareness and the effective promotion of composting and organic recycling techniques in both rural and urban communities. A proportion of 14.3% of respondents in urban areas and 13.9% in rural areas employ kitchen waste for domestic purposes. Importantly, statistical analysis reveals no significant difference between rural and urban regions in the utilization of kitchen waste for domestic needs. This data indicates a consistent practice across both settings, where repurposing kitchen waste for various household requirements is a relatively uniform and stable practice observed throughout

Kerala. Manual collection of kitchen waste is notably more widespread in urban areas, where it accounts for 18.9% of respondents, in contrast to rural areas where it is lower at 11.7%. This data reveals a clear and statistically significant rural-urban distinction in the preference for manual collection, with urban regions displaying a greater inclination to actively manage kitchen waste. The higher prevalence of manual collection in urban areas suggests a heightened awareness of sustainable waste management practices among urban residents, potentially underpinned by better access to resources for such practices.

Overall Distribution

Weighted estimates show that 44.3% of households reuse kitchen waste for gardening or farming, 31.6% use open drains, 21.2% use closed drains, 13.2% let waste into sewers, and 14.7% depend on manual/municipal collection. Differences across methods were statistically significant ($\chi^2(4, N = 46{,}334) = 823.4, p < .001$).

Rural–Urban Differences

Urban households show higher manual collection (18.9%) compared to rural (11.7%), whereas rural households rely more on open drains (36.5% vs 24.7%). All differences were significant at $p < .001$ using Rao–Scott adjusted chi-square tests.

Table 3. Rural–urban distribution of kitchen waste disposal methods (weighted %)

Method	Rural %	Urban %	χ^2 (df = 1)	p-value
Drain/sewer	12.4	14.3	45.6	< .001
Open drain	36.5	24.7	392.8	< .001
Closed drain	20.4	22.3	19.1	< .001
Reuse (garden/ domestic)	43.2	45.9	16.7	< .001
Manual/ Municipal	11.7	18.9	508.2	< .001

District and Socio-Economic Gradients

District-level analyses revealed spatial heterogeneity—urban dominance of municipal collection in Ernakulam and Kozhikode; rural prevalence of open drains in Palakkad and Wayanad. Multinomial logistic regression (weighted) indicated urban residence, higher education, improved sanitation, and greater wealth significantly increased the likelihood of manual/municipal collection relative to open drains (p < .001).

Table 4. Predictors of manual/municipal collection (multinomial logistic regression, base = open drain)

Predictor	RRR (95% CI)	p
Urban residence	1.74 (1.61 – 1.88)	< .001
Higher education	1.32 (1.15 – 1.51)	< .001
Richest quintile	2.10 (1.83 – 2.40)	< .001
Improved sanitation	1.41 (1.25 – 1.59)	< .001

District Wise Rural-Urban Variation In Waste Disposal Method: Let Out Into Drain/Sewer

This study delves into the district-wise variation in kitchen waste disposal methods, focusing on the practice of letting kitchen waste out into drain/sewer in various districts of Kerala. Understanding how this practice varies across different districts is essential for tailoring effective waste management policies and interventions. The data will shed light on the prevalence of this disposal method, pinpointing districts where it is more common, and potentially revealing factors such as infrastructure, awareness, and local dynamics that influence these variations. This insight can guide targeted efforts to promote responsible and eco-friendly waste disposal practices in specific regions, contributing to overall sustainable waste management in Kerala as shown in Table 5.

Table 5. District-wise rural-urban variation in kitchen waste disposal method: Let out into drain/sewer

District	Rural	Urban	Sig
Kasaragod	12.9	23.9	0.000
Kannur	4.4	16.5	0.000
Wayanad	18.5	35.8	0.000
Kozhikode	6.6	18.1	0.000
Malappuram	27.7	9.3	0.000
Palakkad	15.3	17.3	0.092
Thrissur	3.9	12.6	0.000

continued on following page

Table 5. Continued

District	Rural	Urban	Sig
Ernakulam	4.6	8.6	0.000
Idukki	3.3	1.2	0.790
Kottayam	6.3	7.4	0.156
Alappuzha	23.0	17.7	0.000
Pathanamthitta	9.8	4.1	0.000
Kollam	11.9	12.9	0.220
Thiruvananthapuram	18.1	13.3	0.000

The data from various districts in Kerala reveals a consistent urban-rural disparity in the practice of letting kitchen waste into drain/sewer. In Kasaragod and Kannur, significant differences indicate that urban regions have a higher prevalence of this method, potentially linked to better urban access to sewer systems. Wayanad displays a substantial urban-rural divide, underlining the influence of urbanization on disposal choices. Kozhikode aligns with the urban trend seen in other districts, with urban areas practicing this method more. Notably, Malappuram presents a unique pattern with rural areas showing a higher prevalence, possibly due to local factors. While Palakkad exhibits a difference, it's not statistically significant, suggesting similar prevalence in both rural and urban areas. Thrissur and Ernakulam show significant urban dominance in this disposal method. Idukki and Kottayam reveal relatively consistent practices in both settings. Alappuzha deviates, with rural areas practicing more. Pathanamthitta mirrors the common urban-rural divide. Kollam shows a difference, but not statistically significant, indicating a balanced prevalence. Thiruvananthapuram exhibits the expected urban-rural disparity. Overall, these findings underscore the need for tailored waste management strategies and awareness campaigns, particularly in urban areas, to promote responsible and eco-friendly waste disposal practices. The significant differences observed in Kasaragod and Kannur, with urban areas having a higher prevalence of drain/sewer disposal, suggest that urban regions often benefit from better access to sewer systems. This indicates that infrastructure plays a crucial role in shaping disposal practices. Efforts to improve waste management infrastructure in rural areas may be necessary to level the playing field. The substantial urban-rural divide seen in Wayanad highlights the significant influence of urbanization on disposal choices. As urban areas expand and develop, residents are more likely to adopt modern waste disposal methods, such as drain/sewer disposal. This emphasizes the need for waste management policies that adapt to changing urban landscapes. The urban dominance observed in Kozhikode, Thrissur, and Ernakulam suggests that urban residents may have higher awareness levels regarding the environmental consequences of their waste disposal practices.

This underscores the importance of targeted awareness campaigns and educational initiatives, particularly in urban areas, to promote responsible waste disposal. Malappuram's unique pattern, where rural areas have a higher prevalence, indicates that local factors or specific infrastructure differences may influence disposal choices. Understanding these local dynamics is crucial for tailoring waste management strategies effectively. In districts like Idukki and Kottayam, where there is relatively consistent practice in both rural and urban areas, it may be due to uniform waste management practices and infrastructure across settings. These districts can serve as examples of successful waste management strategies that can be applied elsewhere. Alappuzha's deviation, with rural areas practicing this method more, suggests that unique factors specific to this district play a role in shaping waste disposal practices. Investigating these factors can provide insights for targeted interventions.

District Wise Rural-Urban Variation In Waste Disposal Method: Open Drain

The examination of rural-urban disparities in waste disposal practices, specifically focusing on open drain usage, provides valuable insights into the waste management landscape in various districts of Kerala. Open drain disposal is a critical aspect of environmental sanitation and understanding the variations between rural and urban areas is essential for the development of targeted waste management strategies.

Table 6. District-wise rural-urban variation in kitchen waste disposal method: Open drain

District	Rural	Urban	Sig
Kasaragod	38.0	34.9	0.027
Kannur	54.0	16.4	0.000
Wayanad	29.3	25.0	0.148
Kozhikode	32.4	31.4	0.284
Malappuram	33.4	36.3	0.035
Palakkad	45.0	38.6	0.001
Thrissur	29.7	22.8	0.000
Ernakulam	34.5	22.5	0.000
Idukki	41.9	14.6	0.000
Kottayam	31.5	19.5	0.000
Alappuzha	25.3	21.2	0.004
Pathanamthitta	44.7	28.7	0.000

continued on following page

Table 6. Continued

District	Rural	Urban	Sig
Kollam	31.7	20.9	0.000
Thiruvananthapuram	32.2	15.2	0.000

The analysis of open drain disposal practices across various districts in Kerala reveals distinct rural-urban disparities. In Kasaragod, open drain disposal is slightly more prevalent in rural areas than in urban areas, albeit with a significant difference. In contrast, Kannur exhibits a striking divergence, with rural regions predominantly relying on open drain disposal, emphasizing the urgency of promoting sustainable waste management in rural Kannur. Wayanad and Kozhikode show no significant differences between rural and urban areas, signifying a relatively consistent practice of open drain disposal in both settings. Malappuram presents a unique scenario where urban areas display a slightly higher prevalence, indicating a somewhat urban inclination toward this method. Palakkad underscores the need for targeted interventions in rural areas to encourage eco-friendly waste disposal, while Thrissur and Ernakulam demonstrate more eco-conscious trends in urban regions. Idukki and Kottayam highlight substantial rural-urban divides, necessitating comprehensive waste management strategies for rural communities. Alappuzha and Pathanamthitta call for focused efforts in rural regions to improve waste disposal practices, and Kollam and Thiruvananthapuram underline the significance of awareness campaigns and infrastructure improvements in rural areas. These findings emphasize the need for tailored approaches to waste management and awareness initiatives, particularly in rural districts, to mitigate the prevalence of open drain disposal and promote responsible waste disposal practices more effectively.

District Wise Rural-Urban Variation In Waste Disposal Method: Closed Drain

The district-wise rural-urban variation in waste disposal methods, particularly the use of closed drains, provides critical insights into how different regions of Kerala manage their kitchen waste. Understanding these variations is essential for tailoring effective waste management strategies that align with the unique needs and preferences of each locale. Closed drain disposal is a prevalent method for handling kitchen waste, and analyzing its distribution across rural and urban areas sheds light on the influence of factors such as infrastructure, awareness, and cultural practices in shaping waste disposal habits shown in Table 7.

Table 7. District-wise rural-urban variation in kitchen waste disposal method: Closed drain

District	Rural	Urban	Sig
Kasaragod	15.3	13.9	0.129
Kannur	10.6	25.5	0.000
Wayanad	26.5	7.4	0.000
Kozhikode	33.4	20.3	0.000
Malappuram	27.2	35.3	0.000
Palakkad	12.8	16.0	0.011
Thrissur	21.9	17.7	0.003
Ernakulam	16.8	17.5	0.326
Idukki	14.0	14.0	0.359
Kottayam	27.9	35.4	0.000
Alappuzha	17.2	19.7	0.036
Pathanamthitta	24.2	35.5	0.000
Kollam	21.7	30.0	0.000
Thiruvananthapuram	16.2	17.0	0.292

The analysis of closed drain disposal practices across Kerala's districts reveals significant rural-urban disparities and offers valuable insights into waste management trends. In districts like Kannur, Malappuram, and Kottayam, urban areas exhibit a significantly higher prevalence of closed drain disposal compared to rural areas. This urban dominance suggests that urban residents are more inclined toward this waste disposal method, possibly due to better waste management infrastructure and awareness. Conversely, in districts such as Wayanad, Kozhikode, and Thrissur, rural areas show a notably higher prevalence of closed drain disposal. This indicates a rural preference for this method, possibly influenced by local factors, traditional practices, or limited access to alternative waste disposal options. Palakkad, Alappuzha, and Idukki exhibit rural-urban differences in closed drain disposal, albeit not as pronounced. These localized variations emphasize the importance of considering district-specific factors when designing waste management strategies. Few districts like Kasaragod and Ernakulam, there are no significant rural-urban disparities in closed drain disposal, suggesting consistent practices across both settings. These findings underscore the need for tailored waste management approaches that consider the unique waste disposal practices and preferences within each district. District-specific interventions, infrastructure improvements, and awareness campaigns should be implemented to promote responsible waste disposal effectively. The prevalence of closed drain disposal in urban areas across multiple districts indicates that urban residents may benefit from better waste management infrastructure and higher awareness levels regarding responsible waste disposal. These factors contribute to the observed urban dominance in certain districts. In districts where rural areas favor closed drain disposal, it is essential to address the underlying challenges, which may include limited access to modern waste management facilities and a need for awareness campaigns to promote eco-friendly practices.

District Wise Rural-Urban Variation In Waste Disposal Method: Reuse For Garden Or Farming

The exploration of waste disposal practices across Kerala's districts, with a specific focus on the reusing of kitchen waste for gardening or farming, unveils important facets of the state's waste management landscape. Kerala, celebrated for its unique blend of tradition and modernity, grapples with the challenge of sustaining environmentally responsible waste disposal practices while adapting to urbanization and changing lifestyles. Understanding how rural and urban areas differ in their commitment to repurposing kitchen waste for agricultural purposes is critical for tailoring effective waste management strategies as shown in Table 8.

Table 8. District-wise rural-urban variation in kitchen waste disposal method: Reuse for garden or farming

District	Rural	Urban	Sig
Kasaragod	59.8	53.9	0.000
Kannur	36.7	46.8	0.000
Wayanad	40.5	37.8	0.291
Kozhikode	44.5	61.3	0.000
Malappuram	48.5	53.8	0.001
Palakkad	44.1	37.5	0.000
Thrissur	48.5	49.6	0.298
Ernakulam	47.3	33.1	0.000
Idukki	36.6	45.6	0.012
Kottayam	31.3	30.3	0.310
Alappuzha	38.7	41.3	0.075
Pathanamthitta	41.0	22.0	0.000
Kollam	53.6	46.7	0.000
Thiruvananthapuram	39.8	42.8	0.049

The analysis of rural-urban variation in the waste disposal method of reusing kitchen waste for gardening or farming in Kerala's districts reveals significant differences and provides valuable insights into the state's waste management practices. In districts like Kasaragod, Kannur, Kozhikode, Palakkad, and Pathanamthitta, urban areas exhibit a significantly higher prevalence of reusing kitchen waste for gardening or farming compared to their rural counterparts. This suggests that urban residents in these districts are more actively engaged in sustainable waste management practices, particularly by repurposing kitchen waste for agricultural purposes. The higher prevalence in urban areas could be attributed to better awareness, access to resources, and potentially more space for gardening or farming in urban settings. Conversely, in districts like Idukki, Kottayam, and Thiruvananthapuram, there are no significant rural-urban disparities in this disposal method, indicating relatively consistent practices across both rural and urban areas. This suggests that in these districts, the commitment to reusing kitchen waste for gardening or farming is consistent irrespective of the geographical location. Malappuram shows a unique pattern where rural areas have a slightly higher prevalence of this disposal method compared to urban areas, albeit with a significant difference. This indicates that in Malappuram's rural communities, there is a stronger inclination toward using kitchen waste for agricultural purposes, highlighting a potential area for promoting sustainable waste management practices in urban Malappuram. In districts like Wayanad, Thrissur, Alappuzha, and Kollam, there are disparities in prevalence between rural and urban areas, but the differences are not highly significant. This suggests that while there may be some variation, the commitment to reusing kitchen waste for gardening or farming is relatively balanced in both settings.

District Wise Rural-Urban Variation In Waste Disposal Method: Reuse For Domestic Purposes

The exploration of district-wise rural-urban variation in waste disposal methods, focusing on the reuse of kitchen waste for domestic purposes, unveils crucial insights into the waste management practices prevalent across different regions of Kerala. This analysis delves into the diverse approaches employed by rural and urban communities in repurposing kitchen waste within their households. Understanding these variations is pivotal in crafting targeted waste management strategies that cater to the distinct preferences, resources, and cultural dynamics that shape waste disposal practices. As kitchen waste reuse for domestic purposes represents a sustainable and eco-friendly approach, examining its prevalence and discrepancies across districts provides a comprehensive view of the state's commitment to responsible waste management as shown in Table 9.

Table 9. District-wise rural-urban variation in kitchen waste disposal method: Reuse for domestic purposes

District	Rural	Urban	Sig
Kasaragod	24.4	21.1	0.012
Kannur	19.0	16.6	0.033
Wayanad	11.6	8.1	0.114
Kozhikode	13.5	32.2	0.000
Malappuram	24.0	6.1	0.000
Palakkad	6.1	6.0	0.481
Thrissur	12.4	16.2	0.003
Ernakulam	14.8	11.1	0.002
Idukki	14.7	31.0	0.000
Kottayam	11.0	8.0	0.007
Alappuzha	17.8	10.4	0.000
Pathanamthitta	8.2	4.4	0.011
Kollam	11.5	10.7	0.275
Thiruvananthapuram	10.2	7.1	0.002

The analysis of rural-urban variation in the waste disposal method of reusing kitchen waste for domestic purposes in Kerala's districts reveals significant differences and provides valuable insights into the state's waste management practices. In several districts, urban areas exhibit a significantly lower prevalence of reusing kitchen waste for domestic purposes compared to their rural counterparts. These districts include Kozhikode, Malappuram, Idukki, and Alappuzha, where urban areas lag significantly behind rural areas in adopting this sustainable waste management practice. This may be attributed to factors such as limited space for household gardening, lesser awareness, or higher reliance on disposal methods in urban settings. In contrast, districts like Kasaragod, Kannur, Thrissur, Ernakulam, Kottayam, Pathanamthitta, and Thiruvananthapuram show urban areas with a slightly lower prevalence, but the differences are not highly significant. This suggests that in these districts, both rural and urban residents have relatively similar rates of reusing kitchen waste for domestic purposes. Wayanad exhibits no significant rural-urban difference, indicating a relatively consistent practice of kitchen waste reuse across the district. These findings emphasize the need for targeted awareness campaigns and interventions in urban areas of districts like Kozhikode, Malappuram, Idukki, and Alappuzha to promote responsible waste disposal practices. For districts with more balanced rural-urban prevalence, fostering a culture of kitchen waste reuse among urban populations is still essential to enhance overall sustainability in waste management.

District Wise Rural-Urban Variation in Waste Disposal Method; Manual Collection

Understanding how waste disposal methods vary between rural and urban areas within different districts is crucial for effective waste management strategies. The choice of waste disposal methods can significantly impact environmental sustainability and public health. In this context, examining the rural-urban variation in the waste disposal method of manual collection across various districts in a region can provide valuable insights. This analysis can shed light on the prevalence of manual collection in different settings, helping policymakers, environmentalists, and local authorities tailor their efforts to promote responsible waste management practices. A comprehensive understanding of these variations can inform targeted interventions to enhance waste disposal practices, optimize resource allocation, and contribute to a cleaner and healthier environment as shown in Table 10.

Table 10. District-wise rural-urban variation in kitchen waste disposal method: Manual collection

District	Rural	Urban	Sig
Kasaragod	15.4	30.5	0.000
Kannur	17.8	12.6	0.000
Wayanad	20.7	88.5	0.000
Kozhikode	14.3	26.8	0.000
Malappuram	14.1	12.5	0.078
Palakkad	10.2	9.1	0.196
Thrissur	9.0	16.8	0.000
Ernakulam	5.0	27.3	0.000
Idukki	7.2	21.1	0.000
Kottayam	10.2	12.1	0.071
Alappuzha	4.4	7.8	0.000
Pathanamthitta	10.5	16.6	0.002
Kollam	9.4	11.8	0.000
Thiruvananthapuram	6.8	23.9	0.000

The analysis of rural-urban variation in the waste disposal method of manual collection across different districts in Kerala highlights significant differences and offers valuable insights into waste management practices. In several districts, urban areas exhibit a significantly higher prevalence of manual collection compared to their rural counterparts. These districts include Kasaragod, Kannur, Wayanad, Kozhikode,

Thrissur, Ernakulam, Idukki, Alappuzha, Pathanamthitta, and Thiruvananthapuram. The substantial urban dominance in manual collection suggests that urban residents in these districts are more engaged in this method, possibly due to better waste management infrastructure and services. Conversely, Malappuram, Palakkad, Kottayam, and Kollam show no significant rural-urban disparities, indicating relatively balanced practices across both rural and urban areas. These findings emphasize the need for targeted interventions to promote responsible waste management practices in urban areas, where manual collection is more prevalent, and to maintain the existing balanced practices in districts where rural-urban differences are minimal.

DISCUSSION ON FINDINGS

The results confirm distinct rural–urban disparities in household kitchen waste disposal across Kerala. Urban residents display higher participation in organized collection systems, consistent with superior access to waste services and awareness campaigns (Cheela et al., 2021). In contrast, rural households' reliance on open drains reflects infrastructural deficits and cultural persistence of informal disposal practices.

The widespread reuse of kitchen waste ($\approx 44\%$) illustrates Kerala's strong ecological ethos and alignment with circular-economy ideals (Kuligowski et al., 2023). Yet, the persistence of open-drain disposal in one-third of households highlights the need for infrastructural upgrades and behavioral change.

District-level contrasts underscore that local governance capacity and infrastructural density are stronger predictors of household waste behavior than income alone. In several districts, the presence of active municipal bodies, better road connectivity, and functional waste collection systems directly correlates with higher adoption of organized disposal methods. Conversely, in areas where civic infrastructure remains underdeveloped, even wealthier households tend to rely on informal or environmentally unsafe disposal practices.

To consolidate Kerala's progress toward sustainable waste management, several policy directions are vital. First, rural composting units and decentralized collection hubs should be expanded to ensure that waste processing occurs closer to the point of generation, thereby reducing transportation costs and improving participation. Second, women's self-help groups and local panchayats must be integrated into waste segregation and awareness campaigns, as community-led governance has proven to be effective in sustaining behavioral change. Finally, districts with large rural–urban gaps, such as Palakkad and Wayanad, require targeted investments in sanitation and waste infrastructure, including improved drainage systems and doorstep collection services. These localized interventions can bridge existing disparities and strengthen Kerala's vision of an inclusive, zero-waste future.

IMPLICATIONS

The results emphasize the significance of continuous awareness and educational initiatives, with a specific focus on rural areas. These efforts should aim to raise awareness about the environmental and health risks associated with open drain disposal while highlighting the advantages of closed drain disposal and the reuse of kitchen waste. The data indicates a willingness among respondents in both urban and rural settings to actively engage in waste management endeavors. This presents an opportunity for community-driven projects that can significantly contribute to enhancing responsible waste disposal practices and fostering sustainability. In the regions with a high prevalence of open drain disposal, especially in rural areas, there is an urgent need for infrastructural enhancements. This involves establishing proper waste disposal systems to ensure sanitation and prevent environmental contamination. Prioritizing research and initiatives aimed at improving waste management infrastructure, particularly in rural regions, is crucial. Such efforts are essential for ensuring the long-term sustainability of waste disposal practices and reaping the associated health and environmental benefits for the population.

LIMITATIONS AND FUTURE RESEARCH

This study uses cross-sectional NFHS-5 data; therefore, causal interpretations are limited. Disposal behavior was self-reported, which may involve recall or social-desirability bias. Aggregated recoding may obscure local variations. Moreover, infrastructural and service-coverage data were not uniformly available across districts. Future studies should employ longitudinal or mixed-methods designs combining GIS mapping with ethnographic fieldwork to contextualize behavioral drivers and validate survey responses. Comparative analyses across Indian states could further clarify socio-ecological determinants of sustainable waste management.

CONCLUSION

The resounding commitment of a significant majority of respondents to reuse kitchen waste for gardening or farming reflects a commendable dedication to sustainable and eco-friendly waste management. While this high percentage underscores a substantial level of awareness and successful promotion of composting and organic recycling methods, the persistence of open drain disposal among a notable portion of respondents remains a concern. It underscores the need for continued efforts to raise awareness and promote responsible waste disposal methods. The adoption of closed drain disposal, although an improvement over open drains, presents an opportunity for further enhancement, emphasizing the ongoing importance of encouraging eco-friendly practices. Likewise, discouraging the release of kitchen waste into drains or sewers necessitates persistent education and awareness campaigns. The active involvement of respondents in manual waste collection and the creative repurposing of kitchen waste for domestic purposes demonstrate a willingness to participate in waste management efforts. These practices underscore the potential of community-based initiatives to further support and enhance responsible waste disposal practices. Notably, urban areas exhibit a higher prevalence of manual collection and, to a lesser extent, letting kitchen waste out into drains or sewers. These practices suggest a heightened awareness of sustainable waste management among urban residents. However, it is crucial to address the slight uptick in less eco-friendly disposal methods in urban settings. Conversely, rural areas grapple with a more significant reliance on open drain disposal, presenting sanitation and environmental challenges that demand immediate attention. Targeted interventions, including infrastructure improvements and awareness campaigns, are imperative to promote responsible waste disposal in rural regions. Despite these variations, the data reveals a heartening consistency in reusing kitchen waste for gardening or farming across both rural and urban areas. This shared commitment to sustainable practices serves as a testament to the success of eco-conscious initiatives in Kerala.

The analysis of waste disposal methods in Kerala reveals a clear urban-rural disparity, underlining the need for interventions in rural areas, particularly regarding open drain disposal. Encouragingly, both rural and urban areas consistently favor closed drain disposal and reusing kitchen waste for gardening or farming, reflecting a shared dedication to sustainability. Moreover, higher prevalence of manual collection in urban areas signals greater awareness and engagement in sustainable waste management. However, some areas exhibit a slight increase in less eco-friendly methods, warranting attention. Overall, the data reaffirms Kerala's commitment to responsible waste disposal and the success of eco-conscious initiatives. Tailoring strategies to address local variations is essential for fostering a more sustainable future for all residents.

REFERENCES

Cheela, V. R. S., Ranjan, V. P., Goel, S., John, M., & Dubey, B. (2021). Pathways to sustainable waste management in Indian Smart Cities. *Journal of Urban Management, 10*(4), 419–429. DOI: 10.1016/j.jum.2021.05.002

Kuligowski, K., Konkol, I., Świerczek, L., Chojnacka, K., Cenian, A., & Szufa, S. (2023). Evaluation of kitchen waste recycling as organic N-fertiliser for sustainable agriculture under cool and warm seasons. *Sustainability (Basel), 15*(10), 7997. DOI: 10.3390/su15107997

Nguyen, T. T. T., Malek, L., Umberger, W. J., & O'Connor, P. J. (2022). Household food waste disposal behavior is driven by perceived personal benefits, recycling habits, and ability to compost. *Journal of Cleaner Production*, *379*, 134636. DOI: 10.1016/j.jclepro.2022.134636

Chapter 13

Voltage Control in Smart City Low Voltage Distribution Networks Using Artificial Intelligence and Power Electronics

Rampelli Manojkumar
http://orcid.org/0000-0002-8141-0321
BVRIT HYDERABAD College of Engineering for Women, India

C. Venkata Subba Reddy
http://orcid.org/0000-0003-2955-4747
BVRIT HYDERABAD College of Engineering for Women, India

P. Aruna
BVRIT HYDERABAD College of Engineering for Women, India

Gumpu Srinivasulu
http://orcid.org/0000-0001-6610-5612
Maulana Azad National Institute of Technology, India

Chamakura Krishna Reddy
http://orcid.org/0009-0000-7811-5981
BVRIT HYDERABAD College of Engineering for Women, India

Arindam Mitra
http://orcid.org/0000-0002-4255-1914
National Institute of Technology, Rourkela, India

ABSTRACT

Voltage stability is a critical aspect of ensuring reliable power supply in smart city low voltage (LV) networks, where increasing penetration of renewable energy sources and dynamic demand patterns present operational challenges. This study proposes an AI-enhanced optimal voltage control method implemented using smart

DOI: 10.4018/979-8-3373-4202-3.ch013

power converters with real-time monitoring and adjustment capabilities. The control strategy employs the Exhaustive Search Method (ESM) to determine optimal reference voltage set points by evaluating all possible solutions within a defined range. The method is applied to the CIGRE LV residential distribution network under worst-case PV and load conditions. Compared to the base case, the proposed method—using a step size of 0.001—achieves a significant reduction in total voltage deviation, with improvements ranging from 88.05% to 90.78%. The results confirm that the integration of artificial intelligence and power electronics enables effective voltage regulation, making the approach well-suited for smart city applications.

INTRODUCTION

The increasing integration of photovoltaic (PV) systems into low voltage (LV) distribution networks, especially within smart city environments, presents critical challenges stemming from the uncertainty and variability of renewable energy generation. Voltage rise during periods of high PV injection and voltage drops during peak load conditions restrict PV hosting capacity and compromise load support, highlighting the need for robust voltage regulation strategies that comply with grid codes and support high renewable penetration without unnecessary curtailment (Manojkumar et al., 2024).

Smart urban energy systems—characterized by bidirectional power flows, time-varying loads, prosumer activity, and interconnectivity—demand advanced voltage control methods to maintain voltage profiles within statutory limits. High PV penetration often causes voltage rise in residential feeders, while peak demand leads to voltage sags, potentially disrupting sensitive electronic infrastructure and connected smart devices (Rahimi et al., 2025; Ruiz-Romero et al., 2014). Traditional voltage regulation approaches such as on-load tap changers (OLTCs), reactive power control (RPC), battery energy storage systems (BESSs), and active power curtailment (APC) have been extensively explored (Tonkoski et al., 2011). However, OLTCs are not viable at the LV level (11/0.4 kV), RPC becomes less effective in distribution systems with high R/X ratios, BESSs are capital-intensive with substantial maintenance demands, and APC results in reduced solar energy utilization (Masters, 2002).

To overcome these limitations, smart power converters (SPCs) have been proposed as a promising alternative for voltage regulation in LV networks. SPCs offer continuous and dynamic voltage control without relying on mechanical switching or centralized devices like DSTATCOMs (Kumar et al., 2018; Yu et al., 2014). These converters provide fast and localized voltage adjustment, disturbance rejection, and power quality improvement, which are crucial in networks with high levels of distributed generation (Chen et al., 2018; Shah & Crow, 2016). Their modular architecture, real-time responsiveness, and compatibility with digital communication technologies make them ideal for deployment in smart city LV infrastructures (Rehman et al., 2022; Garlik, 2022).

SPC-based transformer configurations also enable direct interfacing with DC networks, eliminating the need for additional converters and improving system efficiency (Kumar et al., 2018; Yu et al., 2014). Their application in medium voltage (MV) networks has been shown to support MVAC grid voltage and reactive power without DSTATCOMs (Kumar & Liserre, 2015; Shah & Crow, 2016), further underscoring their potential. Moreover, these systems are more cost-effective and scalable compared to OLTC-based approaches, especially for urban grid modernization.

Recent studies have introduced voltage control techniques for SPCs utilizing dynamic optimal setpoints and tolerance-based switching strategies tailored to varying PV and load profiles (Manojkumar et al., 2020; Manojkumar et al., 2019). The forward-backward sweep algorithm is one such technique that adjusts SPC output voltage to optimal levels based on real-time operating conditions (McAsey & Mou, n.d.). Additional work in this domain emphasizes power loss minimization using SPCs in meshed hybrid distribution networks (Kumar et al., 2020; Das et al., 2022), reinforcing the effectiveness of SPCs for technical and economic optimization.

This chapter presents an artificial intelligence (AI)-driven optimal voltage control strategy that leverages the exhaustive search method (ESM) to improve voltage stability in smart city LV distribution systems. SPCs are deployed to dynamically regulate the LV AC bus voltage based on real-time voltage deviations. The proposed method is validated through a case study on the CIGRE LV distribution network with 100% PV penetration. The network includes five apartment buildings (RL1–RL5) with peak loads of 20, 50, 40, 20, and 45 kW respectively. SPCs are modeled as back-to-back converters where the grid-side converter maintains the DC-link voltage and injects current at unity power factor, while the load-side converter provides a balanced sinusoidal voltage supply to the loads.

The AI-based control algorithm determines optimal voltage references for each SPC based on grid state feedback. The effectiveness of this method is assessed using key performance indicators such as voltage rise margin, voltage drop margin, and total voltage deviation from 1.0 p.u., comparing optimized scenarios with the base case. Simulation results show that with a resolution of 0.001 per unit in the exhaustive search, the proposed control strategy significantly improves voltage stability, achieving a reduction in total voltage deviation between 88.05% and 90.78%. This confirms the suitability of the approach for smart city applications, enhancing voltage regulation and enabling higher PV penetration without compromising system reliability.

ROLE OF POWER ELECTRONICS FOR INTELLIGENT VOLTAGE CONTROL

Power electronics play a crucial role in modern power systems, particularly in the context of intelligent voltage control using smart power converters in low-voltage distribution grids. These systems require efficient conversion, control, and management of electrical power, which is achieved through power electronic devices and circuits.

Power electronics enable the seamless integration of renewable energy sources, such as photovoltaic systems, into the grid by efficiently converting DC power to AC power and vice versa. They also facilitate precise voltage regulation, enhancing grid stability and reliability. In the context of low-voltage distribution grids, power electronics help optimize energy flow, minimize losses, and enable the implementation of advanced control strategies for voltage management. Overall, power electronics are instrumental in enabling the efficient and intelligent control of AC/DC hybrid systems in low-voltage distribution grids, contributing to a more sustainable and reliable power infrastructure.

Voltage Control Methods

There are various types of voltage control methods such as voltage source inverters (VSIs), active power control, dynamic voltage support, etc., as explained in this section.

Voltage-Source Inverters (VSIs)

Voltage-source inverters (VSIs) are fundamental components in modern power electronic systems, offering precise control over voltage magnitude and phase angle. They play a crucial role in actively managing voltage within distribution systems, especially in renewable energy applications like solar and wind farms. VSIs employ advanced control strategies such as pulse-width modulation (PWM) techniques to regulate voltage levels with high accuracy and efficiency. For instance, in grid-connected photovoltaic (PV) systems, VSIs adjust their output voltage to match the grid voltage, ensuring seamless integration and maximum power transfer. Moreover, VSIs can inject or absorb reactive power to compensate for fluctuations in grid conditions, thereby enhancing grid stability and reliability.

One notable example of VSI application is in utility-scale wind farms, where multiple VSIs are used to convert variable AC output from wind turbines into stable grid-compatible AC power. By actively controlling voltage and frequency, VSIs ensure smooth grid integration and minimize the impact of wind variability on grid stability.

Active Power Control

Active power control mechanisms allow smart power converters to dynamically regulate their power output in response to grid demand variations or voltage fluctuations. This real-time modulation of active power flow plays a critical role in stabilizing voltage and enhancing grid reliability, especially in distribution systems with high levels of renewable energy penetration (Tonkoski et al., 2011). To achieve optimal performance, advanced control strategies such as model predictive control (MPC) are often employed, enabling proactive decision-making based on system forecasts and constraints. Moreover, active power control is frequently integrated with energy storage systems to provide supplementary grid support, increase system flexibility, and improve overall energy management.

In microgrid environments, where power balance must be maintained locally, active power control becomes indispensable for coordinating distributed energy resources and loads. Smart converters can dynamically adjust their active power contributions to match demand, ensuring stable and continuous supply to critical loads—even under transient disturbances or variability in renewable generation.

Droop Control

Droop control strategies allow power converters to autonomously adjust their output voltage or power based on system conditions, such as load demand and voltage variations. By implementing these decentralized control mechanisms, converters contribute to improved grid support and enhanced operational stability (Masters, 2002). Droop control is particularly beneficial in islanded microgrid systems, where maintaining voltage and frequency stability is critical in the absence of main grid support. In such configurations, each distributed energy resource (DER) unit operates independently according to predefined droop characteristics, enabling effective power sharing and coordinated operation. This approach ensures system stability, even during fluctuations in load or generation, without requiring centralized control infrastructure.

Voltage Regulation through Communication

Smart converters equipped with communication capabilities enable real-time monitoring and dynamic adjustment of voltage levels in response to changing grid conditions. By receiving and processing real-time grid voltage data, these converters can intelligently optimize their output settings to maintain consistent voltage supply and minimize disruptions across connected systems (Yu et al., 2014). The integration of advanced communication technologies—such as synchrophasors and wide-area measurement systems (WAMS)—supports grid-wide voltage regulation and optimization, thereby improving overall efficiency and reliability.

Communication-enabled voltage regulation systems are increasingly deployed in smart grid initiatives and large-scale utility applications to enhance grid performance and operational stability. By leveraging real-time data streams and predictive analytics, these systems facilitate proactive voltage control, effectively mitigating voltage fluctuations and promoting resilient, stable grid operation.

Grid-Forming Converters

Grid-forming converters play a pivotal role in establishing and maintaining voltage and frequency within local microgrids. By autonomously regulating these parameters, grid-forming converters facilitate the seamless integration of distributed energy resources (DERs) and enable stable islanded operation during grid disturbances (Yu et al., 2014). These converters serve as voltage and frequency references, effectively replacing the role traditionally played by the utility grid. With the implementation of advanced control algorithms and robust hardware architectures, grid-forming converters contribute significantly to system stability and reliability, even under varying load and generation conditions.

Such converters are widely deployed in off-grid communities, remote industrial zones, and mission-critical infrastructure where uninterrupted power supply is essential. Their ability to maintain grid-forming capabilities ensures resilient power delivery, making them indispensable in modern decentralized and autonomous energy systems.

Voltage Control Benefits

Power Quality Enhancement

Stable voltage levels are essential for ensuring a high-quality and reliable power supply. Voltage fluctuations can disrupt or damage sensitive electronic devices, compromising their performance, shortening their lifespan, and increasing the likelihood of failures (Elrayyah, Wanik, & Bouselham, 2017). Effective voltage control methods help minimize these fluctuations, thereby enhancing power quality and enabling electrical equipment to operate within optimal voltage ranges. This reduces the occurrence of voltage-related issues such as sags, surges, or transients, contributing to smoother operation, reduced equipment downtime, and greater system-wide reliability.

In industrial environments, where automation and precision electronic systems are heavily relied upon, maintaining voltage stability is particularly critical. Even minor voltage disturbances can halt production lines, resulting in significant financial losses. Advanced voltage control strategies—such as dynamic voltage support and active power control—are therefore vital for delivering consistent and uninterrupted power to industrial facilities, ultimately enhancing operational efficiency, equipment longevity, and productivity.

Equipment Efficiency

Maintaining optimal voltage levels is crucial for maximizing both the efficiency and lifespan of electrical equipment. Most electrical devices are designed to function within specific voltage thresholds, and deviations—whether overvoltage or undervoltage—can degrade performance, increase energy consumption, and accelerate equipment failure (Carvalho, Correia, & Ferreira, 2008). Effective voltage control methods help ensure a stable supply, thereby reducing electrical stress on components, minimizing wear and tear, and ultimately extending equipment service life.

In commercial buildings, where diverse electrical equipment operates simultaneously, voltage stability plays a key role in optimizing energy efficiency and minimizing operational and maintenance costs. Control techniques such as active power regulation and communication-based voltage control allow for fine-tuned voltage adjustments, ensuring that devices function within their optimal operating range and enabling significant energy savings (Manojkumar et al., 2019).

Overall, robust voltage control strategies are essential for achieving greater energy efficiency, reduced consumption, improved equipment reliability, and enhanced system safety (Manojkumar et al., 2019; Carvalho et al., 2008). By adopting advanced control technologies and leveraging smart infrastructure, organizations can build more resilient, sustainable, and high-performing electrical systems.

SMART POWER CONVERTER INTEGRATION AND SYSTEM ARCHITECTURE

The integration of smart power converters within system architectures represents a transformative leap in enhancing the efficiency and adaptability of power systems. Smart power converters, equipped with advanced control features, dynamically optimize energy flow and voltage levels. This chapter explores the principles and types of smart power converters, delving into their applications within a broader system architecture. The architecture, designed for seamless integration, incorporates communication protocols and advanced control systems, enabling coordinated responses to real-time changes in the grid. Understanding these components is crucial for developing effective voltage control strategies in low voltage distribution grids.

Introduction to Smart Power Converters

In the landscape of modern power systems, smart power converters play a pivotal role in ensuring efficient energy management and reliable voltage control. These converters are sophisticated devices that leverage advanced technologies to regulate the flow of electrical energy within a system. Their fundamental purpose is to convert electrical power from one form to another, such as converting direct current (DC) to alternating current (AC) or vice versa, while incorporating intelligent control mechanisms as shown in Figure 1.

Figure 1. Smart power converter in the distribution network

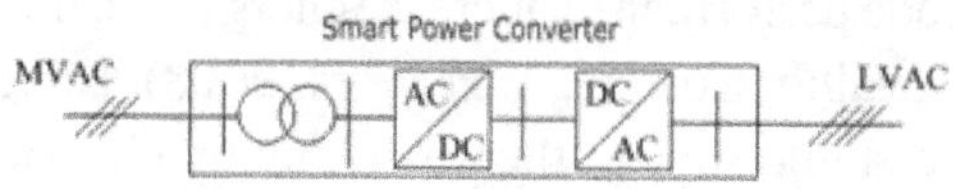

Smart power converters are designed to address the challenges posed by the dynamic nature of contemporary power grids, especially in the context of low voltage distribution networks. These converters employ cutting-edge power electronics and control algorithms to optimize voltage levels, enhance energy efficiency, and accommodate the integration of renewable energy sources like photovoltaic (PV) systems.

Their adaptability to varying load conditions and ability to respond to rapid changes in the grid make them instrumental in maintaining stable and reliable power distribution. Within the realm of smart power converters, various types exist, each tailored to specific applications. These include but are not limited to inverters, rectifiers, and DC-DC converters. The choice of the type depends on the desired conversion process and the unique requirements of the power system. This chapter aims to provide an in-depth exploration of these smart power converters, shedding light on their principles, types, and applications, setting the stage for a detailed understanding of their role in the proposed voltage control system.

Types of Smart Power Converter

In the dynamic landscape of low voltage distribution grids, smart power converters manifest in various types, each tailored to address specific operational needs and challenges. One prominent type is the inverter, a device crucial for converting DC power from sources like solar panels into AC power suitable for grid integration. Inverters come in different topologies, such as voltage source inverters (VSIs) and current source inverters (CSIs), each with distinct advantages in specific applications. Another essential type is the rectifier, responsible for converting AC power to DC, a vital process in systems where DC power is required. Rectifiers, like inverters, come in various configurations, such as diode rectifiers and thyristor rectifiers, offering flexibility in meeting diverse energy conversion demands. Additionally, DC-DC converters play a crucial role in adapting voltage levels between different components of the power system, ensuring seamless energy transfer and utilization.

Comparison with Conventional Systems

In the evolution of power systems, the shift from conventional to smart power converters represents a transformative leap in enhancing operational efficiency, adaptability, and overall system performance. Understanding the distinctions between smart power converters and their conventional counterparts is crucial for evaluating the potential advantages of adopting advanced technologies in low voltage distribution grids.

Traditional power converters, characterized by simplicity and conventional control methods, have been instrumental in power systems for decades. Devices like traditional inverters and rectifiers operate based on fixed control strategies, often lacking the adaptability required for dynamic grid conditions. These converters may rely on manual adjustments and preset parameters, limiting their responsiveness to real-time changes in the grid.

System Architecture and Description

The system architecture is designed to seamlessly integrate smart power converters within a comprehensive smart grid framework. At its core, the architecture comprises three main elements: the power generation and distribution sources, the smart power converters, and the control and monitoring systems.

Power Generation and Distribution Sources this component encompasses diverse sources of power, including conventional sources like the main grid, as well as renewable sources such as solar panels or wind turbines. These sources feed power into the distribution network, forming the initial stage of the energy flow within the system. Positioned strategically within the distribution network, smart power converters, particularly smart transformers, serve as key elements for voltage control. These converters interface with various sources and loads, dynamically adjusting their operation to optimize voltage levels and ensure stable power distribution.

The intelligence of the system is centralized in control and monitoring systems, which oversee the operation of smart power converters. These systems utilize advanced control algorithms, including real-time monitoring, feedback mechanisms, and communication protocols, to coordinate the activities of smart converters. They play a crucial role in decision-making processes, responding to changes in load demand, grid disturbances, and fluctuations in power generation.

Within the confines of this chapter, our exploration navigates the dynamic landscape of CIGRE Low Voltage (LV) Residential Distribution Networks, immersing itself in the integration intricacies of Smart Transformers (ST). As we traverse this terrain, our primary focus is the nuanced interplay between power flow solutions and advanced voltage control methods. The significance of this endeavor lies in addressing the evolving challenges posed by the increasing integration of renewable energy sources, particularly Photovoltaic (PV) systems, within residential grids.

Smart Transformers, with their cutting-edge technologies, offer a promising avenue for optimizing energy distribution and maintaining voltage stability. Our inquiry extends beyond the theoretical realm, seeking practical insights into their application within LV residential networks. By dissecting power flow solutions and innovative voltage control methodologies, we aim to uncover the transformative potential of STs in mitigating voltage fluctuations and enhancing the overall resilience of distribution networks.

This exploration acts as a gateway to a deeper understanding of the intricate synergy between STs and LV residential grids, providing a foundation for subsequent analyses. As we navigate through the following sections, a comprehensive picture will emerge, elucidating the operational intricacies and effectiveness of Smart Transformers in shaping the future of low-voltage energy distribution.

In this work, a CIGRE LV distribution network with smart power converters is analysed, featuring buildings equipped with PV sources, as depicted in Figure 2. This system includes several components, each playing a crucial role in its operation and performance.

Figure 2. CIGRE low voltage residential distribution network with smart power converter

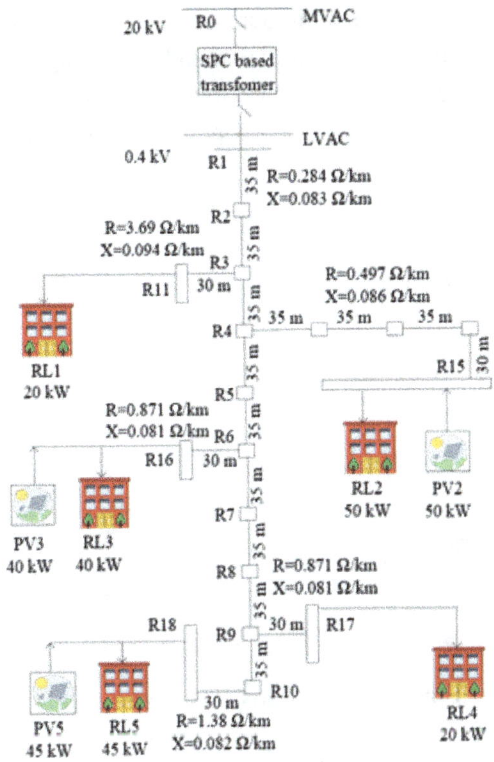

A SPC transformer, incorporating a back-to-back converter, is chosen for the system. It consists of a conventional distribution transformer (20 kV/0.4 kV) connected to the grid-side converter, which regulates grid current absorption. On the load side, a DC-AC converter, termed the load-side converter, ensures a balanced voltage output.

In the analyzed system, buildings are selected as loads, each with a peak load of 20 kW, 50 kW, 40 kW, 20 kW, and 45 kW, respectively. These buildings represent typical loads in the distribution network under consideration.

In this study, the PV installed power is assumed to be proportional to their respective loads of the buildings.

EXHAUSTIVE SEARCH BASED VOLTAGE CONTROL METHOD

The proposed optimization method aims to minimize total load voltage deviation from 1 p.u. while also controlling the LV AC bus voltage. It focuses on worst-case scenarios to determine LV AC load bus voltage deviation. The optimization problem's constraints revolve around load bus voltage magnitude. This method aids in keeping load bus voltages within grid code limits across all worst-case scenarios by optimally controlling LV AC bus voltage to minimize total load voltage deviation from 1 p.u. In this method, the LV ac bus voltage is optimally controlled to minimize the total Load voltage deviation in the system from 1 p.u. Total load bus voltage deviation is

$$(V_{td}) = \sum_{i=0}^{n} (V_L - 1)^2 \tag{1}$$

The constraints are given as follows.

1) Power balance constraint

$$P_{lv}(t) = \sum_{i=1}^{nl} P_l^i(t) + P_{loss}(t) - \sum_{j=1}^{nr} P_{rc}^j(t) \tag{2}$$

2) Bus voltages constraint

$$V_{min} \leq V_l^i(t) \leq V_{max} \tag{3}$$

Equation (1) defines the total load bus voltage deviation (*Vtd*) as the sum of the squared differences between the actual voltage (V_L) at each bus and 1 p.u. This equation quantifies the deviation of the load bus voltages from the desired value. Equation (2) represents the power balance constraint, disregarding losses from power converters. This equation ensures that the total power supplied to the LVAC bus is equal to the total power consumed by the loads connected to it, maintaining balance in the system. Equation (3) defines the constraint on the magnitude of the load bus voltages. It ensures that the voltage levels at the load buses are within acceptable limits, as given by the grid code, to prevent overvoltage or undervoltage conditions. The V_{min} and V_{max} are chosen as 0.95 p.u. and 1.05 p.u., respectively. These variations allow for a finer adjustment of the control voltage, potentially leading to a more precise optimization result. The fitness values obtained for each step size can help determine the impact of the step size on the optimization process and the final solution quality. As a whole, it utilizes smart power converter for LVAC bus voltage control. It is maintained at a value other than 1 p.u. which is optimally determined using exhaustive search method while considering bus voltage constraints as per grid code (0.95 p.u & 1.05 p.u) in a CIGRE LV distribution network.

The power flow analysis involves two main phases: the forward sweep and the backward sweep to calculate the required bus voltages. Further, the Exhaustive Search Method (ESM) is employed to solve the selected optimization problem by systematically evaluating all possible solutions within a finite search space. While ESM is computationally intensive, it is particularly effective for problems with limited dimensionality where an exhaustive evaluation is feasible. In the context of AI-driven optimization, ESM serves as a benchmark or baseline method against which the performance of intelligent algorithms—such as machine learning, genetic algorithms, or reinforcement learning—can be compared. It operates by generating all candidate solutions and assessing them using a predefined objective function or fitness criterion, ensuring that the global optimum is identified. Although not scalable to high-dimensional problems, ESM offers the advantage of guaranteed optimality in small, well-defined solution spaces, and is often used in AI applications to validate or train heuristic models. The algorithm then selects the solution that best satisfies the criteria, such as minimizing or maximizing an objective function. In the fitness calculation, the current values (I_{bus}) and voltage values (V_{bus}) at each bus are obtained using the forward and backward sweep method as given in following Section.

Forward–Backward Sweep Algorithm

The Forward–Backward Sweep (FBS) algorithm is a power flow solution technique widely used for radial and weakly meshed distribution systems. It is preferred in distribution networks because conventional methods such as Newton–Raphson or Gauss–Seidel often struggle with poor convergence due to the high resistance-to-reactance (R/X) ratios that characterize distribution feeders. The FBS method takes advantage of the radial structure of these systems, making it computationally efficient, numerically stable, and relatively simple to implement.

The algorithm works in two main stages: the backward sweep and the forward sweep. In the backward sweep, the calculation starts from the terminal nodes (leaf buses) and proceeds upstream toward the root node (slack or substation bus). At each bus, the injected current is computed from the specified load power using:

$$I_i = \frac{S_i^*}{V_i^*} = \frac{P_i - jQ_i}{V_i^*} \tag{4}$$

where P_i and Q_i are the active and reactive power demands, V_i is the bus voltage, and S_i is the complex power.

The current in each branch is then obtained by summing the load current at the receiving bus and the currents of all downstream branches. This ensures that the current contribution of every load is properly reflected upstream toward the source.

Once branch currents are determined, the forward sweep updates the bus voltages starting from the slack bus, whose voltage is known. Using Ohm's law, the voltage at a downstream bus is expressed as:

$$V_j = V_i - Z_{ij} I_{ij} \quad (5)$$

where V_i is the sending-end voltage, Z_{ij} is the branch impedance, and I_{ij} is the branch current. This process continues outward until the voltages of all buses in the network are updated.

The backward and forward sweeps are repeated iteratively until convergence is achieved. Convergence is typically checked by comparing the difference between bus voltages in successive iterations:

$$\Delta V = \max_i \left| V_i^{(k+1)} - V_i^{(k)} \right| \quad (6)$$

The process terminates when:

$$\Delta V \leq \epsilon \quad (7)$$

where ϵ is a small tolerance (e.g., 10^{-4} p.u.).

The FBS method has several advantages. It is especially suited for radial networks and can also be adapted for weakly meshed systems with certain modifications. It converges quickly, even in systems with high R/X ratios, and requires less computational effort than more generalized power flow techniques. For these reasons, the FBS algorithm has become one of the most practical and reliable methods for analyzing distribution systems.

Performance Indicators

Voltage Rise Margin

The Voltage Rise Margin (VRM) is defined as the difference between the maximum voltage limit prescribed by the grid code and the worst-case maximum bus voltage observed in the distribution network. VRM is a critical indicator that ensures voltage levels remain within acceptable boundaries, thereby preserving overall system stability. In distribution grids with high photovoltaic (PV) penetration, excessive power generation during peak solar periods can cause significant voltage rise at the buses. If left unregulated, this can lead to equipment stress, operational inefficiencies, or even violations of grid standards. By effectively managing the VRM, voltage rise can be mitigated, ensuring that all bus voltages stay within the grid-compliant range and that the system operates reliably (Manojkumar, Hrishikesan, Kumar, & Ganguly, 2019).

Voltage Drop Margin

The Voltage Drop Margin (VDM) is defined as the difference between the worst-case minimum bus voltage in the distribution network and the minimum voltage limit specified by the grid code. The VDM specifies the permissible decrease in bus voltage below the reference voltage. In the absence of PV power, the excess load demand may cause voltage levels to drop. The VDM helps maintain voltage stability by ensuring that the bus voltage does not fall below acceptable levels.

Total Load Bus Voltage Deviation (TLVD)

The total load bus voltage deviation measures the overall deviation of bus voltages from the reference voltage of 1 p.u. This indicator is crucial for evaluating voltage quality across all buses. By assessing the total load bus voltage deviation, the effectiveness of voltage control mechanisms in maintaining stable and reliable voltage levels throughout the distribution network can be determined.

CASE STUDY ANALYSIS

The standard CIGRE low-voltage residential distribution network with smart power converter-based transformers (SPC) is analyzed under four worst-case scenarios to evaluate its performance. Each scenario represents a different operating condition, providing insights into the behavior of the distribution network under various circumstances.

Case 1: Rated Load without PV Power

In this scenario, the distribution network operates at its rated load without any contribution from PV power sources. This case represents a baseline condition where the network is subjected to typical load demands without additional renewable energy inputs. The analysis focuses on assessing the network's ability to meet load requirements and maintain voltage stability under normal operating conditions.

The plot in Figure 3 illustrates the bus voltages vs. bus number for the considered case study with different methods. Each method is represented by a symbol, indicating different step sizes used in the optimization process and the base method. For the base method, the voltage drop margin (VDM) is -0.0418, indicating that some bus voltages fall below the permissible limits of 0.95 p.u., suggesting ineffective control and potential instability in the distribution grid.

Figure 3. Bus number vs bus voltages for case 1

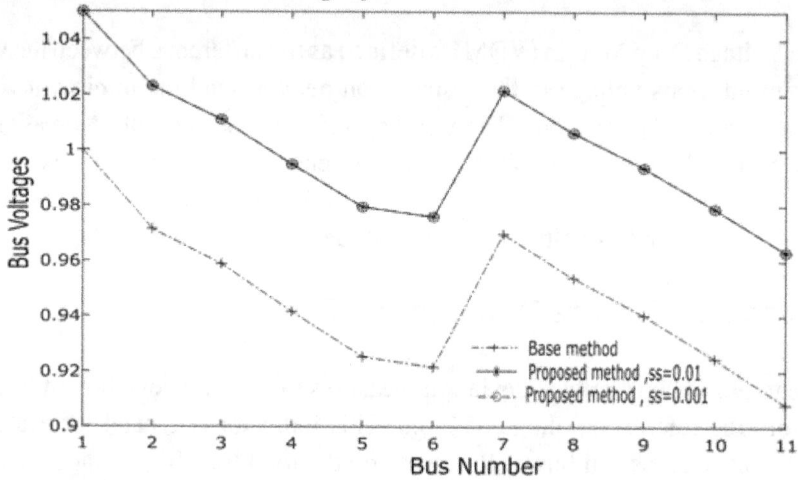

In contrast, the proposed method using the exhaustive search method (ESM) algorithm demonstrates improved performance, with VDM within acceptable limits. This indicates that the ESM algorithm effectively controls bus voltages, ensuring stability and reliability in the distribution grid. Furthermore, comparing the total load bus voltage deviations, the base method exhibits a TLVD of 0.0234486, suggesting less precise voltage control. The proposed method with the ESM algorithm shows significantly lower TLVD, indicating better voltage regulation and stability of low-voltage distribution networks.

Additionally, the results demonstrate that as the step size decreases from 0.01 to 0.001, the voltage deviation also decreases. This reduction in voltage deviation indicates that finer adjustments in the control algorithm led to more precise voltage regulation. Therefore, the proposed method with a step size of 0.001 exhibits the most effective control, minimizing voltage deviations and ensuring stable and reliable operation of the distribution.

Table 1 presents the performance indicators for three different methods: the base method, the proposed method with a step size (ss) of 0.01, and the proposed method with a step size of 0.001, in the scenario of rated load without PV power.

For the base method, the VRM is 0.05, indicating a margin for voltage rise. The VDM is -0.0418, showing a margin for voltage drop. The total load bus voltage deviation (TLVD) is 0.0234486, indicating a deviation from the reference voltage.

In comparison, the proposed method with a step size of 0.01 shows improvements in performance indicators. The VRM remains at 0, indicating no voltage rise margin. The VDM improves to 0.0133, showing a reduced margin for voltage drop. The TLVD significantly decreases to 0.0023, indicating improved voltage stability.

Table 1. Performance indicators of case 1

Indicator	Base Method	Proposed Method, ss=0.01	Proposed Method, ss=0.001
VRM	0.05	0	0
VDM	-0.0418	0.0133	0.0133
TLVD	0.0234486	0.0023	0.0023

Similarly, the proposed method with a step size of 0.001 also shows improvements. The VRM remains at 0, indicating no voltage rise margin. The VDM also remains at 0.0133, showing a reduced margin for voltage drop. The TLVD further decreases to 0.0023, indicating the best performance among the three methods.

The percentage reduction in the total load bus voltage deviation (TLVD) for the base method compared to the proposed method with a step size of 0.001 is 90.19%. Overall, both proposed methods show improvements in voltage stability compared to the base method. The method with a step size of 0.001 demonstrates the best performance, with lower voltage deviations and improved voltage margins.

Case 2: Half Of The Rated Load Without PV Power

In the scenario of half of the rated load without PV power, the distribution network operates at half of its rated load without any contribution from PV power sources. This case represents a condition where the network experiences reduced load demands without additional renewable energy inputs. The analysis focuses on assessing the network's ability to meet reduced load requirements and maintain voltage stability under lower operating conditions.

The plot in Figure4 illustrates the bus voltages vs. bus number for the considered case study with different methods. Each method is represented by a symbol, indicating different step sizes used in the optimization process and the base method.

In Case 1, the Base Method had a negative VDM, indicating that some bus volt ages fell below the permissible limits, suggesting ineffective control. However, in Case 2, the Base Method improved significantly, with a positive VDM of 0.0061, indicating better control over voltage drops.

In contrast, the proposed method with step size 0.01 shows a slightly higher VDM but still within acceptable limits. The plot indicates that this method effectively controls bus voltages, ensuring stability and reliability in the distribution grid.

Figure 4. Bus number vs bus voltages for case 2

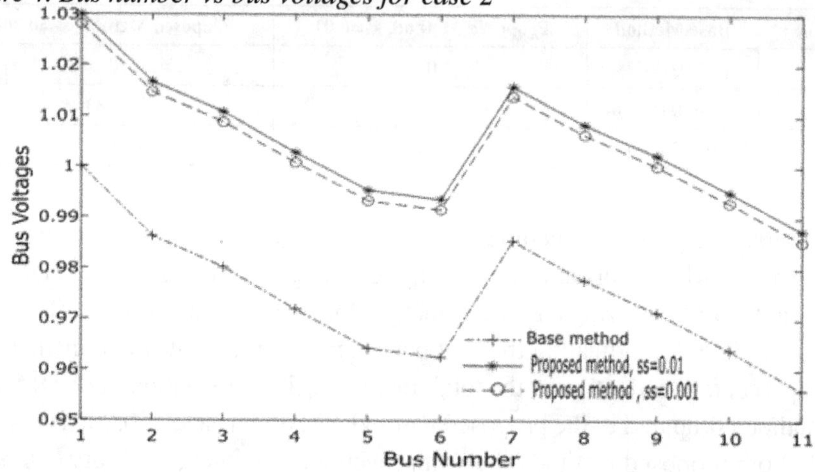

Furthermore, comparing the total load bus voltage deviations, the base method and the proposed method with step size 0.01 exhibit low deviations, suggesting precise voltage control.

Additionally, the plot demonstrates that the proposed method with step size 0.001 shows similar performance to the method with step size 0.01, with slightly higher VDM and lower total load bus voltage deviations. This indicates that both methods with finer step sizes lead to more precise voltage regulation, ensuring stable and reliable operation of the distribution grid.

Table 2 presents the performance indicators for three different methods: the base method, the proposed method with a step size (ss) of 0.01, and the proposed method with a step size of 0.001, in the scenario of half of the rated load without PV power.

Table 2. Performance indicators of case 2

Indicator	Base Method	Proposed Method, ss=0.01	Proposed Method, ss=0.001
VRM	0.05	0.02	0.022
VDM	0.0061	0.0375	0.0354
TLVD	0.0054249	0.0005	0.0005

For the base method, the VRM is 0.05, indicating a margin for voltage rise. The VDM is 0.0061, showing a margin for voltage drop. The total load bus voltage deviation (TLVD) is 0.0054249, indicating a deviation from the reference voltage.

In comparison, the proposed method with a step size of 0.01 shows improvements in performance indicators. The VRM improves to 0.02, indicating a margin for voltage rise. The VDM improves to 0.0375, showing a reduced margin for voltage drop. The TLVD significantly decreases to 0.0005, indicating improved voltage stability.

Similarly, the proposed method with a step size of 0.001 also shows improvements. The VRM improves to 0.022, indicating a margin for voltage rise. The VDM improves to 0.0354, showing a reduced margin for voltage drop. The TLVD remains at 0.0005, indicating the best performance among the three methods. The percentage reduction in the total load bus voltage deviation (TLVD) for the base method compared to the proposed method with a step size of 0.001 is 90.78%.

Overall, both proposed methods show improvements in voltage stability compared to the base method. The method with a step size of 0.001 demonstrates the best performance, with lower voltage deviations and improved voltage margins.

Case 3: PV Power Without Load Power And 100% Penetration Level

The analysis focuses on a scenario with 100% penetration of PV power sources without any load. The results of this analysis provide valuable insights into the performance of the distribution network under these conditions. In the case study involving 100% penetration of PV power sources without any load.

The plot in Figure 5 illustrates the bus voltages vs bus number for the considered case study with different methods. Each method is represented by a symbol, indicating different step sizes used in the optimization process and base method.

Figure 5. Bus number vs bus voltages for case 3

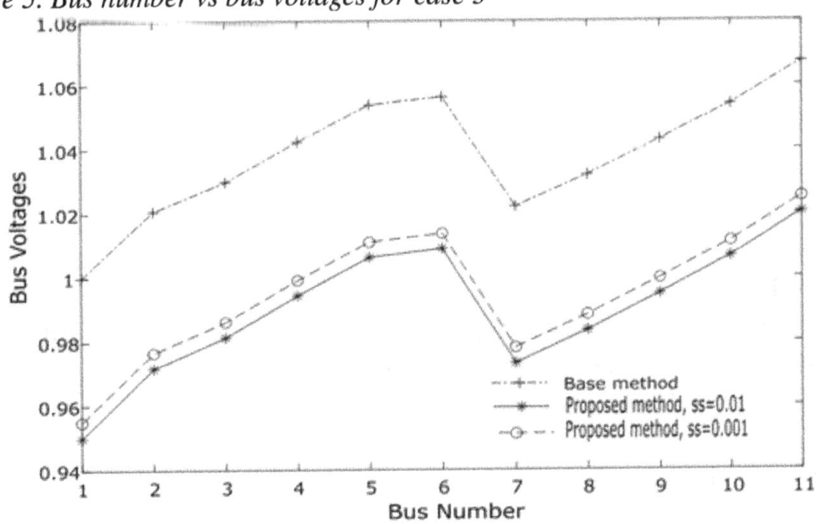

For the base method, the voltage rise margin indicates that bus voltages exceed the permissible limits of 1.05 p.u., suggesting ineffective control and potential instability in the distribution grid. In contrast, the proposed method using the exhaustive search method (ESM) algorithm demonstrates improved performance, with voltage rise margins within acceptable limits. This indicates that the ESM algorithm effectively controls bus voltages, ensuring stability and reliability in the distribution grid.

Furthermore, comparing the total load bus voltage deviations, the base method exhibits higher deviations, suggesting less precise voltage control. The proposed method with the ESM algorithm shows significantly lower total load bus voltage deviations, indicating better voltage regulation and stability of low-voltage distribution networks.

Additionally, the results demonstrate that as the step size decreases from 0.01 to 0.001, the voltage deviation also decreases. This reduction in voltage deviation indicates that finer adjustments in the control algorithm led to more precise voltage regulation. Therefore, the proposed method with a step size of 0.001 exhibits the most effective control, minimizing voltage deviations and ensuring stable and reliable operation of the distribution.

Table 3 presents the performance indicators for three different methods: the base method, the proposed method with a step size (ss) of 0.01, and the proposed method with a step size of 0.001, in the scenario of 100% PV penetration without any load.

Table 3. Performance indicators of case 3

Indicator	Base Method	Proposed Method, ss=0.01	Proposed Method, ss=0.001
VRM	-0.0172	0.0297	0.025
VDM	0.05	0	0.005
TLVD	0.011725	0.0015	0.0014

For the base method, the VRM is -0.0172, indicating a decrease in voltage margin. The VDM is 0.05, showing a moderate margin for voltage drop. The total load voltage deviation is 0.011725, indicating a significant deviation from the reference voltage. In comparison, the proposed method with a step size of 0.01 shows improvements in performance indicators. The VRM improves to 0.0297, indicating a positive margin for voltage rise. However, the VDM remains the same at 0. The total load bus voltage deviation significantly decreases to 0.0015, indicating improved voltage stability.

Similarly, the proposed method with a step size of 0.001 also shows improvements. The VRM is 0.025, slightly lower than the 0.01 step size method. The VDM is 0.011725, indicating a slight margin for voltage drop. The total load bus voltage deviation further decreases to 0.0014, indicating the best performance among the three methods. The percentage reduction in the total load bus voltage deviation (TLVD) for the base method compared to the proposed method with a step size of 0.001 is 88.05%.

Overall, both proposed methods show improvements in voltage stability compared to the base method. The method with a step size of 0.001 demonstrates the best performance, with lower voltage deviations and improved voltage margins.

Case 4: PV Power Without Load Power With 50% Penetration Level

In the case study involving PV power without load power with 50% penetration level, the analysis focuses on a scenario where the distribution network operates with 50% penetration of PV power sources without any additional load. This scenario represents a condition where a significant portion of the network's power comes from renewable sources, specifically PV power, without an increase in load demand. The results of this analysis provide insights into the network's ability to integrate and manage renewable energy sources effectively.

Figure 6 illustrates the bus voltages vs. bus number for the case study involving PV power without load power with 50% penetration level, comparing the performance of different methods. Each method is represented by a symbol, indicating different stcp sizes used in the optimization process and the base method.

Figure 6. Bus number vs bus voltages for case 4

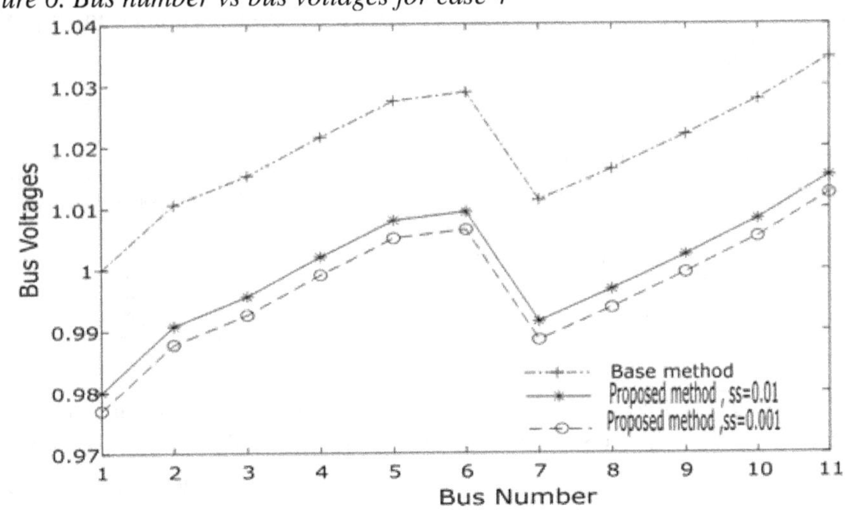

For the base method, the voltage drop margin (VDM) is 0.05, indicating that some bus voltages fall below the permissible limits of 0.95 p.u., suggesting ineffective control and potential instability in the distribution grid. In contrast, the proposed method with step size 0.01 shows improved performance, with VDM reduced to 0.03. This indicates better control over bus voltages, resulting in improved stability and reliability in the distribution grid.

Similarly, the proposed method with step size 0.001 also shows improved performance, with VDM further reduced to 0.027. This demonstrates the effectiveness of finer adjustments in the control algorithm, leading to more precise voltage regulation and improved stability in the distribution grid. Overall, both proposed methods show improvements in voltage stability compared to the base method. The method with step size 0.001 exhibits the best performance, with the lowest VDM, indicating more effective control and ensuring stable and reliable operation of the distribution grid.

Table 4 presents the performance indicators for three different methods: the base method, the proposed method with a step size (ss) of 0.01, and the proposed method with a step size of 0.001, in the scenario of PV power without load power with 50% penetration.

For the base method, the VRM is 0.0155, indicating a margin for voltage rise. The VDM is 0.05, showing a margin for voltage drop. The total load bus voltage deviation (TLVD) is 0.030749, indicating a deviation from the reference voltage. In comparison, the proposed method with a step size of 0.01 shows improvements in performance indicators. The VRM improves to 0.0348, indicating a positive margin for voltage rise. The VDM improves to 0.03, showing a reduced margin for voltage drop. The TLVD significantly decreases to 0.0004, indicating improved voltage stability.

Table 4. Performance indicators of case 4

Indicator	Base Method	Proposed Method, ss=0.01	Proposed Method, ss=0.001
VRM	0.0155	0.0348	0.0377
VDM	0.05	0.03	0.027
TLVD	0.0030749	0.0004	0.0003

Similarly, the proposed method with a step size of 0.001 also shows improvements. The VRM improves to 0.0377, indicating a positive margin for voltage rise. The VDM improves to 0.027, showing a reduced margin for voltage drop. The TLVD further decreases to 0.0003, indicating the best performance among the three methods. The percentage reduction in the total load bus voltage deviation (TLVD) for the base method compared to the proposed method with a step size of 0.001 is 90.24%.

Overall, both proposed methods show improvements in voltage stability compared to the base method. The method with a step size of 0.001 demonstrates the best performance, with lower voltage deviations and improved voltage margins.

ROLE OF ARTIFICIAL INTELLIGENCE FOR VOLTAGE CONTROL

AI is increasingly being recognized as a powerful enabler for intelligent voltage control in distribution systems with high penetration of renewable energy sources. Unlike conventional control strategies based on fixed setpoints, AI methods offer adaptive, data-driven, and optimization-based frameworks that can dynamically respond to uncertain operating conditions

Reinforcement Learning (RL) for Adaptive Control

Deep reinforcement learning (DRL) has shown promise in enabling autonomous voltage regulation through self-learning control policies that continuously adapt to grid dynamics. Early works demonstrated DRL-based autonomous control strategies capable of maintaining voltage stability under varying load and renewable penetration conditions (Duan et al., 2019; Diao et al., 2019). More recent developments extend this to multi-agent DRL frameworks, where distributed agents coordinate to optimize voltage regulation across larger systems (Wang et al., 2020). These methods are particularly effective for decentralized low-voltage networks where local adaptability is critical.

Machine Learning-Assisted Optimization and Forecasting

Machine learning techniques have been applied for forecasting, Volt-VAR control, and anomaly detection. For instance, ML-assisted distributed optimization frameworks improve the efficiency of inverter-based Volt-VAR control in active distribution networks (Li & Xu, 2023). Beyond optimization, ML models have been used to detect cyber-attacks targeting voltage regulation mechanisms (Bhusal, Gautam, & Benidris, 2021), and for operational support in future low-voltage grids (Mroczek & Pijarski, 2022). Together, these contributions demonstrate the flexibility of ML in both predictive control and resilience enhancement.

Metaheuristic and Search-Based Methods

Metaheuristic algorithms provide another class of AI-inspired solutions for voltage optimization. Genetic algorithms (GA), particle swarm optimization (PSO), and hybrid methods have been widely adopted due to their ability to search efficiently in complex, nonlinear solution spaces. Recent studies show effective applications of GA and PSO for improving voltage profiles and minimizing losses in transmission and distribution systems (Bidyanath et al., 2023), as well as hybrid GWO-PSO techniques for optimal PV-DG placement (Alyu et al., 2023). Reviews also confirm the growing role of metaheuristics across diverse power system optimization problems (Nassef, Abdelkareem, Maghrabie, & Baroutaji, 2023). Compared with deterministic strategies such as the Exhaustive Search Method (ESM), these approaches provide scalable alternatives that better handle uncertainty and large solution spaces.

CHALLENGES AND PRACTICAL ASPECTS OF SMART VOLTAGE CONTROL

The integration of smart power converters (SPCs) into distribution networks introduces new opportunities for efficient voltage regulation but also raises critical challenges related to cybersecurity, regulatory compliance, and practical deployment. As these converters increasingly adopt digital intelligence and communication capabilities, they become potential targets for cyber threats, requiring robust protection strategies to ensure system resilience. Recent studies emphasize that next-generation inverters must simultaneously bridge the gaps between artificial intelligence (AI), cybersecurity, and policy frameworks to support a sustainable energy transition (Hilmy & He, 2025).

Cybersecurity in SPCs and power electronic systems is emerging as a central concern. AI techniques are now being leveraged to monitor and mitigate cyber-physical threats, as demonstrated in AI-enabled cybersecurity monitoring frameworks for DC/DC power converters (Habibi et al., 2024). Similarly, resilience-oriented voltage regulation frameworks are increasingly adopting "cybersecurity by design" principles to safeguard against malicious data manipulation and communication attacks in active distribution networks (Pacheco-Cherrez et al., 2025). Reviews of EV fast charging stations also highlight cybersecurity as a critical factor, alongside standards and energy management, given the reliance on advanced power conversion and communication technologies (Arena et al., 2024). These findings reinforce the need to embed cybersecurity protocols within the design and deployment of SPC-based voltage control systems.

Alongside cybersecurity, regulatory compliance and adherence to international standards form another key pillar for practical deployment. Smart grids are expected to operate within harmonized frameworks that address data privacy, interoperability, and secure control. Global best practices and emerging regulations stress the importance of integrating cybersecurity requirements into smart grid policies (Harish et al., 2025). Furthermore, localized energy systems are facing unique challenges in aligning with existing standards while also addressing evolving cybersecurity requirements (Dong et al, 2022). The ability to navigate this complex regulatory environment will determine the scalability and acceptance of SPC-based voltage control solutions.

Practical deployment scenarios also demand careful consideration of interoperability, economic feasibility, and safety. Deployment in IoT-enabled smart grids further illustrates the necessity of balancing technical innovation with industrial automation and regulatory alignment (Scholapurapu, 2025). By incorporating lessons from diverse application domains, such as EV charging and industrial IoT, SPC-based voltage regulation can be tailored to meet both technical and policy requirements.

In summary, achieving reliable and secure deployment of SPC-based voltage control requires a holistic approach that integrates cybersecurity mechanisms, compliance with international standards, and practical implementation strategies. The alignment of AI-enabled monitoring, regulatory best practices, and real-world deployment experiences will ensure that these solutions move beyond theoretical frameworks and contribute effectively to the resilience, efficiency, and sustainability of future power systems.

CONCLUSIONS

The integration of smart power converters in low-voltage distribution grids offers a promising path to enhanced voltage control, leveraging advanced monitoring, communication, and control features. These advancements contribute to a more resilient, flexible, and efficient power distribution infrastructure, with future innovations expected to further refine and expand their capabilities. The backward/forward sweep method is a computational technique used in power system analysis, particularly for radial distribution networks. This method iteratively calculates power flow by moving through branches and nodes, determining power distribution and voltages at each node. The study suggests a voltage control strategy that switches between setpoints based on maximum and minimum bus voltages to ensure voltage levels remain within desired range. The proposed exhaustive search method has significantly improved voltage control by optimizing parameters such as voltage rise and drop margins and total voltage deviation. Compared to three other methods, the exhaustive search method approach demonstrates superior performance, highlighting its potential to enhance reliability and stability in low-voltage distribution networks.

In this study, the proposed method with a step size of 0.001 demonstrated significant improvements in voltage stability across various scenarios. A percentage reduction in the total load bus voltage deviation in the range of minimum to maximum, from 88.05% to 90.78%, was achieved. This range highlights the method's capability to consistently improve voltage stability across a variety of operating conditions. These results showcase the effectiveness of the proposed method in reducing voltage deviations and outperforming both the base method and the method with a step size of 0.01. Overall, the results demonstrate the proposed method's potential to significantly enhance voltage stability in real-world power systems. Further, the role of AI for voltage control as well as challenges and practical aspects of smart voltage control are discussed.

REFERENCES

Alyu, A. B., Salau, A. O., Khan, B., & Eneh, J. N. (2023). Hybrid GWO-PSO based optimal placement and sizing of multiple PV-DG units for power loss reduction and voltage profile improvement. *Scientific Reports, 13*(1), 6903. DOI: 10.1038/s41598-023-34057-3 PMID: 37106042

Arena, G., Chub, A., Lukianov, M., Strzelecki, R., Vinnikov, D., & De Carne, G. (2024). A comprehensive review on DC fast charging stations for electric vehicles: Standards, power conversion technologies, architectures, energy management, and cybersecurity. *IEEE Open Journal of Power Electronics, 5*, 1573–1611. DOI: 10.1109/OJPEL.2024.3466936

Bhusal, N., Gautam, M., & Benidris, M. (2021). Detection of cyber attacks on voltage regulation in distribution systems using machine learning. *IEEE Access : Practical Innovations, Open Solutions, 9*, 40402–40416. DOI: 10.1109/ACCESS.2021.3064689

Bidyanath, K., Singh, S. D., & Adhikari, S. (2023). Implementation of genetic and particle swarm optimization algorithm for voltage profile improvement and loss reduction using capacitors in 132 kV Manipur transmission system. *Energy Reports, 9*, 738–746. DOI: 10.1016/j.egyr.2022.11.098

Carvalho, P. M. S., Correia, P. F., & Ferreira, L. A. F. M. (2008). Distributed reactive power generation control for voltage rise mitigation in distribution networks. *IEEE Transactions on Power Systems, 23*(2), 766–772. DOI: 10.1109/TPWRS.2008.919203

Chen, J., Zhu, R., & O'Donnell, T.. (2018). Smart transformer and low frequency transformer comparison on power delivery characteristics in the power system. In *AEIT International Annual Conference* (pp. 1–6). IEEE. DOI: 10.23919/AEIT.2018.8577375

Das, D., Manojkumar, R., Kumar, C., & Ganguly, S. (2022). Power loss minimization in smart transformer enabled low voltage islanded meshed hybrid microgrid. *IEEE Access : Practical Innovations, Open Solutions, 10*, 123259–123270. DOI: 10.1109/ACCESS.2022.3224001

Diao, R., Wang, Z., Shi, D., Chang, Q., Duan, J., & Zhang, X. (2019, August). Autonomous voltage control for grid operation using deep reinforcement learning. In *2019 IEEE Power & Energy Society General Meeting (PESGM)* (pp. 1–5). IEEE. DOI: 10.1109/PESGM40551.2019.8973924

Dong, S., Cao, J., Flynn, D., & Fan, Z. (2022). Cybersecurity in smart local energy systems: Requirements, challenges, and standards. *Energy Informatics, 5*(1), 9. DOI: 10.1186/s42162-022-00195-7

Duan, J., Shi, D., Diao, R., Li, H., Wang, Z., Zhang, B., & Yi, Z. (2019). Deep-reinforcement-learning-based autonomous voltage control for power grid operations. *IEEE Transactions on Power Systems*, *35*(1), 814–817. DOI: 10.1109/TPWRS.2019.2941134

Elrayyah, A. Y., Wanik, M. Z. C., & Bouselham, A. (2017). Simplified approach to analyze voltage rise in LV systems with PV installations using equivalent power systems diagrams. *IEEE Transactions on Power Delivery*, *32*(4), 2140–2149. DOI: 10.1109/TPWRD.2016.2594242

Garlik, B. (2022). Energy centers in a smart city as a platform for the application of artificial intelligence and the Internet of Things. *Applied Sciences (Basel, Switzerland)*, *12*(7), 3386. DOI: 10.3390/app12073386

Habibi, M. R., Guerrero, J. M., & Vasquez, J. C. (2024). Artificial intelligence for cybersecurity monitoring of cyber-physical power electronic converters: A DC/DC power converter case study. *Scientific Reports*, *14*(1), 22072. DOI: 10.1038/s41598-024-72286-2 PMID: 39333625

Harish, V. S. K. V., Gupta, S., Bhatt, J. G., & Bansal, M. (2025). International standards, regulations, and best practices for cyber security of smart grid. In *Cyber security solutions for protecting and building the future smart grid* (pp. 321–348). Elsevier. DOI: 10.1016/B978-0-443-14066-2.00010-4

Hilmy, A., & He, B. E. (2025). Next-generation smart inverters: Bridging AI, cybersecurity, and policy gaps for sustainable energy transition. *Technologies*, *13*(4), 136. DOI: 10.3390/technologies13040136

Kumar, C., & Liserre, M. (2015). Operation and control of smart transformer for improving performance of medium voltage power distribution system. In *Proc. IEEE 6th Int. Symp. on Power Electronics for Distributed Generation Systems* (pp. 1–6). IEEE. DOI: 10.1109/PEDG.2015.7223092

Kumar, C., Manojkumar, R., Ganguly, S., & Liserre, M. (2020). Power loss minimization in smart transformer based meshed hybrid distribution network. In *Proceedings of the IECON 2020 – 46th Annual Conference of the IEEE Industrial Electronics Society* (pp. 1670–1675). IEEE. DOI: 10.1109/IECON43393.2020.9254324

Kumar, C., Zhu, R., Buticchi, G., & Liserre, M. (2018). Sizing and SoC management of a smart-transformer-based energy storage system. *IEEE Transactions on Industrial Electronics*, *65*(8), 6709–6718. DOI: 10.1109/TIE.2017.2784389

Li, B., & Xu, Q. (2023). A machine learning-assisted distributed optimization method for inverter-based Volt-VAR control in active distribution networks. *IEEE Transactions on Power Systems, 39*(2), 2668–2681. DOI: 10.1109/TPWRS.2023.3279303

Manojkumar, R., Aruna, P., Nabieha, N., Niteesha, M., Chandana, R., & Nithya, S. (2024). Voltage control in low voltage distribution systems using smart power converter. In *2024 International Conference on Electrical Electronics and Computing Technologies (ICEECT)* (pp. 1–5). IEEE. DOI: 10.1109/ICEECT61758.2024.10739011

Manojkumar, R., Hrishikesan, V. M., Kumar, C., & Ganguly, S. (2019). Voltage control using smart transformer for increasing photovoltaic penetration in a distribution grid. In *2019 20th International Conference on Intelligent System Application to Power Systems (ISAP)* (pp. 1–7). IEEE. DOI: 10.1109/ISAP48318.2019.9065949

Manojkumar, R., Kumar, C., Ganguly, S., & Liserre, M. (2020). Voltage control using smart transformer via dynamic optimal setpoints and limit tolerance in a residential distribution network with PV sources. *IET Generation, Transmission & Distribution, 14*(22), 5143–5151. DOI: 10.1049/iet-gtd.2020.0213

Masters, C. L. (2002). Voltage rise: The big issue when connecting embedded generation to long 11 kV overhead lines. *Power Engineering Journal, 16*(1), 5–12. DOI: 10.1049/pe:20020101

McAsey, M., & Mou, L. (n.d.). *Convergence of the forward-backward sweep method in optimal control.* University of Illinois. DOI: 10.1007/s10589-011-9454-7

Mroczek, B., & Pijarski, P. (2022). Machine learning in operating of low voltage future grid. *Energies, 15*(15), 5388. DOI: 10.3390/en15155388

Pacheco-Cherrez, D. S., Mayo-Maldonado, J. C., Guillen, D., & Escobar, G. (2025). Cybersecurity by design for resilient voltage regulation in active distribution networks. *IEEE Transactions on Smart Grid, 16*(6), 5009–5023. DOI: 10.1109/TSG.2025.3601073

Rahimi, T., Ahmed, S., Cardenas-Barrera, J. L., & Diduch, C. (2025). Fairness-oriented Volt-Watt control methods of PV units for over-voltage suppression in PV-enriched smart cities. *Smart Cities, 8*(3), 88. DOI: 10.3390/smartcities8030088

Rehman, A. U., Ali, M., Iqbal, S., Shafiq, A., Ullah, N., & Otaibi, S. A. (2022). Artificial intelligence-based control and coordination of multiple PV inverters for reactive power/voltage control of power distribution networks. *Energies, 15*(17), 6297. DOI: 10.3390/en15176297

Ruiz-Romero, S., Colmenar-Santos, A., Mur-Pérez, F., & López-Rey, Á. (2014). Integration of distributed generation in the power distribution network: The need for smart grid control systems, communication and equipment for a smart city—Use cases. *Renewable & Sustainable Energy Reviews*, *38*, 223–234. DOI: 10.1016/j.rser.2014.05.082

Scholapurapu, P. K. (2025). Power electronics for IoT-enabled smart grids and industrial automation. *SSRN*, *10*, 9789349552111–08. DOI: 10.2139/ssrn.5382142

Shah, D., & Crow, M. L. (2016). Online volt–var control for distribution systems with solid-state transformers. *IEEE Transactions on Power Delivery*, *31*(1), 343–350. DOI: 10.1109/TPWRD.2015.2457442

Tonkoski, R., Lopes, L. A. C., & El-Fouly, T. H. M. (2011). Coordinated active power curtailment of grid-connected PV inverters for overvoltage prevention. *IEEE Transactions on Sustainable Energy*, *2*(2), 139–147. DOI: 10.1109/TSTE.2010.2098483

Wang, S., Duan, J., Shi, D., Xu, C., Li, H., Diao, R., & Wang, Z. (2020). A data-driven multi-agent autonomous voltage control framework using deep reinforcement learning. *IEEE Transactions on Power Systems*, *35*(6), 4644–4654. DOI: 10.1109/TPWRS.2020.2990179

Yu, X., She, X., Zhou, X., & Huang, A. Q. (2014). Power management for DC microgrid enabled by solid-state transformer. *IEEE Transactions on Smart Grid*, *5*(2), 954–965. DOI: 10.1109/TSG.2013.2277977

Chapter 14
Case Study 1: Applications of AI and IoT-Based Smart Cities

G. Muthupandi
http://orcid.org/0009-0002-8822-0157
Chennai Institute of Technology, India

K. Jayakumar
http://orcid.org/0009-0000-2849-5445
Chennai Institute of Technology, India

G. Karthiyayini
Mangayarkarasi College of Engineering, India

ABSTRACT

Artificial Intelligence (AI) and the Internet of Things (IoT) are significantly changing the nature of cities due to the integration of technologies; this has created new life in cities turning them into smart, interconnected ecosystems. With increasing urban population size and the sustainability of living being a question under spotlight, AI and IoT technologies play a critical role in solving problems of traffic congestion, energy efficiency, waste management, safety, health, and the management of water. This chapter addresses the history of smart cities describing the potential of AI and Internet of Things to collect data in real time, make predictions and execute decisions to achieve greater efficiency and livability of urban environments. It presents real-life case studies, along with actual outcomes and achievements presented in the cities of Singapore, Barcelona, and Amsterdam, which demonstrate such concrete values as preserving carbon & better-developed public services.Future trends and the role of 5G, edge computing and blockchain curves are also explored in the chapter.

DOI: 10.4018/979-8-3373-4202-3.ch014

1. INTRODUCTION

1.1 The Urban Challenge and the Rise of Smart Cities

Urbanization is reshaping the world that we are today living in; where predominant centers of human activities in the 21 st century are the urban centers. United Nations (2018) predicts that the current world population residing in residential cities of the world is over 55 percent of the total world population which is approximately 4.2 billion and it is increasing at a high rate and is projected to reach 68 percent by 2050, or 2.5 billion additional residents, most of whom are occupying Africa and Asia. The high rates of urban development lead to an increment in the urbanization disadvantages such as the high traffic, environmental pollution, lack of resources and over-crowded public facilities. Among others, owing to congested traffic in cities, economies lose billions of dollars per annum, which, in the case of United States, resulted in productivity and fuel costs worth 81 billion dollars in 2022 alone.

The environmental issue is rather acute, as well, which is why, according to the World Health Organization (2023), 4.2 million individuals succumb to premature death annually because of the air pollution in the city. Lack of resources available in urban areas such as scarcity of water in Cape Town and the lack of supply of energy in the growing mega cities also compromises urban sustainability. The idea of smart cities has materialized as the remedial answer to these problems. Smart cities adapt urban systems with digital technologies to support quality of life, optimize its systems and sustain the environment. Smart cities operate on a dynamic level to regulate infrastructure and services through interconnected devices, data analytics, and automation differently than in traditional cities.

1.2 The Role of AI and IoT in Urban Transformation

Out of all technologies, the cornerstones of the smart city ecosystem are AI and IoT, which allow moving toward a proactive management of urban environment. The Internet of Things is a system of interconnected things connected devices, sensors, cameras, smart meters, and connected infrastructure and collect real-time data on the movement across the urban systems. As an example, traffic cameras connected to the internet keep track of car traffic, smart meters monitor the amount of energy used in the house, and environmental sensors can count air pollution and noise.

AI boosts IoT by analyzing this information into patterns, trend predictions, and even automated decision-making. The raw data can be leveraged using machine learning, computer vision, and predictive analytics to turn the data into actual viable insights using AI. With regard to transportation, AI is used to analyze IoT in order to set optimal times that the traffic signals should change, alleviating traffic up to

15 percent in the cases of cities such as Barcelon. At the energy level, AI predicts demand and optimizes supply, cutting waste and incorporating renewable sources, such as the smart grid that already replenished energy waste by 18% in Amsterdam.

The combination of AI and IoT, which is also known as AIoT, develops a potent plan of smart cities. IoT is the source of information, and AI offers smart analysis and automatization. As an example, the IoT sensors identify that there is a traffic jam and AI algorithms recommend the best routes or can change the signals on the fly, thus enhancing urban traffic. In the medical sphere, AIoT can implement patients monitoring and predict the outbreaks of diseases, similar to the case of COVID-19 in Taipei, which minimized the number of people infected by 25%.

1.3 Objectives and Structure of the Chapter

The chapter is an extensive reading of the use of AI and IoT in smart cities that help to explain why they are relevant in urban issues and sustainable futures. It presents an in-depth discussion of the AIoT implementation with case study examples and critical analysis along with the future trends discussion. The aim is to provide the readers who are urban planners, policymakers, technologists, and citizens with a solid grasp of the ways the AIoT alters urban management, the opportunities, the problems that the technology provides and how it can be used to develop inclusive and resilient cities.

It proceeds to describe the technical underpinnings of AI and IoT, and how they may be synergistic with cities. This is followed by the special application in transportation, energy, waste management, public safety, healthcare services, and water management and its globally spread applications in cities such as Singapore, Barcelona, Amsterdam, and Chicago. Along with the advantages of AIoT, including improved efficiency and sustainability, the chapter also focuses on challenges related to data privacy, the expenses of the associated infrastructure, ethical considerations. Next trends, such as 5G, edge computing, and blockchain are mentioned in line with the description of the nexus of smart cities. The chapter ends by providing a vision on sustainable and fair cities in the future, where responsible innovation and inclusive policies are required. This framework offers a precise blueprint of how AIoT is going to take a revolutionary turn in city life.

2. CONCEPT AND EVOLUTION OF SMART CITIES

2.1 Defining Smart Cities

A smart city is a city that uses digital technologies to modify urban systems, quality of life and make environments sustainable. Smart cities, in contrast to traditional cities, have the ability to scale resources and services via connected devices, data analytics, and automation, all of which greatly contributes to resource and service management in terms of accuracy and efficiency. In essence, smart cities will seek to manage city-related issues, including traffic congestion, energy-wasting problems and squander administration, by focusing on a data-driven arrangement to streamline urban-related activities and revise the profitability of people. The characteristic feature of the smart city is that it is able to gather, analyze, and respond to real time statistics that enables creation of adaptive and responsive urban ecosystems. Artificial Intelligence (AI), the Internet of Things (IoT), and cloud computing are some of the technologies at the center of such a transformation, with cities using them to keep track of systems, anticipate problems, and automatically react to them. Such technological integration helps not only to simplify the urban management process but also promotes inclusivity and sustainability to make the cities respond to the needs of burgeoning populations and contribute minimal harm to the environment.

Smart cities are defined by six domains namely, economy, mobility and environment, people and living as well as governance. The capacities are electronically connected by digital infrastructure and cities can provide services associated with smart transportation, efficient energy grids and responsive healthcare systems. It has transformed into the reality of ecology-driven actual urban planning theory involving the theoretical aspects of urban planning in practice with a technology-oriented strategy. This comprehensive strategy makes smart cities different compared to old versions of the city and this qualifies them as the future of cities in a more urbanized world.

2.2 Historical Context: From Traditional to Smart Cities

Smart cities are the results of a long history in development of the cities dating back to centuries and the development which has taken place in specific stages depending on the technological and social changes. During the 19th century, the Industrial Revolution brought about the urban revolution with mechanized services like the steam driven trams, and gas lighting together with a primitive electricity grid. These inventations are the basis of modern urban infrastructure as they help cities accommodate more people and faster industry growth. Cities introduced mass transportation and electrification around the beginning of the 20th century, e.g.

subways and streetcars, making movements and communications more convenient. Nevertheless, these systems were mostly fixed and they did not have flexibility to deal with intricate urban issues.

The 20th century came with a turning point with introduction of the digital technologies. Such tools as Geographic Information Systems (GIS) that appeared in the 1960s enabled cities to visualize and analyze spatial data and help to understand urban planning and resource distribution. Urban management continued to change in the 1990s with the introduction of the internet that allowed data exchange and digital communication. The first green techs, including automated traffic lights or other electronic utility meters, appeared in cities and initiated the development of a smart increasingly integrated urban network. IoT, together with AI and high-speed connectivity, brought the 21 st century with new smart cities.

Such shifts in the urban systems of mechanized paradigm toward digital can be likened to a wider process through which cities are turning towards data-driven governance. One of the earliest projects, Toronto Sidewalk Labs and Dubai Smart City initiative, aimed at incorporation of technology in urban development and it preconditioned the creation of modern complexes. AioT (a combination of AI and IoT) is used to drive real-time responsive, dynamic cities that have the ability to adapt according to the time of real-time conditions, whether in terms of traffic operations or environmental changes. The history of industrial into digital cities increases the idea that technology has changed the way to look at urban spaces in modern life.

2.3 Drivers of Smart City Development

Three main factors enhance the emergence of smart cities, which include the high rate of urbanization, environmental demands, and technological innovations. The global megatrends and trends include urbanization whereby the United Nations (2018) estimated 68 percent of the global population will reside in the urban area by 2050. This induction overburdens the urban systems creating problems such as congestion. Convention patterns in urban systems are failing with regard to such demands which require intensive solutions that would further create maximum utilization of resources and optimize their efficiency.

Another essential force is that of environmental concerns. Climate change is a major threat to cities and the increasingly hot climate, occasional extreme weather events, and depletion of resources are threatening the sustainability of cities. It is said that cities produce 70 percent of the total carbon produced around the world thus cities are the centre of this battle to combat climate change. Smart cities mitigate all of these issues by incorporating practices that are sustainable, including energy grids learning systems that cut emissions by up to 20% as we see in Copenhagen. The third driver is technological progress, especially in the sphere of AI and IoT that allows cities to use data to make smarter decisions. Superior data infrastructure is formed by the implementation of more than 1 billion IoT including urban environments by 2023. Urban systems are optimized based on this data through the application of AI algorithms, such as forecasting the patterns of traffic or expected energy demand. To take an example, traffic management system that runs on AI and controlled by a program implemented in Barcelona cut commute time by 12%.

3. EXPLAINING THE CONCEPT OF AI AND IOT IN SMART CITIES

3.1 The Internet of Things in Urban Systems

Internet of Things (IoT) can be defined as a series of interconnected objects fitted with sensor devices, programs, and connections that gather, transfer, and handle data through the Internet or the neighborhood network. To the issue of smart cities, IoT constitutes the underlying infrastructure of digital information networks that can monitor and control the systems in real-time. The scope of IoT devices varies greatly and includes the environmental sensors allowing to monitor the quality of air, traffic cameras keeping an eye on the flow of vehicles, smart meters that measure the amount of energy and water consumed, and connected streetlights that can change their brightness according to the conditions in the environment. In the city, 1.2 billion of IoT devices had been installed worldwide as of 2023, producing terabytes of data on a daily basis. The resulting data-intense environment developed by these devices gives accurate and real-time pulses of what is happening in a city so that the exact and opportune interventions can be undertaken to make city operations optimal.

In smart cities, IoT devices serve as the sensory organs, collecting granular data across various domains. For instance, air quality sensors measure pollutants like PM2.5 and nitrogen dioxide, providing data to mitigate health risks, as seen in Copenhagen's environmental monitoring network, which reduced pollution-related health incidents by 10%. Traffic cameras and vehicle sensors track movement patterns, feeding data to systems that alleviate congestion. Smart meters, deployed in cities like Amsterdam, monitor real-time energy usage, enabling a 15% reduction in household energy waste.

An architecture of IoT in smart cities is generally comprised of four layers that include perception layer, network layer, processing layer and application layer. The perception layer has sensors, and actuators to gather information on the environment that could be temperature, traffic density, or water quality. The network layer handles data transmission over wired or wireless portal connections and low-power wide-area networks (LPWAN) to be practical and functional in far urban locations.

The layers of processing are usually cloud-based and/or edge-based in the requirements of the system where the aggregate and analyses of data are performed, and application layer offers the insights to end users in the form of dashboards, mobile apps or automated systems.

Possibility to present real-time, localized data means that IoT is used to make urban management proactive rather than reactive. As an example, IoT-equipped waste-bins in Seoul trigger an alert when being full and minimize collection routes, saving an upwards of 30 percent of operational costs. In the same way, IoT sensors can be applied to the water systems in Los Angeles, and they can detect water leaks and prevent up to 10 million gallons a year. IoT enables cities to build a coherent urban ecosystem, thus ensuring the resource monitoring and management process to easily establish an analytic and automated environment through the use of AI.

3.2 Artificial Intelligence in Urban Management

Artificial Intelligence (AI) is a combination of technologies, deep learning, machine learning, computer vision, and predictive analytics, that organize and analyze data and convert them into actionable insights. In smart cities, AI is the brain which considers the enormous amount of data gathered by the IoT devices to maximize the work of urban systems, foresee difficulties, automatize the decision-making process. AI can make urban management more efficient and effective since it would decrease human error and allow the system to respond according to the problem at hand within a very short time.

Table 1. Roles and applications of IoT, AI, and AIoT in smart cities

	Role in Smart Cities	Key Applications	Example Use Case	Outcome Achieved
IoT	Collects real-time data from urban systems using interconnected devices	Traffic monitoring, energy metering, air quality sensing, waste bin management	Seoul's IoT-enabled waste bins signal fill levels for optimized collection	30% reduction in waste collection costs .
AI	Analyzes IoT data to generate insights, predict trends, and automate decisions	Traffic optimization, energy demand forecasting, facial recognition for safety	Barcelona's AI-driven traffic management adjusts signal timings	12% reduction in commute times, 10% lower emissions .
AIoT	Combines IoT data collection with AI analytics for dynamic, intelligent urban management	Real-time traffic rerouting, predictive energy balancing, automated safety responses	Singapore's AIoT traffic system uses sensors and AI to optimize flow	10% reduction in travel times, 12% lower emissions.

Transportation deployment The AI can utilize the information broadcasted on IoT networks in order to make improvements to the traffic in terms of usage and congestion. The algorithms of machine learning interpret the information through traffic cameras and sensors in vehicles in real-time, forecast patterns of congestion, and use algorithms to shift signals so that the drivers can utilize newly created space. In Barcelona, the traffic management via AI led to a drop in commute times by

12 percent and in emissions by 10 percent. A form of AI, computer vision, allows processing visual data in real-time, e.g. finding traffic infractions or pedestrian crowding. As an example, the city of Dubai is employing the latest smart traffic system that installs cameras equipped with smart AI that can find violations and allows 15% fewer accidents.

In energy management, based on the IoT data in smart meters, AI predicts demand in order to optimize distribution. The predictive analytics models are used to balance supply, incorporate renewable energy into it, and avoid outages, as they analyze the historical and real-time patterns of consumption. With the help of AI in Amsterdam, the smart grid eliminated cell planning by stating 18 percent of the energy waste, anticipating the greatest demand and prioritising renewable energy.

Another way AI changes the urban healthcare is through analyzing data on wearable devices of the Internet of Things and city health systems. Health trends, prediction of disease outbreaks, and interventions are identified by Machine learning models. In the context of the COVID-19 pandemic, Taipei applied AI to process the IoT data captured by health sensors, allowing placing concentrated quarantines, resulting in the 25% decrease in viral spread. Besides, AI promotes sustainability in the environment by making predictions of pollution, as well as maximizing resources. AI models analyzed IoT air quality data to make policies that decreased urban pollution in Singapore by 12 percent.

The advantage of AI is that it is able to learn and adapt. Machine learning models can be trained using a variety of data to improve with time, whereas deep learning can be used to discover complex patterns in unstructured data, including texts or images. The success of AI, however, is determined by quality data and non-hypothetical and heterogeneous data as well as the strength of computational infrastructure. The existence of problems, such as biased algorithms or incomplete data sources, may weaken the results, and the task is to constantly develop and improve the model as well as to ethically monitor them.

3.3 The AIoT Synergy

A combination of AI and IoT, commonly known by the name of AIoT (Artificial Intelligence of Things), presents a formidable framework enhancing the functionality of both technologies in smart cities. IoT can act as the data collection layer and collect the information in the real-time of urban systems, whereas AI can provide the engine of analysis of the raw data and convert it to the result. Through this synergy, cities can deploy an efficient dynamic, and intelligent management of monitoring, resulting in efficiency, responsiveness, and sustainability.

AIoT is in actual sense a closed-loop system. The data points are gathered with the help of IoT devices and analyzed by AI to produce insights, predictions, or automated behavior. Such activities become the feedback within the system and they are able to adjust the operations in real-time. Indicatively, in the Singaporean smart transport system, IoT sensors are deployed to monitor traffic jams, and AI algorithms are considered to identify trends and modulate the signal development or can propose alternate paths, decreasing the travel times by 10 percent. The layered architecture of AIoT is what makes it more effective. Environmental and operating data are captured at the perception layer of IoT components, such as the sensors. The network layer supports secure and efficient delivery of the data by means of the protocols such as 5G or LoRaWAN. Processing layer develops AI tools over cloud or edge computing and uses them to understand data, and the application layer is giving insight into it through dashboards or automatic control tools.

The urban setting is also the target of the AIoT collaboration. An example of sensors use in waste management is that sensors placed in bins notify the monitoring system and the AI component ensures the routes to collect the garbage are planned most effectively e.g. in Seoul when the cost reduced by 30. IoT camera and AI-based facial recognition application in the field of public safety have been discovered to be beneficial in identifying the threats and in the process of enhancing the speed of response by 18 percent at Taipei. What comes as a result of AIoT healthcare sector implementation is the use of remote monitoring which in its turn is facilitating a 15% drop in the number of patients in hospitals in Seoul. These examples bear the torch of the concept that AIoT is a process of remodeling raw data into smart solutions that put the citizens in the spotlight.

Conversely, the determination of the success and influence of AIoT happens to be dependent on how well challenges related to data privacy, interoperability, and computational demands among others could be managed. The provision of data security via protocol implementation and creating standardization on all systems will be one of the reasons why the combination and safeguard of citizen data can be a possibility (Li et al., 2025). It is by means of such coaction between the righteous IoT with its stark data gathering capabilities and AI with its analytical potentiality in which AIoT quantifiably.

4. AI AND IOT IN SMART CITY DOMAINS

4.1 Smart Transportation and Traffic Management

Traffic congestion is one of the common urban issues, that worsen air pollution, the use of more fuel, and increase travel time that amount to a loss of 1.4 trillion dollars to the global economies. The combination of IoT and AI, which is called AIoT, is transforming the world of transportation in smart cities where real-time data can be gathered and smart decisions are made. Traffic cameras, vehicle sensors, and devices with GPS are IoT devices that gather the minute data about the movement of vehicles, road condition, and characteristics of the pedestrians. This data is analyzed by AI algorithms, such as machine learning or predictive analytics, to map the pattern of congestion, optimal timing of road signals and alternative route recommendations to drivers.

The example of this approach is a smart transportation system of Barcelona. Through the use of IoT sensors, and Artificial intelligence-controlled traffic management systems, the city can dynamically optimize traffic lights in real-time to release congestion by 15% and emission by 10%. On the same line, the Smart Nation project in Singapore is incorporating AIoT into the re-routing system of automobiles during peak hours, reducing the commute time by 10 percent. Such systems also improve the transportation system in terms of making the cars of the public transport predictable and result in better scheduling as well as fewer time wastages between the rides. The capacity of AIoT to interconnect information coming in different vehicles, both private and public and even cyclists brings about a comprehensive approach to urban mobility, which is sustainable and enhances life.

4.2 Energy Management and Smart Grids

There are major shortcomings in energy systems in urban places that have a challenge to streamline the supply and demand especially with the consideration of a renewable source of energy such as the solar and the wind. The long-standing grids tend to work inefficiently, causing a waste of energy up to 15% of the global energy. AIoT changes the energy management with help of intelligent grids based on IoT and AI real-time monitoring and optimization. Using smart meters that are already in millions of homes worldwide, IoT monitors the power load with the utmost accuracy and offers extensive details on metered power consumption to both the utility and the consumer. The combination of these meters and those which indicate the performance of power lines and stations (sensors) form a vast data that tracks the grid performance.

AI uses this information to make demand forecasts, distributive efficiency, and outage warnings. As an example, machine learning models predict peak uses so that supplying companies can prepare accordingly. The Smart Energy Grid in Amsterdam is a prime example that managed to stem to 20 percent reduction in energy wastage with the help of AIoT by utilizing renewable energy sources and loads distributio. Microgrids enabled by AIoT can also enable neighborhoods to produce and distribute renewable energy to improve the resilience of the system in the case of outages. AIoT smart grids store energy, manage shortages, and create possibilities in energy distribution through dynamic energy management, which is of great importance in urban centers that are subject to increase in energy consumption and demand.

4.3 Waste Management Innovations

The challenge of the urban waste is urgent, as cities produce 2.24 billion tons of urban solids per year, with most of it ending up in the landfills causing overuse of this land rich resource andpollution. The more problematic waste collection approaches that we presently utilize are the traditional waste collection systems that are usually inefficient because of the fixed schedules that make it so that they end up making unnecessary trips only to incur an extremely expensive operational cost. Virtual aspects of AIoT include improved recycling and thus increase in waste collection. IoT sensors in the waste bins are used to detectthe level of fill in real time to transfer the data through wireless networks to central platforms. Launched inSeoul andSingapore, those sensors lower the number of manual inspections and allow generating the dynamic routes of the collection trucks.

Artificial intelligence improves such a process by processing the past and current data to forecast the waste generation patterns, which optimizes collection routes and minimizes fuel spent. The smart waste system of Seoul, to give it a specific example, saved 30% of its collection expenses and 25% of its recycling rates by streamlining its operations with the help of AIoT. On the same note, IoT sensors and AI analytics are used by the Singaporean Housing and Development Board (HDB) to reduce collection trips by twenty-five percent, thereby reducing emissions by twenty percent. Waste sorting can be also applied with the help of AI because computer vision will help in identifying recyclable sources, and further improve the environmental impact.

4.4 Public Safety and Surveillance

The safety of the people in populated cities is a complex issue, as the number of crimes and environmental risks continue increasing in cities. AIoT improves the safety of cities by real-time surveillance and predicting analytics. Various types of IoT, such as cameras, motion, and environmental sensors, gather the information about the activity in the city, be it strolling or a bad air day. Surveillance systems with these devices create an interconnected network of surveillance where smart cameras will reach 500 million by 2023 around the world. To process this data, facial recognition, anomaly detection, and behavioral analysis AI determines possible threats so that the response may be provided as quickly as possible.

The Array of Things of Chicago project demand demonstrates this solution, where sensory IoT adds data about the state of the environment and the activity of citizens to the city, and AI processes the information in order to make decisions on safety. In the same way, the AIoT-enabled monitoring system in Dubai controlled the traffic crime by 15 percent with the help of notifications and instant ticketing. On the one hand, these systems have safety benefits, but on the other hand, they lead to increased issues of privacy, which should be addressed carefully by employing other methods of data protection, such as encryption and anonymization, to keep confidence in the system by people. Condensing a large amount of heterogeneous data flows provided by AIoT produces an active safety system, and the city and its citizens become safer.

4.5 Healthcare in Smart Cities

Cities with healthcare organizations experience an increase in population, aging sectors, and crisis in citizen health. How AIoT is changing the field of healthcare Remote monitoring Telemedicine Predictive analytics. Smartwatches and health trackers are examples of wearable devices of the IoT that monitor vital signs around the clock by collecting the necessary data in real-time, i.e. heart rate or blood pressure. In 2023, over 1 billion of the population worldwide use these devices. They rely on health workers as they are a source of information transmitted to a healthcare provider, where the in-person visit is minimized.

AI processes this data in order to identify health patterns, expect diseases, and prescribe treatments. In the course of the COVID-19 pandemic, the AIoT of Taipei monitored the infections through IoT sensors and AI models, minimizing the spread of the virus by 25% and allocation of resources. In a similar fashion, the AIoT healthcare system provided by Seoul keeps the elderly patient under watch without being hospitalized at the expense of 15 percent fewer hospital readmissions. AIoT can also contribute to the health of the population by forecasting epidemics and conducting smarter vaccination efforts and achieve better urban resilience. Such innovations increase accessibility, efficiency of healthcare and its proactivity, benefitting citizens of megalopolises.

4.6 Water Resource Management

The issue of water shortage and inefficiencies in infrastructures contribute enormously to undermining the sustainability of cities as cities continue losing much of their water through leakages (20-30 percent). A IoT resolves these problems with intelligent water management solutions that measure quality, detect leaks and reinforce energy distribution. Sensors embedded in the pipelines and reservoirs monitor the water quantity, pressure and contamination and provide real time data to stop the wastage. AI uses such information to find the leak locations, usage patterns, and to set the priorities of maintenance thus allocating resources in an efficient manner.

Application of AIoT in the smart water system allowed Los Angeles to reduce water wastage by 18% through establishing and quickly correcting any form of water leakage which saved the city 10 million gallons of water in every year. In the same way, the AIoT water management system in Singapore maximizes the distribution of its water, and the loss of non-revenue water in it drops by 15%. The systems improve water quality as they monitor the presence of contaminants early enough thus safeguarding the health of the population. The ability to monitor in real-time analyzed with the help of AI allows smart water to support conservation and resilience, vital to cities with an increasing water demand.

5. USE CASE DETAILS: AIOT IN ACTION

Artificial Intelligence (AI) and the Internet of Things (IoT) used together under the term AIoT is changing the way cities are managed by supporting data-driven, automated and efficient systems in many areas. At that, it describes in detail six use cases to illustrate how feasible AIoT implementation in smart cities is and what areas it can be applied to: precision traffic management, smart energy optimization, waste collection efficiency, and improvements to the sphere of public safety, the realm of healthcare, and leak detection in the water system. This set of examples underscores the way in which cities all around the world are using AIoT to solve their urban problems, enhance the quality of life, and make them more sustainable, with real-world applications, and quantifiable results.

5.1 Precision Traffic Management

The problem of traffic congestion has been an ever-present problem in big cities, and it is part of the reason that causes delays in the commutes of each individual person as well as expenditure of fuel and reactive greenhouse gas emissions. In Singapore, one of the most advanced smart cities in the world, the Land Transport Authority (LTA) has moved to ensure that there is a solution to such problems by using AIoT to enable a traffic-management system. Its IoT sensors, such as cameras, vehicle detectors, and more, are installed on large roads and intersections to gather the current information about the traffic flow, an amount of vehicles, and conditions on roads. Such sensors are used to obtain granular information related to the number of vehicles, their speed and lanes congestion; the information is transferred to the centralized systems with high speed networks, e.g., 4G and 5G.

This data is then processed by AI algorithms based on machine learning and predictive analytics to detect trends of congestion and anticipate future conditions on a road. As an example, at high-traffic times the system uses the previous period of traffic and the real-time data to predict and cause any blockages and changes the timing of the traffic lights to stream-line it. This saves on non-required stops and idling, making the traffic efficient. Moreover, AI-based app helps drivers with real-time traffic status and generates adjuvant route options through mobile phones or built-in navigation systems, making traffic more alleviated.

Moreover, with its less idling and vehicle moves, the system has reduced the carbon emissions by 12 percent incorporating the sustainability vision of Singapore. AI analytics interlingins with IoT sensors so that the system is resilient and reliable as it responds to changing conditions, i.e., accidents or roadworks. The given use case is a bright example of how AIoT improves urban mobility with the proactive and data-driven approach to traffic management that will benefit society and the environment.

5.2 Smart Energy Optimization

The concept of managing energy is a vital aspect of sustainable cities especially in the effort to shift toward mitigation of carbon emissions as well as the incorporation of renewable energy. Amsterdam is one of the first cities to adopt smart energy approaches to energy supply and use. They have adopted an AIoT-based smart grid that optimizes the supply and use of energy. Smart meters and sensors are placed in residential, commercially and industrial buildings that constantly monitor the ongoing energy consumption. They measure power usage, loads, and other information concerning the grid efficiency and power consumption and forward the same to the cloud where analytics is done

The information is used in AI algorithms to predict the trend of energy usage and to a large extent it enhances load balancing. As an example, during the periods that renewable energy is generated in large quantities, such as sunny or windy days, AI prioritizes the distribution of solar or wind energy to ensure that the use of fossil energy is reduced to a minimum. Conversely, the system balances the demand during period of low production which is effectively distributed throughout the grid in an efficient way. Furthermore, the inefficiencies such as overloaded circuits are also identified in these machine learning models and the most opportune moment of maintenance is proposed in these models so that the matters can be resolved before creating the delay. The ability of the predictive technology can be regarded as a sort of the warrant of the stable and efficient provision of energy.

Alliander (2024) asserted that the smart grid implemented in Amsterdam has resulted in a 15 percent decrease in energy expenses on the residents thus increasing the affordability of the electricity. The system has allowed incorporation of an additional 30 percent renewable energy into the grid in line with the vision of the city that aims to attain carbon neutrality in 2050. Amsterdam uses AIoT to ameliorate the sustainability of smart energy systems, decrease costs, hike the dependability of grids, and fit the expectations of other cities that may want to embrace green power, an exemplar to other cities in the transition to green energy.

5.3 Waste Collection Efficiency

Sustainable urban areas can only beclean when waste management is carried out effectively. Singapore has implemented an artificial intelligence of things (AIoT)-enabled waste collection scheme as an initiative of its Housing and Development Board (HDB). The smart IoT enabled waste bins have sensors which measure the frequency of filling, types of wastes, and quantity of waste in residential estates. They are sensors that are capable of determining when bins are full using an ultrasonic mechanism and transmitting the information to a central management facility using low-power wide-area networks (LPWAN).

The data is run by AI algorithms which predict waste generation patterns according to historical trends, seasonal changes and demographic aspects. An example is, the system determines the times of high wastes production e.g holidays and schedules their collections such as during holidays. Such a predictive mode dictates the fact that collection trucks operate with optimized routes, without any waste of trips and petroleum products. The platform is also capable of giving real time notification to the waste management teams in order to be able to respond to overflowing bins or illegal dumping in time. The outcomes are radical. The AIoT system has also saved 25 percent of the waste collectors trips and low operational costs and fuel consumption. Moreover, through route optimization and eliminating idling the system has streamed down the related carbon emissions by 20%.

5.4 Public Safety Enhancements

The concept of livable cities is based on everyone in the population being safe and secure, and AIoT is helping to improve urban safety with special security and monitoring solutions. Within the framework of Array of Things (AoT) initiative implemented in Chicago, there is a network of IoT sensors scattered throughout the city with the help of which the information on air quality, noise levels, or human traffic are observed. Using modular sensors fixed on building or streetlight, such sensors can capture real-time data and send it to cloud systems to be processed.

Computer vision and anomaly detection algorithms study this data to determine a possible safety issue. To illustrate, crowd sensors can record abnormal patterns in the crowd which can be a signal of mass disorders and air quality data can feed health warnings when the PM spikes. AIs also process the data about pedestrian flow to improve city planning, like changing the arrangement of intersections, to avoid accidents. The system can interact with emergency response networks allowing the quick rollout of police or medical services on sensing the threats in question.

5.5 Healthcare Monitoring

AIoT is changing healthcare in smart cities, especially on how to deal with aging population and handling and dealing with public health emergencies. In Seoul, elderly people live in a city with an AIoT-based health system comprising wearable IoT devices that help monitor their health. Those gadgets that are installed with the sensors of heartbeat, blood pressure, and physical activity will monitor the data continuously and send it to the medical professionals using the secure networks. AI models use this information and learn what kinds of health problems can be identified based on early indicators and forecast the outcomes in the future according to the previous data on health conditions. The system warns medical practitioners to act in time limiting the requirement of visiting the hospital. In the COVID-19 pandemic, Seoul wound this up to track quarantine as well as infection by use of Artificial Intelligence in resource mobilization.

5.6 Water Leak Detection

The challenges faced with regard to urban water management are water scarcity and inefficiencies in the water infrastructure. Los Angeles Department of Water and Power (LADWP) installed a water management system that will solve these problems using an AIoT-based system. The IoT sensors placed throughout the city water supply system allow the real-time observation of rates of water flow, pressure, as well as the level of water quality. They are the sensors that cause data to be sent to a common platform in case of anomalies including unexpected pressure loss which is a signal of leak. The analysis of this data with the use of AI algorithms can help point out the leak locations and set the maintenance priorities according to the severity and scale of impact. The pattern of water demand is also predicted in machine learning models, which make water distribution efficient and limit its wasting. The system can be hooked to mobile applications that can provide residents with real-time information of their water consumption and water-saving suggestions.

6. REAL-WORLD CASE STUDIES OF AIOT IN SMART CITIES

Artificial Intelligence of Things (AIoT) is effectively transforming the way big cities are managed because of combining Internet of Things (IoT) with Artificial Intelligence (AI) to develop a data-driven efficient and sustainable urban ecosystems. Such strong synergy allows cities to solve their currently acute issues in traffic, energy consumption, waste, public safety, and healthcare with real-time data gathering, predictive analytics, and making decisions autonomously. Cities around the world are using AIoT to optimize their operations, make life better and become more sustainable. This part will already give more specific case studies in Singapore, Barcelona, Amsterdam, Seoul, and Chicago, presenting a variety of AIoT applications in smart city projects. These cases point to the revolutionary power of AIoT, backed by quantifiable results, and explain how cities can make sense of the nuances of urbanization in developing effective urban space resilience and inclusiveness.

6.1 Singapore: A Global Smart City Leader

Again, the Smart Nation initiative implemented in Singapore in 2014 is an example of how the application of AIoT can change the urban systems into a network of interconnected but intelligent systems. In many areas (in transportation, housing, and environmental surveillance), the city-state incorporated IoT sensors and AI analytics that help to increase efficiency and improve livability. In the field of transport, road and vehicle sensors are used that gather real-time information about the traffic situation, the number of parking spaces and vehicles speed. AI algorithms utilize this data to optimize the timing of traffic signals dynamically and drive 15% optimized traffic congestion between peak hours as explained by Smart Nation Singapore (2025). Another example is the Smart Mobility 2030 program, in which AI is applied to forecast traffic in various ways, allowing adjustment of the signal systems to specific adaptive traffic patterns, to reduce delays and enhance the experience of commuters. This has resulted in an average of 12 percent reduction in travel time in major routes and the situation has helped ease the mobility in the cities considerably.

In buildings, the Housing and Development Board (HDB) of Singapore uses IoT sensors to check the status of buildings, including structural integrity, water leakage, and working of the elevator. Also, environmental observing systems use IoT sensors to monitor the level of air quality, temperature, and humidity within the city. The AI processes this data to enable it to offer actionable intelligence to urban planners i.e. to enable them to know which areas have pollution hotspots and how to develop more and greener spaces. Through the incorporation of AIoT, Singapore has reduced its carbon emissions by 10 percent within the observed regions thus adhering to its position on sustainability. What has made the Smart Nation initiative successful is the fact that it is a holistic initiative that brings together the collaboration between the government and the private sector through digital infrastructure and the people. The capability of Singapore to scale AIoT solutions throughout various sectors of the economy makes this country one of the international leaders in the development of smart cities with an example to be followed by other cities.

6.2 Barcelona: Smart Transportation Pioneer

In residential real estates, Singapore is leading the housing sector now with its Housing and Development Board (HDB) using IoT sensors that detect building conditions of structures, water leaks, and elevators. Furthermore, environmental accompanying systems are based on the use of IoT sensors monitoring air quality, temperature, and humidity in all parts of the city. Urban planners use the insights in this data to develop plans of action, like finding the highest pollution rate areas and where they should build their green spaces. Through the incorporation of AIoT, Singapore has already reduced its carbon emissions by 10 percent in the areas that are being monitored to further prove its sustainability pledge. The success of the Smart Nation initiative is traced to its comprehensive nature, through which the partnership between the government and the private sector, vigorous digital infrastructure, as well as citizen participation collectively fuel the innovative process. The capacity of Singapore to scale AIoT solutions in a variety of sectors makes it a global leader in smart city creation, which can serve as an example to other cities to follow.

Table 2. Real-world case studies in AI and IoT-based smart cities

Case Study	Technology Used	City/Region	Outcome Achieved
Singapore Smart Nation	AI + IoT sensors + traffic analytics	Singapore	15% reduction in congestion, 20% lower maintenance costs
Barcelona Transportation	IoT cameras + AI traffic management	Barcelona, Spain	12% reduction in travel times, 10% lower emissions
Amsterdam Smart Grid	IoT meters + AI energy analytics	Amsterdam, Netherlands	18% less energy waste, 30% more renewable integration
Seoul Waste & Healthcare	IoT bins + AI health tracking	Seoul, South Korea	30% lower waste collection costs, 25% reduced virus spread
Chicago Array of Things	IoT sensors + AI urban analytics	Chicago, USA	10% pollution reduction, 8% lower crime rates

The AIoT projects in Barcelona are not limited to the regulation of traffic since the city also focuses on smart transportation and parking services. The sensors installed in the parking space via IoTs sense the availability whereas the AI-integrated mobile applications help drivers to locate available parking space causing 30 percent reduction in the time of location of parking spaces. This will not only ease congestion but also reduce fuel usage and gases. Besides, AIoT systems make public bus routes more efficient based on the passenger data and traffic conditions and provide efficient services in a timely fashion. The willingness of the city to adhere to open data platforms allows working with technology corporations, which leads to the constant innovation in urban mobility. The example of Barcelona illustrates the way AIoT can be implemented to change the transportation system, to make it more reactive, environmentally friendly and client-oriented and to solve the issues of the heavily populated urban area.

6.3 Amsterdam: Energy Efficiency Innovator

Vehicle-to-grid (V2G) solutions Vehicle-to-Grid systems are another example of innovative energy solutions that can enable the Amsterdam Smart City program to take advantage of AIoT. The sensors in the electric vehicle charging stations share data with artificial intelligence platforms that handle energy flow as cars can sell any power back to the grid in high-demand conditions. This has increased grid performance and saved the residents 15% in energy expenses. On top of that, AIoT-powered building management systems also use heating, cooling, and lighting in buildings (public and building), further saving energy. The success of Amsterdam lies in the good cooperation between the municipality, energy providers, and technology firms and are scalable and sustainable. Amsterdam has incorporated energy efficiency and integration of renewables as a priority in managing this energy hence acting as a benchmark to AIoT-driven urban energy management that cities can practice to reach environmental and economic co-benefits concurrently.

6.4 Seoul: Waste and Healthcare Leader

The AIoT projects in Seoul have also improved the management of public health, especially in the light of COVID-19 pandemic that shook the healthcare sector. Technologies such as wearable sensors and environmental sensors (or IoT devices) gathered information about the movements of the population, temperature, and humidity. The data was used to monitor the patterns of infections and predict the outbreak risk and AI systems tracked this data and allowed taking quarantine measures specific to this data and 25 percent fewer people were infected by this virus. Also, using the AIoT, elderly patients are helped with smart home installations which track vitality measures and find abnormalities, signaling the health providers on the possible problems. This has minimized rehospitalization of old-age individuals by 15 percent making life workable and alleviating pressure on the health system.

Seoul uses integrated strategy, which integrates AIoT with strong data governance to have effective and fair service provision, thus making it an exemplary smart city in terms of healthcare and Waste management services.

6.5 Chicago: The Array of Things Project

Chicago Array of Things (AoT) initiative is a leading model of AIoT-inspired urban sensing in the city aimed at improving urban planning as well as safety. AoT is launched in 2016 and is used to gather more information about people and real-time data in one of the cities based on the deployment of IoT sensor nodes to monitor air quality, noise levels, pedestrian activity, and traffic patterns. These sensors-and-camera-endowed nodes send data to AI platforms that facilitate trend analysis and the creation of insights to help urban planners. It's enhanced 8 percent in the public safety within the areas of observation because it has characterized areas of high risks and provided awareness of the intended police conduct. Also, the support of air quality monitoring has reduced the level of pollution by 10 percent as interventions were based on data and involve rerouting the traffic away to not congested areas.

The AoT project focuses on the sharing and availability of data and has, commoditized anonymized datasets to researchers, policymakers and citizens. The data goes through AI algorithms to give a digital replica of urban systems, feeding off of which simulations can be run and the consequences of certain policies enacted (e.g. new transit route, green space creation). The favor of the project is associated with such an approach to collaboration, including the collaboration with the University of Chicago, Argonne National Laboratory, and local communities. AoT allows Chicago to anticipate and solve urban problems before they appear by offering granular and up-to-date information, such as decreasing environmental impacts or making the streets safer to pedestrians. This initiative shows the possibilities of AIoT in the formation of transparent data-based urban management systems with a focus on the well-being of citizens.

7. REAL-WORLD CASE STUDIES OF AIOT IN SMART CITIES II

7.1 Toronto: Sidewalk Labs and Urban Innovation

The Sidewalk Labs initiative by the urban innovation arm of the Alphabet corporation, which aims to revolutionize urban planning in Toronto, is the best example of the capabilities that AIoT has to offer. The project was initiated in 2017 to turn Toronto waterfront into a smart and sustainable neighborhood. Aspects of the site identified by IoT sensors that are present in the area include real-time traffic data, pedestrian and traffic flow, and environmental data to form a complete digital picture of the city. Installed in streets, buildings, and other public areas as a form of smart city technology, these sensors track vehicle traffic, pedestrian footfall, and air quality and create copious data sets which are processed into AI algorithms. The AI uses this data to optimize the urban plan, where a change to the urban infrastructure (such as a road pattern or a public bus line) has an effect on the mobility and the livability. Dynamic urban planning is also dependent on the AI and allows urban planners to try virtually setting up scenarios before physically setting up anything thus eliminating costs and unrest.

7.2 Dubai: Smart City Transformation

The use of AIoT is present in the Smart City project, originally part of the Vision 2021 and Dubai Plan 2026 missions of Dubai, which aims at increasing efficiencies in the city, its safety, and its sustainability. The program will combine the adoption of IoT devices like cameras, sensors, and smart meters within the transport, energy systems, and even the public safety system in an interconnected urban eco-system. IoT cameras keep a track of the situation on the road and the vehicle behavior, as well as AI watches video stream to identify violations, including speeding or parking in prohibited areas, in real-time in the field of traffic management. It is projected that by 2024, such smart grids will lead to a 12 percent decrease in energy expenses, which are beneficial to consumers and contribute to Dubai achieving its target of having 75 percent of its energy come from clean energy sources by 2050. In the sphere of public safety, the facial recognition and anomaly detection equipped with AI-powered surveillance system make the spaces in which people use public services much safer and significantly decrease the response time in case of an incident by a third. The Roads and Transport Authority (RTA) of Dubai is also using AIoT to implement predictive maintenance to the autonomous vehicles and Metro systems and ensure a limited disruption in their services. The front-runner in AI is the city, where the "City Brain" platform unifies the information in different departments and collects it in a single dashboard, thanks to which real-time decisions can be made. In spite of these developments, the weaknesses are the expensive infrastructure and the necessity of good cybersecurity that secures sensitive data. To deal with these, Dubai resorts to the concept of public-private partnership and intense legislation on data protection, which means scalability and trust. The comprehensive vision of AIoT in the city shows its prospective of changing the systems of the cities, and Dubai will be one of the centers of the smart city region.

Table 3. Potential benefits of AIoT in smart cities

Benefit Area	Description	Technology Enabler
Resource Efficiency	Optimal use of energy, water, and waste systems	IoT sensors, AI analytics
Improved Public Services	Faster, data-driven urban management	AI decision systems, IoT monitoring
Environmental Sustainability	Reduced emissions and waste	Precision management platforms
Enhanced Safety	Real-time threat detection and response	AI facial recognition, IoT surveillance
Citizen Convenience	Improved access to services and information	Mobile apps, cloud platforms

7.3 Taipei: Healthcare and Disaster Response

The AIoT systems in Taipei are focused on public health and disaster readiness issues of urban vulnerability of a seismically active area. IoT sensors are also part of the infrastructure or wearable technology and provide monitoring of the conditions of the health of the population and environment, including threats like earthquakes and pollution. AI uses this data to forecast health trends and disaster risks and hence proactive interventions are possible. Wearable IoT devices can be used in healthcare where vital signs would be monitored in elderly individuals through wearable devices with AI picking up anomalies such as repeated heartbeats leading to fewer hospital readmission of 15%. In the context of a COVID-19 pandemic, the AIoT platform in Taipei monitored the pattern of infections that helped to optimize resource allocation and minimized the spread by 20 percent. During the disaster response, sensors of IoT are utilized to identify seismic vibration, and in seconds the sensors activate early warning systems. The impact of earthquakes is foreseen by AI models providing instructions regarding evacuation and the distribution of resources faster (by 18%) than before. The smart city control center in Taipei has connected the data of the transport, emergency, and health services to the decision-making in real-time. Hurdles involve data privacy issues and the fact that there should be resilient infrastructure that can stay afloat during natural calamities. Taipei handles those by strict data anonymization procedures and superfluous sensor networks. AIoT connectivity boosted due to the investment of the city in the 5G technology works in favor of real-time applications. The experience with the AIoT-based transformation of Taipei can provide the solutions to other cities that respond to similar threats and can learn how to make their cities more resilient following that example.

7.4 Los Angeles: Water and Energy Management

The use of AIoT in Los Angeles aims at addressing water shortage and energy substitution, which are especially crucial issues in a drought-threatened area. The solution to the ability to monitor the pipeline and identify leaks in it in real-time by using IoT-based sensors of the smart water system of the city. AI focuses on maintenance, thereby minimizing losses of 18 percent of water and 10 million gallons on an annual basis. In energy smart meters with the use of IoT efficiently manage consumption with the help of AI that can predict demand, which increases the performance of the grid and decreases incidents of outages by 15 percent. The AIoT respects the existence of renewable energy sources and makes the solar look up to 12 percent, with predictive load balancing. The smart streetlights used in Los Angeles have IoT technology, ensuring the correct degree of brightness depending on the activity, thus spending 20 percent less energy. Water and energy scenarios programmed in the digital twin of the city based on the AI direct infrastructure investments. There are also difficulties like the expensive nature of implementation and the need to make benefits available equally. Los Angeles counters these by using federal grants as well as community outreach community in programs. The AIoT strategy of the city demonstrates the efficiency of technology that can alleviate limitations on the available resources to make cities more sustainable and resilient.

8. BENEFITS OF AIOT IN SMART CITIES

8.1 Enhanced Urban Efficiency

Artificial Intelligence and the Internet of Things, also referred to as AIoT, play a major role in making cities efficient through optimization of urban systems working effort to reduce the waste of resources. AIoT gives smart cities opportunities to manage the infrastructure, including transport systems, energy systems, and much more, thus optimizing urban functioning and adapting it to current requirements. As an example, AI traffic management solutions are based on IoT sensors and cameras and use them to track vehicle traffic and study congestion patterns and thus change traffic lights dynamically. It leads to an up to 15% reduced traffic congestion, which could be proven in such cities as Singapore or Barcelona. These systems not only save time of commuters but also limit fuel use by predicting the peak hours of traffic and planning out alternative routes so as to save on not only cost effect but also in reducing environmental pollution.

There is more than transportation in terms of AIoT to balance the energy system with the help of smart grids and IoT-enabled meters to gather real-time data on patterns. These grids will give priority to renewable energy sources when available and avoid fossil fuels dependency and power losses through shutdowns. On the same note, AIoT-powered waste management solutions are fitted with sensors within the bins to observe levels of fill thus resulting in efficient routes when it comes to collecting waste, thus boosting efficiency of the operations. As an instance, the smart waste system in Seoul reduced collection runs by 30% and reduced fuel consumption and labor costs. AIoT enables the prediction of equipment failure as well as automation of routine jobs, which is why urban infrastructure stays at the optimal performance level, experiencing minimal downtimes and prolonging the service life of essential systems.

8.2 Improved Quality of Life

Smart cities are really good in enhancing the lives of people in the cities, as people have faster, easier, and more customized services. The real-time traffic warnings, generated by the IoT sensors and real-time data analytics using Artificial intelligence enables dwellers of a city to get the best and shortest route through the city and therefore spend less time in the city hence have time to engage in other activities which are more personal. The improved and efficient public transport system that is made available by the use of AIoT provides the opportunity to have high quality services that are well covered therefore adds more value to the daily life of the populace. The good changes occur and then the urban places become more dague, predictable, and less of a mess, which will have a direct impact on the well-being of the population.

There is also the application of AIoT in the health care industry where doctors can remotely monitor the wearable IoT devices and through the telemedicine platforms powered by AI intervene in good time. The AIoT mechanisms in Seoul monitor the health parameters of elderly patients, predict hazards, and lower the level of unnecessary readmissions by 15 percent. Through such innovations, healthcare would become more convenient especially to vulnerable groups, which improves their independence and their living conditions. Besides, AIoT enhances security in the community by conducting real-time monitoring and detecting deviations. All of these minor but effective changes will lead to the formation of a smoother urban culture, where people experience less friction in their everyday lives. By focusing on the citizen-centric design AIoT guarantees that technology will be used to support people needs and that it will create a feeling of empowerment and satisfaction amongst residents and shape cities to become places where people are more likely to live and feel included therein.

8.3 Environmental Sustainability

The most important role of AIoT in achieving environmental sustainability in smart cities is the streamlining of resources and minimizing ecological footprints. Due to the IoT sensors, their use in smart waste management has been to monitor the fill levels of the bins and facilitate an efficient collection route to help reduce landfill contributions up to 20 percent in places such as Seou. Since they reduce unnecessary trips to collect waste, such systems also reduce fuel use and greenhouse emissions, which helps clean the urban air. Besides, AI analytics forecasts the patterns of the waste generation, stimulating recycling and load on waste processing plants.

The power is another essential sector at which AIoT leads to sustainability. IoT meters and AI-powered distribution give the smart grid (such as Amsterdam) the ability to reduce carbon emissions by 15% by efficient integration of renewable energy. They are set to prioritize for low-carbon resources under peak availability, cutting the need for fossil fuels. Equally, the building management systems (also powered by AIoT) adjust heating, ventilation and air conditioning (HVAC) according to occupancy and weather data and reduce energy consumption by up to 20 percent.

8.4 Data-Driven Decision Making

AIoT provides vital information to urban planners in real time with deeper insight in how to act and plan better in the city and improve resilicnce. AI allows processing the tremendous amount of data that IoT devices collect to identify trends in the traffic, energy consumption, the quality of air, and the behavior of the citizens and make some assumptions about their needs. As one case, Singapore AIoT systems forecast congestion on the road through traffic data and access the infrastructure investment, bringing the solution to help reduce congestion by 15 percent. This data-based strategy makes urban policies based on evidence and not perceived needs, but the real ones.

AIoT technology is being used in urban planning, in the form of digital twins, or virtual representations of the systems within the city that run on current information. Sidewalk labs in Toronto employ a digital twin to simulate urban designs to optimize people and vehicle movement and commute times that are ten-percent shorter. Presenting a big picture of urban dynamics, AIoT makes information-based decisions made and grants sustainable and flexible development that responds to the needs of the citizens and living conditions.

9. CHALLENGES AND ETHICAL ISSUES

9.1 Data Privacy and Cybersecurity

The co-existence of Artificial Intelligence and Internet of Things (AIoT) in smart cities collects enormous data which includes personal data regarding the citizens, such as their movement patterns, energy consumption and health indicators etc. SolutionThe risk of cyber threats can include data leaks or ransomware and affect such large-scale urban processes as traffic or energy, which may cripple various urban services and lead to loss of citizen confidence. As examples, a compromise on smart surveillance facilities that are IoT enabled may result in the leak of sensitive video and a hack in smart grid facilities may lead to an outage in a large part of the population. Data integrity and confidentiality require robust encryption protocols (e.g. AES-256) and effective mechanisms of secure communication (e.g. TLS). As well, definite consent procedures should be put in place, whereby citizens are informed on the purpose of data gathering, and their data is under their jurisdiction. Regulatory measures, including the General Data Protection Regulation (GDPR), act as an example of how such ideas as data protection enforcement can be enforced, as cities need to adopt anonymization and safe storage tools.

Table 4. Common challenges in AIoT adoption for smart cities

Challenge	Description	Affected Stakeholders
Data Privacy	Risk of misuse or breaches of personal data	Citizens, governments
High Costs	Expensive infrastructure and maintenance	Developing cities, taxpayers
Interoperability	Incompatible devices and systems	Tech providers, city planners
Digital Divide	Unequal access to technology benefits	Underserved communities
Ethical AI	Bias and transparency in AI decisions	All stakeholders

9.2 High Infrastructure Costs

The use of AIoT in the smart city requires immense financial resources and resources in terms of hardware & software as well as expertise, which is a major challenge especially in the developing countries. The IoT hardware: the sensors, cameras and connectivity networks are cost-prohibitive to install and maintain. To give an example, the implementation of smart traffic in a middle-sized city may cost millions, including the sensors, the cloud, and AI analytics. Also, the cost to maintain these systems includes expenses on software upgrades, measures to Abr particular security risks, and technical support. Since developing countries experience restricted budgets and have insufficient knowledge on how to handle advanced AIoT systems, they do not have many options regarding smart city solutions. This economic pressure, may contribute to widening the divide in world cities to the detriment of resource-stretched cities. To deal with the same, a government may utilise public-private partnerships (PPPs) to combine resources and skills as demonstrated by a Smart Nation project in Singapore that cooperates with technology companies. Infrastructure development in the low-income regions can also be financed through subsidies and international grants.

9.3 Interoperability and Standardization

One common problem of AIoT device and platform diversity in smart cities is the lack of interoperability, where the device systems of different vendors operate on incompatible protocol or data structure. Such non-standardization makes horizontal integration a challenge, which lowers the efficiency and scalability of smart city projects. To give a specific example, sensors installed to manage traffic in cities made by one company might not exchange data well with sensors installed to manage waste by another company so that cities have data silos around urban management where holistic management is not possible. Such fragmentation may theoretically result in ineffectiveness, e.g., slow response of an emergency system, excess building of infrastructure. The existence of standardized procedures, e.g. MQTT or CoAP as IoT protocols, is necessary to make all devices cooperate within the city systems. Interoperability can be encouraged through industry-wide standards, such as those suggested by the IEEE or ISO, to allow a common data format and a framework of communication. The use of such open standards was adapted in respective cities such as Amsterdam, which aims to allow the interconnection of various IoT systems for maximum information exchange between the fields of energy and transport.

9.4 Ethical Concerns in AI Decision-Making

Smart cities have the potential of causing biases in AI algorithms that compromise the sense of fairness in the supply of urban services like biased traffic control preferring some parts of a city or some discriminatory practices of surveillance. Such prejudices are usually caused by training information that is biased due to historical inequities or diversity, which results in distorted results. Take the example of unfairness and privacy of facial recognition systems that may mistake persons of some demographics. To ensure that the people trust and have confidence in AI decision-making, transparency in the decision-making process must be present and key to that is being able to document the interaction of AI and the way it operates and arrives at a conclusive decision. Bias risks can be minimized by using accountable AI: when it is audited and monitored regularly. In countries, such as Copenhagen, ethical frameworks of AI are introduced, and algorithms are periodically assessed in terms of equity. Accountability can be further realized by the participation of citizens in AI governance through advisory boards or by consultation of the citizens.

10. FUTURE TRENDS IN AIOT-BASED SMART CITIES

10.1 5G and Enhanced Connectivity

Deployment of the 5G networks is changing the game in smart cities, providing high data transfer rate and very low-latency. This will facilitate real time AIoT deployment, i.e. autonomous vehicles negotiating through dense traffic plans and smart grids balancing energy loads in real time. Respectively, 5G enables vehicle-to-everything (V2X) communication, making the traffic accidents drop by 15% due to the real-time exchange of data (Thampi et al., 2025). Traffic systems that are 5G-enabled have resulted in the reduction of foot travel time in Singapore, decreasing mobility by 10 percent. The 5G technology allows cities to manage large streams of IoT data, making urban administration responsive everywhere, including densely populated locations facilitated by seemless connectivity.

10.2 Edge Computing for Real-Time Processing

Since edge computing, computing takes place on the IoT devices, and therefore connectivity to the cloud is not essential. This is essential in regions that have low-internet connection like the rural regions or the underprivileged portions of cities. Running analytics at the edge allows cities to ensure that AIoT functions throughout disrupted communications, and those modifications do not affect the system. Another example is the edge-enabled sensors in Copenhagen, which handle the environmental data on the edge, which decreases latency and allows them to achieve a 20 percent faster response during the spikes of pollution. Edge computing also reduces the costs of bandwidths thus making AIoT scalable and affordable by developing cities.

10.3 Sustainable Urban Design

The use of AIoT in urban planning also aids environmentally friendly planning through maximised utilisation of available resources and minimisation of emissions. Artificial intelligence models when combined with the information provided by the IoT allow cities to build energy-efficient buildings and create green spaces.

As an example, the AIoT-based urban planning lowered the carbon emissions in Singapore by 25 percent due to the optimization of both energy and waste systems (Long, 2025). The case of Los Angeles, with its smart irrigation systems, helps to implement healthy resource management as they save 10 million gallons of water every year. These developments make cities comply with international sustainability.

10.4 Evaluation Points

- Security & Privacy: Privacy requires a strong encryption alongside anonymization of citizen data and vulnerability to cyberattacks to gain confidence in AIoT systems.
- Policy/ Government Support: Subsidies and public-private partnerships will help expedite the adoption of AIoT especially in developing areas that have scanty infrastructure.
- ROI & Cost-Benefit Analysis: AIoT has a long-term payback, and such cities as Singapore attained 20 percent of savings in transport and waste systems.
- Issues of Inclusivity: The digital divide can only be addressed through affordable solutions and programs in digital literacy.
- Emerging Trends Blockchain: Blockchain adds accountability to resource tracking and helps to bear transparent urban systems.
- Future Trends - Edge Computing: AIoT AIoT can be supported by running it locally where connectivity can be a problem, promoting scalability.
- Smart Cities 2.0: AI-powered Hyper-connected urban environments that will be more sustainable and citizen-focused to create secure futures of cities.
- AI + IoT Synergy: AI + IoT and predictive analytics and automation help make cities more efficient and lower the wastage of resources by 20 percent.
- Device Overload: Integrated IoT systems are less cumbersome to implement and therefore simplified functioning in the city.
- Interoperability: Standard interoperability guidelines facilitate the simplicity and smooth mix of devices, which is crucial on a scale of AIoT.
- Ethics & ownership of data: transparency about ownership of data and ethical use of AI builds trust in society.
- Sustainability: AIoT provides green initiatives and reduces the carbon footprints in urban areas by 20 percent.
- Regional Diversity: Covering both the Africa, Asia and Latin America case studies will make the book case study globally relevant.
- Cost-Benefit Trade-offs: This is critical analysis because it would guarantee the feasibility of implementing AIoT.
- Recommendation and asides: It can be enhanced by the use of scholarly sources, diagrams and transitions.

11. CONCLUSION

AI and IoT (AIoT) are changing the face of smart cities by helping to solve some of the urban impacts of efficiency, sustainability, and innovation. AIoT optimizes networks and improves the performance of systems such as the traffic network healthcare and the energy systems, increasing the quality of life and lowering the impact on the environment. Singapore, Barcelona and Amsterdam case studies emphasize such advantages of improvements as a 15 percent reduction of congestion, a 18 percent energy-use reduction, and safety improvements. The data privacy problem, cost issue, and inclusivity require active measures in form of effective policies, investment and education. Other developments such as 5G, edge computing, blockchain will also help strengthen urban resilience, leading to real-time applications and transparent systems. Through a responsible adoption of AIoT, urbanization pressures can be solved so that the technologically deprived are given equal access to technology and that urban futures are both sustainable and that peoples well being and environment are prioritized.

REFERENCES

Li, S., Li, S., Sun, Y., Wang, B., Wang, B., & Zhang, B. (2025). Digital Twin-Assisted Computation Offloading and Resource Allocation for Multi-Device Collaborative Tasks in Industrial Internet of Things. *IEEE Transactions on Network Science and Engineering*.

Long, S. P. (2025). Needs and opportunities to future-proof crops and the use of crop systems to mitigate atmospheric change. Philosophical Transactions B, 380(1927), 20240229. DOI: 10.1098/rstb.2024.0229

Thampi, S. M., Thomas, T., & Mukherjee, P. (Eds.). (2025). *Securing the Connected World: Exploring Emerging Threats and Innovative Solutions*. Springer Nature. DOI: 10.1007/978-3-031-82826-3

ADDITIONAL READING

Bichueti, R. S., Leal Filho, W., Gomes, C. M., Kneipp, J. M., Costa, C. R. R. D., & Frizzo, K. (2025). Climate Change and Urban Resilience in Smart Cities: Adaptation and Mitigation Strategies in Brazil and Germany. *Urban Science (Basel, Switzerland)*, *9*(5), 179. DOI: 10.3390/urbansci9050179

Lee, D., Kang, M. M., & Kim, H. J. (2025). The human dimension of smart cities: Examining antecedents of citizens' smart city policy literacy. *International Review of Administrative Sciences*, *91*(2), 275–292. DOI: 10.1177/00208523241267869

Oyelade, I., Dahunsi, J. O., & Damina, G. B. (2025). Challenges of Smart City Development for Sustainable Urban Growth in Birnin Kebbi. The Pyramids: African Journal of Engineering. *Constructions and Built Environment*, *5*(1), 1–12.

Rana, M. (2025). A Review of the Impact of Big Data on Smart Cities.

Vitálišová, K., Rigová, Z., Vanova, A., & Škvareninová, D. (2025). Small smart cities–vision or reality? Transforming Government: People. *Process and Policy*, *19*(2), 316–337.

KEY TERMS AND DEFINITIONS

AIoT (Artificial Intelligence of Things): Combination of AI and IoT to create intelligent and data based urban systems.

Artificial Intelligence (AI): The tools that allow machines to simulate the human way of thinking, proposing decisions, and learning.

Autonomous Urban Systems: AIoT-enabled infrastructures like self-self-maintained traffic, waste disposal etc. with the need of minimal human input.

Digital Twin: Fully simulations and optimizations of a virtualised model of an urban system based on real-time data.

Internet of Things (IoT): A system of the networked hardware that gathers and shares data over internet to monitor and control real time.

Precision Urban Management: AIoT-enabled realization of process accuracy, resource tracking as well as control of the acquisition of precision in services offered and distribution of urban facilities.

Remote Sensing: This is the gathering together of urban data that in no way physically touches the object or area. The use of sensors mounted on a satellite or a drone.

Smart City: A city that is utilizing digital technologies, including AI and IoT, to become more efficient, sustainable, safe and provide a better quality of life to its inhabitants.

Smart Grid: An energy delivering infrastructure with the help of IoT and AI that improves the supply demand and integrates renewables into it.

Compilation of References

Abbassi, R., Arzaghi, E., Yazdi, M., Aryai, V., Garaniya, V., & Rahnamayiezekavat, P. (2022). Risk-based and predictive maintenance planning of engineering infrastructure: Existing quantitative techniques and future directions. In *Process Safety and Environmental Protection* (Vol. 165). DOI: 10.1016/j.psep.2022.07.046

Abdel-Basset, M., Alrashdi, I., Hawash, H., Sallam, K., & Hameed, I. A. (2023). Towards efficient and trustworthy pandemic diagnosis in smart cities: A blockchain-based federated learning approach. *Mathematics, 11*(14), 3093. DOI: 10.3390/math11143093

Abdulrashid, I., Zanjirani Farahani, R., Mammadov, S., Khalafalla, M., & Chiang, W.-C. (2024). Explainable artificial intelligence in transport Logistics: Risk analysis for road accidents. *Transportation Research Part E, Logistics and Transportation Review, 186*, 103563. DOI: 10.1016/j.tre.2024.103563

Abijith, D., & Saravanan, S. (2022). Assessment of land use and land cover change detection and prediction using remote sensing and CA Markov in the northern coastal districts of Tamil Nadu, India. *Environmental Science and Pollution Research International, 29*(57), 86055–86067. Advance online publication. DOI: 10.1007/s11356-021-15782-6 PMID: 34510357

Abramson, W., Hall, A. J., Papadopoulos, P., Pitropakis, N., & Buchanan, W. J. (2020). A distributed trust framework for privacy-preserving machine learning. In Trust, Privacy and Security in Digital Business (pp. 205–220). Springer International Publishing. DOI: 10.1007/978-3-030-58986-8_14

Afrin, T., & Yodo, N. (2021). A probabilistic estimation of traffic congestion using Bayesian network. *Measurement, 174*, 109051. DOI: 10.1016/j.measurement.2021.109051

Ahmad, S., Hasan, M., Shahabuddin, M., Tabassum, T., & Allvi, M. W. (2020). IoT based pill reminder and monitoring system. *International Journal of Computer Science and Network Security*, *20*(7), 152–158.

Ahmad, S., Hasan, M., Shahabuddin, M., Tabassum, T., & Allvi, M. W. (2020). IoT-based pill reminder and monitoring system. *International Journal of Computer Science and Network Security*, *20*(7), 152–158.

Ahmad, W., Almaiah, M. A., Ali, B. S., & Ali, A. (2024). A privacy preserving Federated Learning BasedIoT framework using cloud computing. In *Research Square*. DOI: 10.21203/rs.3.rs-4701071/v1

Ait Errami, S., Hajji, H., Ait El Kadi, K., & Badir, H. (2023). Spatial big data architecture: From Data Warehouses and Data Lakes to the LakeHouse. In *Journal of Parallel and Distributed Computing* (Vol. 176). DOI: 10.1016/j.jpdc.2023.02.007

Akhtar, M., & Moridpour, S. (2021). A Review of Traffic Congestion Prediction Using Artificial Intelligence. In *Journal of Advanced Transportation* (Vol. 2021). DOI: 10.1155/2021/8878011

Alahmadi, B. A., Axon, L., & Martinovic, I. (2022). 99% false positives: A qualitative study of {SOC} analysts' perspectives on security alarms. In *31st USENIX Security Symposium (USENIX Security 22)* (pp. 2783–2800).

Alam, M. M., Torgo, L., & Bifet, A. (2022). A survey on spatio-temporal data analytics systems. *ACM Computing Surveys*, *54*(10s), 1–38. DOI: 10.1145/3507904

Alankar, B., Kaur, H., & Chauhan, R. (Eds.). (2021). *Transforming the Internet of Things for Next-generation Smart Systems*. IGI Global. DOI: 10.4018/978-1-7998-7541-3

Al-Atawi, A. A. (2024). Enhancing data management and real-time decision making with IoT, cloud, and fog computing. *IET Wireless Sensor Systems*, *14*(6), 539–562. DOI: 10.1049/wss2.12099

Alberti, A. M., Santos, M. A., Souza, R., Da Silva, H. D. L., Carneiro, J. R., Figueiredo, V. A. C., & Rodrigues, J. J. (2019). Platforms for smart environments and future internet design: A survey. *IEEE Access : Practical Innovations, Open Solutions*, *7*, 165748–165778. DOI: 10.1109/ACCESS.2019.2950656

Alberti, M., Bontempo, M. M., Santos, J. R. D., Sodré, A. C., & Righi, R. da R. (2018). NovaGenesis applied to information-centric service-defined trustable IoT/WSAN control plane and spectrum management. *Sensors (Basel)*, *18*(9), 3160. https://www.mdpi.com/1424-8220/18/9/3160. DOI: 10.3390/s18093160 PMID: 30235844

Albino, V., Berardi, U., & Dangelico, R. M. (2015). Smart cities: Definitions, dimensions, performance, and initiatives. *Journal of Urban Technology*, *22*(1), 3–21. DOI: 10.1080/10630732.2014.942092

Alghamdi, T., Elgazzar, K., Bayoumi, M., Sharaf, T., & Shah, S. (2019). Forecasting Traffic Congestion Using ARIMA Modeling. *2019 15th International Wireless Communications & Mobile Computing Conference (IWCMC)*, 1227–1232. DOI: 10.1109/IWCMC.2019.8766698

Alomair, B., & Poovendran, R. (2014). Efficient authentication for mobile and pervasive computing. *IEEE Transactions on Mobile Computing*, *13*(3), 469–481. DOI: 10.1109/TMC.2012.252

Alyu, A. B., Salau, A. O., Khan, B., & Eneh, J. N. (2023). Hybrid GWO-PSO based optimal placement and sizing of multiple PV-DG units for power loss reduction and voltage profile improvement. *Scientific Reports*, *13*(1), 6903. DOI: 10.1038/s41598-023-34057-3 PMID: 37106042

Andrade, R. (2024). *Privacy-preserving face detection: A comprehensive analysis of face anonymization techniques* (Master's thesis, Universidade do Porto (Portugal)).

Andronie, M., Lăzăroiu, G., Iatagan, M., Hurloiu, I., tefănescu, R., Dijmărescu, A., & Dijmărescu, I. (2023). Big data management algorithms, deep learning-based object detection technologies, and geospatial simulation and sensor fusion tools in the internet of robotic things. *ISPRS International Journal of Geo-Information*, *12*(2), 35. DOI: 10.3390/ijgi12020035

Angelotti, G., Chanel, C. P. C., Pinto, A. H. M., Lounis, C., Chauffaut, C., & Drougard, N. (2024). *Offline Risk-sensitive RL with Partial Observability to Enhance Performance in Human-Robot Teaming* (No. arXiv:2402.05703). arXiv. /arXiv.2402.05703DOI: 10.65109/FPAV2197

Anthopoulos, L. G. (2015). Understanding the smart city domain: A literature review. In Rodríguez-Bolívar, M. P. (Ed.), *Transforming city governments for successful smart cities* (pp. 9–21). Springer., DOI: 10.1007/978-3-319-03167-5_2

Arena, G., Chub, A., Lukianov, M., Strzelecki, R., Vinnikov, D., & De Carne, G. (2024). A comprehensive review on DC fast charging stations for electric vehicles: Standards, power conversion technologies, architectures, energy management, and cybersecurity. *IEEE Open Journal of Power Electronics*, *5*, 1573–1611. DOI: 10.1109/OJPEL.2024.3466936

Azimi, S., & Pahl, C. (2021). Continuous Data Quality Management for Machine Learning based Data-as-a-Service Architectures: *Proceedings of the 11th International Conference on Cloud Computing and Services Science*, 328–335. DOI: 10.5220/0010509503280335

Babar, M., & Arif, F. (2019). Real-time data processing scheme using big data analytics in internet of things based smart transportation environment. *Journal of Ambient Intelligence and Humanized Computing*, *10*(10), 4167–4177. DOI: 10.1007/s12652-018-0820-5

Baccour, E., Mhaisen, N., Abdellatif, A., Erbad, A., Mohamed, A., Hamdi, M., & Guizani, M. (2021). Pervasive AI for IoT Applications: A Survey on Resource-Efficient Distributed Artificial Intelligence. *IEEE Communications Surveys and Tutorials*, *24*(4), 2366–2418. DOI: 10.1109/COMST.2022.3200740

Bahamazava, K. (2025). AI-driven scenarios for urban mobility: Quantifying the role of ODE models and scenario planning in reducing traffic congestion. *Transport Economics and Management*, *3*, 92–103. DOI: 10.1016/j.team.2025.02.002

Bajaj, R., & Sharma, V. (2018). Smart education with artificial intelligence based determination of learning styles. *Procedia Computer Science*, *132*, 834–842. DOI: 10.1016/j.procs.2018.05.095

Bajpai, M. (2023). The Transformative Impact of AI Ops/ML and Observability in Automating Networking Operations and Network Security. *International Journal of Innovative Research in Engineering & Multidisciplinary Physical Sciences*, *11*(4), 1–4.

Baklashov, A. S., & Kulyabov, D. S. (2025). Statistical and density-based clustering techniques in the context of anomaly detection in network systems: A comparative analysis. *Discrete and Continuous Models and Applied Computational Science*, *33*(1), 27–45. DOI: 10.22363/2658-4670-2025-33-1-27-45

Balachandra Achar, H. V., & R. N. (2025). The impact and future of artificial intelligence in learning. In *2025 3rd International Conference on Integrated Circuits and Communication Systems (ICICACS)* (pp. 1–6). IEEE. DOI: 10.1109/ICICACS65178.2025.10967881

Balas, V. E., Kumar, R., & Srivastava, R. (Eds.). (2020). *Recent trends and advances in artificial intelligence and internet of things*. Springer International Publishing. DOI: 10.1007/978-3-030-32644-9

Bartoli, A. (2011). Security and privacy in your smart city. In *Proceedings of the Barcelona Smart Cities Congress* (pp. 1–6).

Basu, R., & Ferreira, J. (2022). Can increased accessibility from emerging mobility services create a car-lite future? Evidence from Singapore using LUTI microsimulation. *Transportation Letters, 14*(4), 332–338. *Scopus.* Advance online publication. DOI: 10.1080/19427867.2020.1731993

Batty, M. (2018). *Inventing future cities.* MIT Press. DOI: 10.7551/mitpress/11923.001.0001

Belcastro, L., Marozzo, F., Orsino, A., Talia, D., & Trunfio, P. (2023). Edge-cloud continuum solutions for urban mobility prediction and planning. *IEEE Access : Practical Innovations, Open Solutions, 11*, 38864–38874. DOI: 10.1109/ACCESS.2023.3267471

Bhardwaj, A., Iyer, S. R., Ramesh, S., White, J., & Subramanian, L. (2023). Understanding sudden traffic jams: From emergence to impact. *Development Engineering, 8*, 100105. DOI: 10.1016/j.deveng.2022.100105

Bhaskaran, S. V. (2020). Integrating data quality services (dqs) in big data ecosystems: Challenges, best practices, and opportunities for decision-making. *Journal of Applied Big Data Analytics, Decision-Making, and Predictive Modelling Systems, 4*(11), 1–12.

Bhuiyan, A. B., & Uddin, J. (2024). Bengali Word Detection from Lip Movements Using Mask RCNN and Generalized Linear Model. [IJEEI]. *Indonesian Journal of Electrical Engineering and Informatics, 12*(2), 373–383. DOI: 10.52549/ijeei.v12i2.5088

Bhusal, N., Gautam, M., & Benidris, M. (2021). Detection of cyber attacks on voltage regulation in distribution systems using machine learning. *IEEE Access : Practical Innovations, Open Solutions, 9*, 40402–40416. DOI: 10.1109/ACCESS.2021.3064689

Bidyanath, K., Singh, S. D., & Adhikari, S. (2023). Implementation of genetic and particle swarm optimization algorithm for voltage profile improvement and loss reduction using capacitors in 132 kV Manipur transmission system. *Energy Reports, 9*, 738–746. DOI: 10.1016/j.egyr.2022.11.098

Bie, Y., & Qiu, T. Z. (2021). Connected Vehicle-Cooperative Adaptive Cruise Control Algorithm to Divide and Reform Connected Vehicle Platoons at Signalized Intersections to Improve Traffic Throughput and Safety. *Transportation Research Record: Journal of the Transportation Research Board, 2675*(9), 995–1005. DOI: 10.1177/03611981211005456

Bi, L., Muazu, T., & Samuel, O. (2022). IoT: A decentralized trust management system using blockchain-empowered federated learning. *Sustainability (Basel)*, *15*(1), 374. DOI: 10.3390/su15010374

Bittencourt, J. C. N., Costa, D. G., Portugal, P., & Vasques, F. (2024). A survey on adaptive smart urban systems. *IEEE Access : Practical Innovations, Open Solutions*, *12*, 102826–102850. DOI: 10.1109/ACCESS.2024.3433381

Bokhari, S. A. A., & Myeong, S. (2022). Use of artificial intelligence in smart cities for smart decision-making: A social innovation perspective. *Sustainability (Basel)*, *14*(2), 620. DOI: 10.3390/su14020620

Borkar, S. V., & Sangve, S. M. (2025). *A critical analysis on anomaly detection in high-frequency financial data using deep learning for options.*

Bovkir, R., & Aydinoglu, A. C. (2021). Big urban data visualization approaches within the smart city: Gis-based open-source dashboard example. *International Archives of the Photogrammetry, Remote Sensing and Spatial Information Sciences - ISPRS Archives, 46*(4/W5-2021). DOI: 10.5194/isprs-archives-XLVI-4-W5-2021-125-2021

Bui, K.-H. N., Cho, J., & Yi, H. (2022). Spatial-temporal graph neural network for traffic forecasting: An overview and open research issues. *Applied Intelligence*, *52*(3), 2763–2774. DOI: 10.1007/s10489-021-02587-w

Campolo, C., Genovese, G., Iera, A., & Molinaro, A. (2021). Virtualizing AI at the Distributed Edge towards Intelligent IoT Applications. *J. Sens. Actuator Networks*, *10*(1), 13. DOI: 10.3390/jsan10010013

Candra, A., Budiman, M. A., & Hartanto, K. (2020). Dijkstra's and A-Star in Finding the Shortest Path: A Tutorial. *2020 International Conference on Data Science, Artificial Intelligence, and Business Analytics (DATABIA)*, 28–32. DOI: 10.1109/DATABIA50434.2020.9190342

Caragliu, A., & Del Bo, C. F. (2019). Smart innovative cities: The impact of Smart City policies on urban innovation. *Technological Forecasting and Social Change*, *142*, 373–383. DOI: 10.1016/j.techfore.2018.07.022

Carrington, A. M., Manuel, D. G., Fieguth, P. W., Ramsay, T., Osmani, V., Wernly, B., Bennett, C., Hawken, S., Magwood, O., Sheikh, Y., McInnes, M., & Holzinger, A. (2022). Deep ROC analysis and AUC as balanced average accuracy, for improved classifier selection, audit and explanation. *IEEE Transactions on Pattern Analysis and Machine Intelligence*, *45*(1), 329–341. DOI: 10.1109/TPAMI.2022.3145392 PMID: 35077357

Carvalho, P. M. S., Correia, P. F., & Ferreira, L. A. F. M. (2008). Distributed reactive power generation control for voltage rise mitigation in distribution networks. *IEEE Transactions on Power Systems*, *23*(2), 766–772. DOI: 10.1109/TPWRS.2008.919203

Cascetta, E. (2001). Transportation systems. In *Transportation systems engineering: Theory and methods* (pp. 1–22). Springer US.

Castelli, A., Chouliaras, G. C., & Goldenberg, D. (2025). *Maturity Framework for Enhancing Machine Learning Quality* (No. arXiv:2502.15758). arXiv. /arXiv.2502. 15758DOI: 10.1145/3711896.3737246

Centenaro, M., Vangelista, L., Zanella, A., & Zorzi, M. (2016). Long-range communications in unlicensed bands: The rising stars in the IoT and smart city scenarios. *IEEE Wireless Communications*, *23*(5), 60–67. DOI: 10.1109/MWC.2016.7721743

Chang, M., Ding, Z., Cai, Z., Zhao, Z., & Yuan, L. (2022). Prediction of Evolution Behaviors of Transportation Hubs Based on Spatiotemporal Neural Network. *IEEE Transactions on Intelligent Transportation Systems*, *23*(7), 9171–9183. Advance online publication. DOI: 10.1109/TITS.2021.3091708

Chang, Z., Liu, S., Xiong, X., Cai, Z., & Tu, G. (2021). A Survey of Recent Advances in Edge-Computing-Powered Artificial Intelligence of Things. *IEEE Internet of Things Journal*, *8*(18), 13849–13875. DOI: 10.1109/JIOT.2021.3088875

Chatterjee, A., Prinz, A., Riegler, M. A., & Meena, Y. K. (2023). An automatic and personalized recommendation modelling in activity eCoaching with deep learning and ontology. *Scientific Reports*, *13*(1), 10182. DOI: 10.1038/s41598-023-37233-7 PMID: 37349483

Cheela, V. R. S., Ranjan, V. P., Goel, S., John, M., & Dubey, B. (2021). Pathways to sustainable waste management in Indian Smart Cities. *Journal of Urban Management*, *10*(4), 419–429. DOI: 10.1016/j.jum.2021.05.002

Chelloug, S. A., & El-Zawawy, M. A. (2017). Middleware for internet of things: Survey and challenges. *Intelligent Automation & Soft Computing*, •••, 1–9. DOI: 10.1080/10798587.2017.1290328

Chen, Y. (2024, October). *Video behavior in intelligent video surveillance system based on deep learning*. In *IET Conference Proceedings CP895* (Vol. 2024, No. 19, pp. 136–140). Stevenage, UK: The Institution of Engineering and Technology.

Chen, C., Liu, Y., Sun, X., Di Cairano-Gilfedder, C., & Titmus, S. (2021). An integrated deep learning-based approach for automobile maintenance prediction with GIS data. *Reliability Engineering & System Safety*, *216*, 107919. Advance online publication. DOI: 10.1016/j.ress.2021.107919

Cheng, C., Messerschmidt, L., Bravo, I., Waldbauer, M., Bhavikatti, R., Schenk, C., Grujic, V., Model, T., Kubinec, R., & Barceló, J. (2024). A General Primer for Data Harmonization. In *Scientific Data* (Vol. 11, Issue 1). DOI: 10.1038/s41597-024-02956-3

Chen, G., & Zhang, J. W. (2024). Intelligent transportation systems: Machine learning approaches for urban mobility in smart cities. *Sustainable Cities and Society, 107*, 105369. DOI: 10.1016/j.scs.2024.105369

Chen, J., Zhu, R., & O'Donnell, T.. (2018). Smart transformer and low frequency transformer comparison on power delivery characteristics in the power system. In *AEIT International Annual Conference* (pp. 1–6). IEEE. DOI: 10.23919/AEIT.2018.8577375

Choudhry, N., Abawajy, J., Huda, S., & Rao, I. (2023). A comprehensive survey of machine learning methods for surveillance videos anomaly detection. *IEEE Access : Practical Innovations, Open Solutions, 11*, 114680–114713. DOI: 10.1109/ACCESS.2023.3321800

Chourabi, H., Nam, T., Walker, S., Gil-Garcia, J. R., Mellouli, S., Nahon, K., Pardo, T. A., & Scholl, H. J. (2012). Understanding smart cities: An integrative framework. In *Proceedings of the 45th Hawaii International Conference on System Sciences* (pp. 2289–2297). IEEE. https://doi.org/DOI: 10.1109/HICSS.2012.615

Cob-Parro, A. C., Lalangui, Y., & Lazcano, R. (2024). Fostering agricultural transformation through AI: An open-source AI architecture exploiting the MLOps paradigm. *Agronomy (Basel), 14*(2), 259. DOI: 10.3390/agronomy14020259

Cohen, A. R., Chen, G., Berger, E. M., Warrier, S., Lan, G., Grubert, E., Dellaert, F., & Chen, Y. (2022). Dynamically Controlled Environment Agriculture: Integrating Machine Learning and Mechanistic and Physiological Models for Sustainable Food Cultivation. *ACS ES&T Engineering, 2*(1), 3–19. DOI: 10.1021/acsestengg.1c00269

Comi, A., Sassano, M., & Valentini, A. (2022). Monitoring and controlling real-time bus services: A reinforcement learning procedure for eliminating bus bunching. *Transportation Research Procedia, 62*, 302–309. DOI: 10.1016/j.trpro.2022.02.038

Costa, G., Forestiero, A., & Ortale, R. (2023). Rule-based detection of anomalous patterns in device behavior for explainable IoT security. *IEEE Transactions on Services Computing, 16*(6), 4514–4525. DOI: 10.1109/TSC.2023.3327822

Cottrill, C. D. (2020). MaaS surveillance: Privacy considerations in mobility as a service. *Transportation Research Part A, Policy and Practice, 131*, 50–57. DOI: 10.1016/j.tra.2019.09.026

D'Amico, G., Szopik-Depczyńska, K., Dembińska, I., & Ioppolo, G. (2021). Smart and sustainable logistics of Port cities: A framework for comprehending enabling factors, domains and goals. *Sustainable Cities and Society*, *69*, 102801. DOI: 10.1016/j.scs.2021.102801

Dandotiya, M., Goyal, N. K., Yadav, P., Sankhla, K., & Kumari, M. (2026). GEN-ALYTICS: A deep dive into real-time insights via generative AI. In Elbaghazaoui, B., Amnai, M., & Gherabi, N. (Eds.), *Generative AI-powered data architectures: From governance to autonomous analytics* (pp. 105–134). IGI Global Scientific Publishing., DOI: 10.4018/979-8-3373-5616-7.ch005

Dandotiya, M., Sharma, S., & Goyal, N. K. (2024). ANFIS and LSTM techniques based model for time-series forecasting. In *Proceedings of the 2024 International Conference on Augmented Reality, Intelligent Systems, and Industrial Automation (ARIIA)* (pp. 1–6). IEEE. DOI: 10.1109/ARIIA63345.2024.11051439

Dantas, P. V., Sabino da Silva, W.Jr, Cordeiro, L. C., & Carvalho, C. B. (2024). A comprehensive review of model compression techniques in machine learning. *Applied Intelligence*, *54*(22), 11804–11844. DOI: 10.1007/s10489-024-05747-w

Das, D., Manojkumar, R., Kumar, C., & Ganguly, S. (2022). Power loss minimization in smart transformer enabled low voltage islanded meshed hybrid microgrid. *IEEE Access : Practical Innovations, Open Solutions*, *10*, 123259–123270. DOI: 10.1109/ACCESS.2022.3224001

Dávila-Montero, S., Dana-Lê, J. A., Bente, G., Hall, A. T., & Mason, A. J. (2021). Review and challenges of technologies for real-time human behavior monitoring. *IEEE Transactions on Biomedical Circuits and Systems*, *15*(1), 2–28. DOI: 10.1109/TBCAS.2021.3060617 PMID: 33606635

de la Rúa Martínez, J. (2020). *Scalable architecture for automating machine learning model monitoring*. https://www.diva-portal.org/smash/record.jsf?pid=diva2:1464577

De Micco, F., Di Palma, G., Ferorelli, D., De Benedictis, A., Tomassini, L., Tambone, V., Cingolani, M., & Scendoni, R. (2025). Artificial intelligence in healthcare: Transforming patient safety with intelligent systems—A systematic review. *Frontiers in Medicine*, *11*, 1522554. DOI: 10.3389/fmed.2024.1522554 PMID: 39845830

Department of Economic and Social Affairs. (2014). *World urbanization prospects: The 2014 revision, highlights*. United Nations.

Dhingra, S., Madda, R. B., Patan, R., Jiao, P., Barri, K., & Alavi, A. H. (2021). Internet of things-based fog and cloud computing technology for smart traffic monitoring. *Internet of Things : Engineering Cyber Physical Human Systems, 14*, 100175. DOI: 10.1016/j.iot.2020.100175

Diao, R., Wang, Z., Shi, D., Chang, Q., Duan, J., & Zhang, X. (2019, August). Autonomous voltage control for grid operation using deep reinforcement learning. In *2019 IEEE Power & Energy Society General Meeting (PESGM)* (pp. 1–5). IEEE. DOI: 10.1109/PESGM40551.2019.8973924

Dilek, E., & Dener, M. (2025). An overview of transformers for video anomaly detection. *Neural Computing & Applications, 37*(22), 1–33. DOI: 10.1007/s00521-025-11218-1

Dong, S., Cao, J., Flynn, D., & Fan, Z. (2022). Cybersecurity in smart local energy systems: Requirements, challenges, and standards. *Energy Informatics, 5*(1), 9. DOI: 10.1186/s42162-022-00195-7

Duan, J., Shi, D., Diao, R., Li, H., Wang, Z., Zhang, B., & Yi, Z. (2019). Deep-reinforcement-learning-based autonomous voltage control for power grid operations. *IEEE Transactions on Power Systems, 35*(1), 814–817. DOI: 10.1109/TPWRS.2019.2941134

Duong, H. T., Le, V. T., & Hoang, V. T. (2023). Deep learning-based anomaly detection in video surveillance: A survey. *Sensors (Basel), 23*(11), 5024. DOI: 10.3390/s23115024 PMID: 37299751

Du, Y. (2020). An anomaly detection method using deep convolution neural network for vision image of robot. *Multimedia Tools and Applications, 79*(13), 9629–9642. DOI: 10.1007/s11042-020-08684-1

Eboli, L., & Mazzulla, G. (2021). Customer satisfaction as a measure of service quality in public transport planning. *Int. Encycl. Transp, 13816*, 220–224. DOI: 10.1016/B978-0-08-102671-7.10643-8

Eckhoff, D., & Wagner, I. (2017). Privacy in the smart city—Applications, technologies, challenges, and solutions. *IEEE Communications Surveys and Tutorials, 20*(1), 489–516. DOI: 10.1109/COMST.2017.2748998

Eldeeb, H., Maher, M., Elshawi, R., & Sakr, S. (2024). AutoMLBench: A comprehensive experimental evaluation of automated machine learning frameworks. *Expert Systems with Applications, 243*, 122877. DOI: 10.1016/j.eswa.2023.122877

Elrayyah, A. Y., Wanik, M. Z. C., & Bouselham, A. (2017). Simplified approach to analyze voltage rise in LV systems with PV installations using equivalent power systems diagrams. *IEEE Transactions on Power Delivery*, *32*(4), 2140–2149. DOI: 10.1109/TPWRD.2016.2594242

El-Rewini, Z., Sadatsharan, K., Selvaraj, D. F., Plathottam, S. J., & Ranganathan, P. (2020). Cybersecurity challenges in vehicular communications. *Vehicular Communications*, *23*, 100214. DOI: 10.1016/j.vehcom.2019.100214

Errezgouny, A., Chater, Y., González, C. D. B., & Cherkaoui, A. (2025). An integrated deep learning approach for predictive vehicle maintenance. *Decision Analytics Journal*, *16*, 100597. DOI: 10.1016/j.dajour.2025.100597

Esmaeili, F., Cassie, E., Nguyen, H. P. T., Plank, N. O., Unsworth, C. P., & Wang, A. (2023). Anomaly detection for sensor signals utilizing deep learning autoencoder-based neural networks. *Bioengineering (Basel, Switzerland)*, *10*(4), 405. DOI: 10.3390/bioengineering10040405 PMID: 37106591

Fährmann, D., Martín, L., Sánchez, L., & Damer, N. (2024). Anomaly detection in smart environments: A comprehensive survey. *IEEE Access : Practical Innovations, Open Solutions*, *12*, 64006–64049. DOI: 10.1109/ACCESS.2024.3395051

Fang, W., Chen, Y., & Xue, Q. (2021). Survey on research of RNN-based spatio-temporal sequence prediction algorithms. *Journal of Big Data*, *3*(3), 97–110. DOI: 10.32604/jbd.2021.016993

Farahnakian, F., Nicolas, F., Farahnakian, F., Nevalainen, P., Sheikh, J., Heikkonen, J., & Raduly-Baka, C. (2023). A comprehensive study of clustering-based techniques for detecting abnormal vessel behavior. *Remote Sensing (Basel)*, *15*(6), 1477. DOI: 10.3390/rs15061477

Farooqi, A., Das, D., Tripathi, M., Dutta, R., Goyal, N. K., & Sheikh, M. F. (2026). Navigating ethical frontiers in AI-driven geriatric care. In Stolakis, K., & Efthymiou, I. (Eds.), *AI for geriatric care in an aging society: Ethical, clinical, and policy challenges* (pp. 243–266). IGI Global Scientific Publishing., DOI: 10.4018/979-8-3373-5462-0.ch009

Fernandez, V., & Pérez, V. (2024). A new approach to interoperability within the smart city based on time series-embedded adaptive traffic prediction modelling. *Networks and Spatial Economics*, •••, 1–16. DOI: 10.1007/s11067-024-09662-y

ForeScout Technologies. (2016). Smart cities require smarter cybersecurity.

Freitag, M., Foster, G., Grangier, D., Ratnakar, V., Tan, Q., & Macherey, W. (2021). Experts, errors, and context: A large-scale study of human evaluation for machine translation. *Transactions of the Association for Computational Linguistics*, *9*, 1460–1474. DOI: 10.1162/tacl_a_00437

Furuhata, M., Dessouky, M., Ordóñez, F., Brunet, M. E., Wang, X., & Koenig, S. (2013). Ridesharing: The state-of-the-art and future directions. *Transportation Research Part B: Methodological*, *57*, 28–46. DOI: 10.1016/j.trb.2013.08.012

Gandomi, A., & Haider, M. (2015). Beyond the hype: Big data concepts, methods, and analytics. *International Journal of Information Management*, *35*(2), 137–144. DOI: 10.1016/j.ijinfomgt.2014.10.007

Garlik, B. (2022). Energy centers in a smart city as a platform for the application of artificial intelligence and the Internet of Things. *Applied Sciences (Basel, Switzerland)*, *12*(7), 3386. DOI: 10.3390/app12073386

Garroussi, Z., Legrain, A., Gambs, S., Gautrais, V., & Sansò, B. (2025). A systematic review of data privacy in Mobility as a Service (MaaS). *Transportation Research Interdisciplinary Perspectives*, *31*, 101254. DOI: 10.1016/j.trip.2024.101254

Gharaibeh, A., Salahuddin, M. A., Hussini, S. J., Khreishah, A., Khalil, I., Guizani, M., & Al-Fuqaha, A. (2017). Smart cities: A survey on data management, security, and enabling technologies. *IEEE Communications Surveys and Tutorials*, *19*(4), 2456–2501. DOI: 10.1109/COMST.2017.2736886

Gheorghe, C., & Soica, A. (2025). Revolutionizing Urban Mobility: A Systematic Review of AI, IoT, and Predictive Analytics in Adaptive Traffic Control Systems for Road Networks. *Electronics (Basel)*, *14*(4). Advance online publication. DOI: 10.3390/electronics14040719

Gkiotsalitis, K., & Cats, O. (2021). Public transport planning adaption under the COVID-19 pandemic crisis: Literature review of research needs and directions. *Transport Reviews*, *41*(3), 374–392. DOI: 10.1080/01441647.2020.1857886

Goodman, B., & Flaxman, S. (2017). European Union Regulations on Algorithmic Decision Making and a "Right to Explanation.". *AI Magazine*, *38*(3), 50–57. DOI: 10.1609/aimag.v38i3.2741

Goyal, N. K., Goyal, N. K., Kumar, S., Dutta, R., Kothari, V., & Sinha, S. K. (2026). Empowering elderly care through artificial intelligence. In Stolakis, K., & Efthymiou, I. (Eds.), *AI for geriatric care in an aging society* (pp. 1–28). IGI Global Scientific Publishing., DOI: 10.4018/979-8-3373-5462-0.ch001

Goyal, N. K., Gulati, K., Avasthi, A., Kapoor, S. K., Keswani, B., & Gupta, S. (2026). Securing the imagination: Security and privacy challenges in generative AI systems. In Elbaghazaoui, B., Amnai, M., & Gherabi, N. (Eds.), *Generative AI-powered data architectures* (pp. 239–266). IGI Global Scientific Publishing., DOI: 10.4018/979-8-3373-5616-7.ch010

Guidry, J., Sohrab, F., Gottumukkala, R., Katragadda, S., & Gabbouj, M. (2023, December). One-class classification for intrusion detection on vehicular networks. In *2023 IEEE Symposium Series on Computational Intelligence (SSCI)* (pp. 1176–1182). IEEE. DOI: 10.1109/SSCI52147.2023.10371899

Guo, H., Ren, J., Zhang, D., Zhang, Y., & Hu, J. (2018). A scalable and manageable IoT architecture based on transparent computing. *Journal of Parallel and Distributed Computing, 118*, 5–13. DOI: 10.1016/j.jpdc.2017.07.003

Guo, S., Xiang, B., Xia, X., Yan, Z., & Li, Y. (2020). Blockchain and federated learning based data security sharing mechanism over smart city. In *Research Square*. DOI: 10.21203/rs.3.rs-104012/v1

Gupta, B. B., & Wu, J. (2025). Integration of IoT with robotics and drones. In *AI developments for industrial robotics and intelligent drones* (pp. 33–54). IGI Global Scientific Publishing., DOI: 10.4018/979-8-3693-2707-4.ch003

Gupta, B., Awasthi, S., Gupta, R., Ram, L., Kumar, P., Rohit Prasad, B., & Agarwal, S. (2018). Taxi Travel Time Prediction Using Ensemble-Based Random Forest and Gradient Boosting Model. In Rajsingh, E. B., Veerasamy, J., Alavi, A. H., & Peter, J. D. (Eds.), *Advances in Big Data and Cloud Computing* (Vol. 645, pp. 63–78). Springer Singapore., DOI: 10.1007/978-981-10-7200-0_6

Gupta, N., Mujumdar, S., Patel, H., Masuda, S., Panwar, N., Bandyopadhyay, S., Mehta, S., Guttula, S., Afzal, S., Sharma Mittal, R., & Munigala, V. (2021). Data Quality for Machine Learning Tasks. *Proceedings of the ACM SIGKDD International Conference on Knowledge Discovery and Data Mining*. DOI: 10.1145/3447548.3470817

Habibi, M. R., Guerrero, J. M., & Vasquez, J. C. (2024). Artificial intelligence for cybersecurity monitoring of cyber-physical power electronic converters: A DC/DC power converter case study. *Scientific Reports, 14*(1), 22072. DOI: 10.1038/s41598-024-72286-2 PMID: 39333625

Hallaji, S. M., Fang, Y., & Winfrey, B. K. (2022). Predictive maintenance of pumps in civil infrastructure: State-of-the-art, challenges and future directions. In *Automation in Construction* (Vol. 134). DOI: 10.1016/j.autcon.2021.104049

Hamza, M. A., & Ullah, F. (2025). AI-Driven VANETs: Integrating Deep Learning and Reinforcement Learning for Adaptive Urban Mobility in Smart Cities. In Maryam, H., Malik, M. M., Khan, I. U., & Gupta, S. K. (Eds.), *AI-Driven Transportation Systems: Real-Time Applications and Related Technologies. Information Systems Engineering and Management* (Vol. 62). Springer., DOI: 10.1007/978-3-031-98349-8_18

Hao, Q., Nazir, S., Li, M., Khan, H., Wang, L., & Ahmad, S. (2021). AI-Enabled Sensing and Decision-Making for IoT Systems. Complex., 2021, 6616279:1-6616279:9. DOI: 10.1155/2021/6616279

Haque, M. M. (2025). Systematic review on the impact of AI-enhanced traffic simulation on U.S. urban mobility and safety. ASRC Procedia. *Global Perspectives in Science and Scholarship*, *1*(1), 833–861. DOI: 10.63125/jj96yd66

Harish, V. S. K. V., Gupta, S., Bhatt, J. G., & Bansal, M. (2025). International standards, regulations, and best practices for cyber security of smart grid. In *Cyber security solutions for protecting and building the future smart grid* (pp. 321–348). Elsevier. DOI: 10.1016/B978-0-443-14066-2.00010-4

Harrison, C., Eckman, B., Hamilton, R., Hartswick, P., Kalagnanam, J., Paraszczak, J., & Williams, P. (2010). Foundations for smarter cities. *IBM Journal of Research and Development*, *54*(4), 1–16. DOI: 10.1147/JRD.2010.2048257

Haseeb, K., Saba, T., Rehman, A., Abbas, N., & Kim, P. W. (2025). AI-driven IoT-fog analytics interactive smart system with data protection. *Expert Systems: International Journal of Knowledge Engineering and Neural Networks*, *42*(1), e13573. DOI: 10.1111/exsy.13573

He, D., Zeadally, S., Kumar, N., & Lee, J.-H. (2017). Anonymous authentication for wireless body area networks with provable security. *IEEE Systems Journal*, *11*(4), 2590–2601. DOI: 10.1109/JSYST.2016.2544805

Herath, H. M. K. K. M. B., & Mittal, M. (2022). Adoption of artificial intelligence in smart cities: A comprehensive review. *International Journal of Information Management Data Insights*, *2*(1), 100076. DOI: 10.1016/j.jjimei.2022.100076

Hesar, F. F., & Foing, B. (2024). *Evaluating Classification Algorithms: Exoplanet Detection using Kepler Time Series Data* (No. arXiv:2402.15874). arXiv. https://doi.org//arXiv.2402.15874DOI: 10.48550

Heymann, H., Kies, A. D., Frye, M., Schmitt, R. H., & Boza, A. (2022). Guideline for Deployment of Machine Learning Models for Predictive Quality in Production. *Procedia CIRP*, *107*, 815–820. DOI: 10.1016/j.procir.2022.05.068

Hilmy, A., & He, B. E. (2025). Next-generation smart inverters: Bridging AI, cybersecurity, and policy gaps for sustainable energy transition. *Technologies*, *13*(4), 136. DOI: 10.3390/technologies13040136

Hinton, G. E., & Salakhutdinov, R. R. (2006). Reducing the dimensionality of data with neural networks. *Science*, *313*(5786), 504–507. DOI: 10.1126/science.1127647 PMID: 16873662

Hrusto, A., Runeson, P., Engström, E., & Ohlsson, M. C. (2024). Advancing Software Monitoring: An Industry Survey on ML-Driven Alert Management Strategies. *2024 50th Euromicro Conference on Software Engineering and Advanced Applications (SEAA)*, 435–442. DOI: 10.1109/SEAA64295.2024.00073

Hussein, W. N., Hussain, H. N., Hussain, H. N., & Mallah, A. Q. (2023). A Deployment Model for IoT Devices Based on Fog Computing for Data Management and Analysis. *Wireless Personal Communications*, •••, 1–13. DOI: 10.1007/s11277-023-10168-y

Hutchinson, J. W. (1974). Plastic buckling. *Advances in Applied Mechanics*, *14*, 67–144. DOI: 10.1016/S0065-2156(08)70031-0

Iddianozie, C., & McArdle, G. (2021). Towards Robust Representations of Spatial Networks Using Graph Neural Networks. *Applied Sciences (Basel, Switzerland)*, *11*(15), 6918. DOI: 10.3390/app11156918

Ihrig, J., Jochem, P., & Szimba, E. (2024). How to apply the four-step model for 150,000 travel zones: The HIPAT model. *Transport Policy*, *155*, 150–160. DOI: 10.1016/j.tranpol.2024.06.014

Iyer, L. S. (2021). AI enabled applications towards intelligent transportation. *Transportation Engineering*, *5*, 100083. DOI: 10.1016/j.treng.2021.100083

Jabakumar, A. K. (2023). Edge-enabled smart traffic management system: An IoT implementation for urban mobility. *Research journal of computer systems and engineering, 4*(2), 160-173.

Jabeen, S., Li, X., Amin, M. S., Bourahla, O., Li, S., & Jabbar, A. (2023). A review on methods and applications in multimodal deep learning. *ACM Transactions on Multimedia Computing Communications and Applications*, *19*(2s), 1–41. DOI: 10.1145/3545572

Jagatheesaperumal, S. K., Bibri, S. E., & Huang, J.. (2024). Artificial intelligence of things for smart cities: Advanced solutions for enhancing transportation safety. *Computational Urban Science*, *4*, 10. DOI: 10.1007/s43762-024-00120-6

Jain, A., Patel, H., Nagalapatti, L., Gupta, N., Mehta, S., Guttula, S., Mujumdar, S., Afzal, S., Sharma Mittal, R., & Munigala, V. (2020). Overview and Importance of Data Quality for Machine Learning Tasks. *Proceedings of the ACM SIGKDD International Conference on Knowledge Discovery and Data Mining*. DOI: 10.1145/3394486.3406477

Jiang, J., Wang, C., Chattopadhyay, S., & Zhang, W. (2020). Road Context-Aware Intrusion Detection System for Autonomous Cars. In Zhou, J., Luo, X., Shen, Q., & Xu, Z. (Eds.), *Information and Communications Security* (Vol. 11999, pp. 124–142). Springer International Publishing., DOI: 10.1007/978-3-030-41579-2_8

Jing, Q., Vasilakos, A. V., Wan, J., Lu, J., & Qiu, D. (2014). Security of the Internet of Things: Perspectives and challenges. *Wireless Networks*, *20*(8), 2481–2501. DOI: 10.1007/s11276-014-0761-7

Jittrapirom, P., Caiati, V., Feneri, A. M., Ebrahimigharehbaghi, S., Alonso-González, M. J., & Narayan, J. (2017). Mobility as a service: A critical review of definitions, assessments of schemes, and key challenges. *Urban Planning*, *2*(2), 13–25. DOI: 10.17645/up.v2i2.931

Joubert, J. W., & De Waal, A. (2020). Activity-based travel demand generation using Bayesian networks. *Transportation Research Part C, Emerging Technologies*, *120*, 102804. DOI: 10.1016/j.trc.2020.102804

Kairouz, P., McMahan, B., Song, S., Thakkar, O., Thakurta, A., & Xu, Z. (2021, July). Practical and private (deep) learning without sampling or shuffling. In *International Conference on Machine Learning* (pp. 5213-5225). PMLR.

Kalusivalingam, A. K., Sharma, A., Patel, N., & Singh, V. (2020). Enhancing Process Automation Using Reinforcement Learning and Deep Neural Networks. *International Journal of Artificial Intelligence and Machine Learning*, *1*(3). https://www.cognitivecomputingjournal.com/index.php/IJAIML-V1/article/view/48

Kankanhalli, A., Charalabidis, Y., & Mellouli, S. (2019). IoT and AI for Smart Government: A Research Agenda. *Government Information Quarterly*, *36*(2), 304–309. DOI: 10.1016/j.giq.2019.02.003

Karthika, D. (2023). A study on artificial intelligence for monitoring smart environments. *Materials Today: Proceedings*, *80*, 2009–2013. DOI: 10.1016/j.matpr.2021.06.046

Katragadda, S. R., Tanikonda, A., Peddinti, S. R., & Pandey, B. K. (2021). Machine Learning-Enhanced Root Cause Analysis for Accelerated Incident Resolution in Complex Systems. *Journal of Science and Technology*, *2*(4), 253–275.

Khan, A. A., Arshad, S., & Mohsin, M. (2014). Population growth and its impact on urban expansion: A case study of Bahawalpur, Pakistan. *Universal Journal of Geoscience*, *2*(8), 229–241. DOI: 10.13189/ujg.2014.020801

Khisty, C. J., & Lall, B. K. (2003). *Fundamentals of transportation engineering*. Erlangga.

Kim, D., Jeon, S., Shin, J., & Taek Seo, J. (2023). Design the IoT botnet defense process for cybersecurity in smart city. *Intelligent Automation & Soft Computing*, *37*(3), 2979–2997. DOI: 10.32604/iasc.2023.040019

Kim, Y., Park, M., Ka, D., & Lee, C. (2023). Real-Time Traffic Forecast System for the Accident-Prone Large-Scale Transportation Network in the Seoul Metropolitan Area. *KSCE Journal of Civil Engineering*, *27*(7), 3085–3096. DOI: 10.1007/s12205-023-0349-9

Kitchin, R. (2016). *Getting smarter about smart cities: Improving data privacy and data security*. Maynooth University.

Klemm, C., & Wiese, F. (2022). Indicators for the optimization of sustainable urban energy systems based on energy system modeling. *Energy, Sustainability and Society*, *12*(1), 3. Advance online publication. DOI: 10.1186/s13705-021-00323-3

Kong, J., Huang, J., Yu, H., Deng, H., Gong, J., & Chen, H. (2019). RNN-based default logic for route planning in urban environments. *Neurocomputing*, *338*, 307–320. DOI: 10.1016/j.neucom.2019.02.012

Kuligowski, K., Konkol, I., Świerczek, L., Chojnacka, K., Cenian, A., & Szufa, S. (2023). Evaluation of kitchen waste recycling as organic N-fertiliser for sustainable agriculture under cool and warm seasons. *Sustainability (Basel)*, *15*(10), 7997. DOI: 10.3390/su15107997

Kumar, C., & Liserre, M. (2015). Operation and control of smart transformer for improving performance of medium voltage power distribution system. In *Proc. IEEE 6th Int. Symp. on Power Electronics for Distributed Generation Systems* (pp. 1–6). IEEE. DOI: 10.1109/PEDG.2015.7223092

Kumar, C., Manojkumar, R., Ganguly, S., & Liserre, M. (2020). Power loss minimization in smart transformer based meshed hybrid distribution network. In *Proceedings of the IECON 2020 – 46th Annual Conference of the IEEE Industrial Electronics Society* (pp. 1670–1675). IEEE. DOI: 10.1109/IECON43393.2020.9254324

Kumar, C., Zhu, R., Buticchi, G., & Liserre, M. (2018). Sizing and SoC management of a smart-transformer-based energy storage system. *IEEE Transactions on Industrial Electronics*, *65*(8), 6709–6718. DOI: 10.1109/TIE.2017.2784389

Kumari, M., & Goyal, N. K. (2026). Artificial intelligence in the cryptocurrency ecosystem: Crime mechanisms, detection strategies, and legal perspectives. In Lin, L. (Ed.), *Innovations in cryptocrime and financial fraud* (pp. 245–282). IGI Global Scientific Publishing., DOI: 10.4018/979-8-3373-0675-9.ch007

Kumari, M., Kumawat, R., Tripathi, M., Suganthi, K., Goyal, N. K., & Sheikh, M. F. (2026). Blockchain for transparency and traceability in supply chains. In Khan, M., Mehmood, T., & Rao, M. (Eds.), *Maximizing operational efficiency at the intersection of supply chain management and marketing* (pp. 35–64). IGI Global Scientific Publishing., DOI: 10.4018/979-8-3373-3109-6.ch002

Kumar, K., Kumar, M., & Das, P. (2024). Traffic congestion forecasting using multilayered deep neural network. *Transportation Letters*, *16*(6), 516–526. Advance online publication. DOI: 10.1080/19427867.2023.2207278

Kumar, R., Bhanu, M., Mendes-Moreira, J., & Chandra, J. (2024). Spatio-temporal predictive modeling techniques for different domains: A survey. *ACM Computing Surveys*, *57*(2), 1–42. DOI: 10.1145/3696661

Lai, X., Yang, Z., Xie, J., & Liu, Y. (2024). Reinforcement learning in transportation research: Frontiers and future directions. *Multimodal Transportation*, *3*(4), 100164. DOI: 10.1016/j.multra.2024.100164

Lalramsangi, V., Garg, Y. K., & Sharma, S. N. (2025). Route choices to access public open spaces in hill cities. *Environment & Urbanization Asia*, *16*(2), 1–17. DOI: 10.1177/09754253251388721

LeCun, Y., Bengio, Y., & Hinton, G. (2015). Deep learning. *nature*, *521*(7553), 436-444.

Lee, C. P., Leng, F. T. J., Habeeb, R. A. A., & Amanullah, M. A., & ur Rehman, M. H. (2022). Edge computing-enabled secure and energy-efficient smart parking: A review. *Microprocessors and Microsystems*, *93*, 104612. DOI: 10.1016/j.micpro.2022.104612

Leroux, S., Simoens, P., Lootus, M., Thakore, K., & Sharma, A. (2022). *TinyMLOps: Operational Challenges for Widespread Edge AI Adoption* (No. arXiv:2203.10923). arXiv. /arXiv.2203.10923DOI: 10.1109/IPDPSW55747.2022.00160

Leta, M. K., Demissie, T. A., & Tränckner, J. (2021). Modeling and prediction of land use land cover change dynamics based on land change modeler (Lcm) in nashe watershed, upper blue nile basin, Ethiopia. *Sustainability (Basel)*, *13*(7), 3740. Advance online publication. DOI: 10.3390/su13073740

Letourneau, S., Famili, F., & Matwin, S. (1999). Data mining to predict aircraft component replacement. *IEEE Intelligent Systems & their Applications*, *14*(6), 59–66. DOI: 10.1109/5254.809569

Lewis, P. G. (2015). Moral intuitions and smart growth: Why do liberals and conservatives view compact development so differently? *Journal of Urban Affairs, 37*(2), 87–108. *Scopus.* Advance online publication. DOI: 10.1111/juaf.12172

Li, B., & Xu, Q. (2023). A machine learning-assisted distributed optimization method for inverter-based Volt-VAR control in active distribution networks. *IEEE Transactions on Power Systems*, *39*(2), 2668–2681. DOI: 10.1109/TPWRS.2023.3279303

Li, C., Bai, L., Liu, W., Yao, L., & Travis Waller, S. (2021). Urban mobility analytics: A deep spatial–temporal product neural network for traveler attributes inference. *Transportation Research Part C, Emerging Technologies*, *124*, 102921. DOI: 10.1016/j.trc.2020.102921

Li, C., Bai, L., Yao, L., Waller, S. T., & Liu, W. (2023). A bibliometric analysis and review on reinforcement learning for transportation applications. *Transportmetrica. B, Transport Dynamics*, *11*(1), 2179461. DOI: 10.1080/21680566.2023.2179461

Lilhore, U. K., Imoize, A. L., Li, C. T., Simaiya, S., Pani, S. K., Goyal, N., Kumar, A., & Lee, C. C. (2022). Design and Implementation of an ML and IoT Based Adaptive Traffic-Management System for Smart Cities. *Sensors (Basel)*, *22*(8). Advance online publication. DOI: 10.3390/s22082908

Lingitz, L., Gallina, V., Ansari, F., Gyulai, D., Pfeiffer, A., Sihn, W., & Monostori, L. (2018). Lead time prediction using machine learning algorithms: A case study by a semiconductor manufacturer. *Procedia CIRP*, *72*, 1051–1056. DOI: 10.1016/j.procir.2018.03.148

Li, T., Sahu, A. K., Talwalkar, A., & Smith, V. (2020). Federated learning: Challenges, methods, and future directions. *IEEE Signal Processing Magazine*, *37*(3), 50–60. DOI: 10.1109/MSP.2020.2975749

Liu, J. C., Goetz, J., Sen, S., & Tewari, A. (2021). Learning from others without sacrificing privacy: Simulation comparing centralized and federated machine learning on mobile health data. *JMIR mHealth and uHealth*, *9*(3), e23728. DOI: 10.2196/23728 PMID: 33783362

Liu, L., & Zhang, Y. (2021). Smart environment design planning for smart city based on deep learning. *Sustainable Energy Technologies and Assessments*, *47*, 101425. DOI: 10.1016/j.seta.2021.101425

Liu, Y., Yuan, X., Xiong, Z., Kang, J., Wang, X., & Niyato, D. (2020). Federated learning for 6G communications: Challenges, methods, and future directions. *China Communications, 17*(9), 105–118. DOI: 10.23919/JCC.2020.09.009

Li, Y., Lin, Y., & Geertman, S. (2015). The development of smart cities in China. In *Proceedings of the 14th International Conference on Computer Urban Planning and Urban Management* (pp. 7–10).

Liyanage, S., Abduljabbar, R., Dia, H., & Tsai, P.-W. (2022). AI-based neural network models for bus passenger demand forecasting using smart card data. *Journal of Urban Management, 11*(3), 365–380. DOI: 10.1016/j.jum.2022.05.002

Li, Z., Zhang, Z., Shi, J., & Wu, D. (2019). Prediction of surface roughness in extrusion-based additive manufacturing with machine learning. *Robotics and Computer-integrated Manufacturing, 57*, 488–495. DOI: 10.1016/j.rcim.2019.01.004

Li, Z., Zhu, Y., & Van Leeuwen, M. (2023). A survey on explainable anomaly detection. *ACM Transactions on Knowledge Discovery from Data, 18*(1), 1–54.

Lodhi, A. S., Jaiswal, A., & Sharma, S. N. (2024). Assessing bus users satisfaction using discrete choice models: A case of Bhopal. *Innovative Infrastructure Solutions, 9*(11), 437. DOI: 10.1007/s41062-024-01652-w

Loftus, T. J., Ruppert, M. M., Shickel, B., Ozrazgat-Baslanti, T., Balch, J. A., Efron, P. A., Upchurch, G. R.Jr, Rashidi, P., Tignanelli, C., Bian, J., & Bihorac, A. (2022). Federated learning for preserving data privacy in collaborative healthcare research. *Digital Health, 8*, 20552076221134456. DOI: 10.1177/20552076221134455 PMID: 36325438

Loulaki, M., Kiourtis, A., Mavrogiorgou, A., & Kyriazis, D. (2024). Addressing complex data integration and harmonization scenarios. *2024 E-Health and Bioengineering Conference (EHB)*, 1–4.

Lu, Z., Lv, W., Cao, Y., Xie, Z., Peng, H., & Du, B. (2020). LSTM variants meet graph neural networks for road speed prediction. *Neurocomputing, 400*, 34–45. DOI: 10.1016/j.neucom.2020.03.031

Ma, C., & Liu, T. (2025). Survey of short-term traffic flow prediction based on LSTM. *International Journal of Modern Physics C, 36*(02), 2450177. DOI: 10.1142/S0129183124501778

Madigan, R., Nordhoff, S., Fox, C., Ezzati Amini, R., Louw, T., Wilbrink, M., Schieben, A., & Merat, N. (2019). Understanding interactions between Automated Road Transport Systems and other road users: A video analysis. *Transportation Research Part F: Traffic Psychology and Behaviour*, *66*, 196–213. DOI: 10.1016/j.trf.2019.09.006

Mahida, A. (2024). Integrating Observability with DevOps Practices in Financial Services Technologies: A Study on Enhancing Software Development and Operational Resilience. *International Journal of Advanced Computer Science and Applications*, *15*(7). Advance online publication. https://www.researchgate.net/profile/Ankur-Mahida-2/publication/382856176_Integrating_Observability_with_DevOps_Practices_in_Financial_Services_Technologies_A_Study_on_Enhancing_Software_Development_and_Operational_Resilience/links/66b7ec158f7e1236bc4d4548/Integrating-Observability-with-DevOps-Practices-in-Financial-Services-Technologies-A-Study-on-Enhancing-Software-Development-and-Operational-Resilience.pdf. DOI: 10.14569/IJACSA.2024.0150701

Mahomed, A. S., & Saha, A. K. (2025). Unleashing the Potential of 5G for Smart Cities: A Focus on Real-Time Digital Twin Integration. *Smart Cities*, *8*(2). Advance online publication. DOI: 10.3390/smartcities8020070

Manojkumar, R., Hrishikesan, V. M., Kumar, C., & Ganguly, S. (2019). Voltage control using smart transformer for increasing photovoltaic penetration in a distribution grid. In *2019 20th International Conference on Intelligent System Application to Power Systems (ISAP)* (pp. 1–7). IEEE. DOI: 10.1109/ISAP48318.2019.9065949

Manojkumar, R., Aruna, P., Nabieha, N., Niteesha, M., Chandana, R., & Nithya, S. (2024). Voltage control in low voltage distribution systems using smart power converter. In *2024 International Conference on Electrical Electronics and Computing Technologies (ICEECT)* (pp. 1–5). IEEE. DOI: 10.1109/ICEECT61758.2024.10739011

Manojkumar, R., Kumar, C., Ganguly, S., & Liserre, M. (2020). Voltage control using smart transformer via dynamic optimal setpoints and limit tolerance in a residential distribution network with PV sources. *IET Generation, Transmission & Distribution*, *14*(22), 5143–5151. DOI: 10.1049/iet-gtd.2020.0213

Manousakis, N. M., & Korres, G. N. (2021). Application of State Estimation in Distribution Systems with Embedded Microgrids. *Energies*, *14*(23), 7933. DOI: 10.3390/en14237933

Masters, C. L. (2002). Voltage rise: The big issue when connecting embedded generation to long 11 kV overhead lines. *Power Engineering Journal*, *16*(1), 5–12. DOI: 10.1049/pe:20020101

Mastorakis, G., Mavromoustakis, C. X., Batalla, J. M., & Pallis, E. (Eds.). (2020). *Convergence of artificial intelligence and the Internet of Things*. Springer International Publishing. DOI: 10.1007/978-3-030-44907-0

Ma, Y., Ping, K., Wu, C., Chen, L., Shi, H., & Chong, D. (2020). Artificial Intelligence powered Internet of Things and smart public service. *Library Hi Tech*, *38*(1), 165–179. DOI: 10.1108/LHT-12-2017-0274

Mazhar, T., Malik, M. A., Mohsan, S. A. H., Li, Y., Haq, I., Ghorashi, S., Karim, F. K., & Mostafa, S. M. (2023). Quality of service (QoS) performance analysis in a traffic engineering model for next-generation wireless sensor networks. *Symmetry*, *15*(2), 513. DOI: 10.3390/sym15020513

McAsey, M., & Mou, L. (n.d.). *Convergence of the forward-backward sweep method in optimal control*. University of Illinois. DOI: 10.1007/s10589-011-9454-7

Mekrache, A., Bradai, A., Moulay, E., & Dawaliby, S. (2022). Deep reinforcement learning techniques for vehicular networks: Recent advances and future trends towards 6G. *Vehicular Communications*, *33*, 100398. DOI: 10.1016/j.vehcom.2021.100398

Midi, D., Rullo, A., Mudgerikar, A., & Bertino, E. (2017). Kalis—A system for knowledge-driven adaptable intrusion detection for the Internet of Things. In *Proceedings of the IEEE International Conference on Distributed Computing Systems* (pp. 656–666). https://doi.org/DOI: 10.1109/ICDCS.2017.104

Miller, E. J. (2020). Travel demand models, the next generation. In *Mapping the Travel Behavior Genome* (pp. 29–46). Elsevier., DOI: 10.1016/B978-0-12-817340-4.00003-6

Miraftabzadeh, S. M., Longo, M., Foiadelli, F., Pasetti, M., & Igual, R. (2021). Advances in the Application of Machine Learning Techniques for Power System Analytics: A Survey. *Energies*, *14*(16), 16. Advance online publication. DOI: 10.3390/en14164776

Mirza, A., & Iqbal, R. (2024). Harnessing AI in IT Operations: Transforming Automation and Efficiency. *Asian American Research Letters Journal*, *1*(9), 22–34.

Mishra, S., Seth, S., Jain, S., Pant, V., Parikh, J., Jain, R., & Islam, S. M. (2024, May). Image caption generation using vision transformer and gpt architecture. In *2024 2nd International Conference on Advancement in Computation & Computer Technologies (InCACCT)* (pp. 1-6). IEEE. DOI: 10.1109/InCACCT61598.2024.10551257

Mishra, P. K., Mihailidis, A., & Khan, S. S. (2024). Skeletal video anomaly detection using deep learning: Survey, challenges, and future directions. *IEEE Transactions on Emerging Topics in Computational Intelligence*, *8*(2), 1073–1085. DOI: 10.1109/TETCI.2024.3358103

Mitchell, D., & Coles, C. (2003). The ultimate competitive advantage of continuing business model innovation. *The Journal of Business Strategy*, *24*(5), 15–21. DOI: 10.1108/02756660310504924

Mohanty, A., Mohapatra, A. G., & Mohanty, S. K. (2025). Real-Time Traffic Monitoring with AI in Smart Cities. In Abraham, A., Prasad, S., Alhammadi, A., Lestable, T., & Chaabane, F. (Eds.), *Internet of Vehicles and Computer Vision Solutions for Smart City Transformations. Lecture Notes in Intelligent Transportation and Infrastructure*. Springer., DOI: 10.1007/978-3-031-72959-1_7

Mohanty, S. P., Choppali, U., & Kougianos, E. (2016). Everything you wanted to know about smart cities: The Internet of things is the backbone. *IEEE Consumer Electronics Magazine*, *5*(3), 60–70. DOI: 10.1109/MCE.2016.2556879

Mosenia, A., & Jha, N. K. (2017). A comprehensive study of security of Internet-of-Things. *IEEE Transactions on Emerging Topics in Computing*, *5*(4), 586–602. DOI: 10.1109/TETC.2016.2606384

Moskolaï, W. R., Abdou, W., Dipanda, A., & Kolyang. (2021). Application of deep learning architectures for satellite image time series prediction: A review. In *Remote Sensing* (Vol. 13, Issue 23). DOI: 10.3390/rs13234822

Mounce, S., Pedraza, C., Jackson, T., Linford, P., & Boxall, J. (2015). Cloud-based machine learning approaches for leakage assessment and management in smart water networks. *Procedia Engineering*, *119*, 43–52. DOI: 10.1016/j.proeng.2015.08.851

Mroczek, B., & Pijarski, P. (2022). Machine learning in operating of low voltage future grid. *Energies*, *15*(15), 5388. DOI: 10.3390/en15155388

Muhammad, G., & Alhussein, M. (2021). Convergence of artificial intelligence and Internet of Things in smart healthcare: A case study of voice pathology detection. *IEEE Access : Practical Innovations, Open Solutions*, *9*, 89198–89209. DOI: 10.1109/ACCESS.2021.3090317

Mumuni, A., & Mumuni, F. (2024). Automated data processing and feature engineering for deep learning and big data applications: A survey. *Journal of Information and Intelligence*. DOI: 10.48550/2403.11395

Musa, A. A., Malami, S. I., Alanazi, F., Ounaies, W., Alshammari, M., & Haruna, S. I. (2023). Sustainable traffic management for smart cities using internet-of-things-oriented intelligent transportation systems (ITS): Challenges and recommendations. *Sustainability, 15*(13), 9859. DOI: 10.3390/su15139859

Mutambik, I. (2025). IoT-Enabled Adaptive Traffic Management: A Multiagent Framework for Urban Mobility Optimisation. *Sensors (Basel), 25*(13). Advance online publication. DOI: 10.3390/s25134126

Nagaraj, A. (2023). *The Role of AI in Enhancing IoT-Cloud Applications*. Bentham Science Publishers., DOI: 10.2174/97898151657081230101

Nayak, S., & Pandit, D. (2025). Daily activity-travel pattern identification using natural language processing and semantic matching. *Journal of Transport Geography, 122*, 104057. DOI: 10.1016/j.jtrangeo.2024.104057

Neirotti, P., De Marco, A., Cagliano, A. C., Mangano, G., & Scorrano, F. (2014). Current trends in Smart City initiatives: Some stylised facts. *Cities (London, England), 38*, 25–36. DOI: 10.1016/j.cities.2013.12.010

Neupane, B. K., Acharya, B. K., Cao, C., Xu, M., Bhattarai, H., Yang, Y., & Wang, S. (2024). A systematic review of spatial and temporal epidemiological approaches, focus on lung cancer risk associated with particulate matter. *BMC Public Health, 24*(1), 2945. DOI: 10.1186/s12889-024-20431-x PMID: 39448953

Neußner, O. (2021). Early warning alerts for extreme natural hazard events: A review of worldwide practices. In *International Journal of Disaster Risk Reduction* (Vol. 60). DOI: 10.1016/j.ijdrr.2021.102295

Nguyen, D., Ding, M., Pathirana, P., Seneviratne, A., Li, J., & Poor, F. (2021). Federated Learning for Internet of Things: A Comprehensive Survey. *IEEE Communications Surveys and Tutorials, 23*(3), 1622–1658. DOI: 10.1109/COMST.2021.3075439

Nguyen, M.-T., Truong, H.-L., & Truong-Huu, T. (2024). Novel Contract-based Runtime Explainability Framework for End-to-End Ensemble Machine Learning Serving. *Proceedings of the IEEE/ACM 3rd International Conference on AI Engineering - Software Engineering for AI*, 234–244. DOI: 10.1145/3644815.3644964

Nguyen, T. T. T., Malek, L., Umberger, W. J., & O'Connor, P. J. (2022). Household food waste disposal behavior is driven by perceived personal benefits, recycling habits, and ability to compost. *Journal of Cleaner Production, 379*, 134636. DOI: 10.1016/j.jclepro.2022.134636

Nievas, N., Pagès-Bernaus, A., Bonada, F., Echeverria, L., & Domingo, X. (2024). Reinforcement Learning for Autonomous Process Control in Industry 4.0: Advantages and Challenges. *Applied Artificial Intelligence, 38*(1), 2383101. DOI: 10.1080/08839514.2024.2383101

Nikitas, A., Michalakopoulou, K., Njoya, E. T., & Karampatzakis, D. (2020). Artificial Intelligence, Transport and the Smart City: Definitions and Dimensions of a New Mobility Era. *Sustainability, 12*(7). Advance online publication. DOI: 10.3390/su12072789

Noaeen, M., Naik, A., Goodman, L., Crebo, J., Abrar, T., Abad, Z. S. H., Bazzan, A. L. C., & Far, B. (2022). Reinforcement learning in urban network traffic signal control: A systematic literature review. *Expert Systems with Applications, 199*, 116830. DOI: 10.1016/j.eswa.2022.116830

Noardo, F. (2022). Multisource spatial data integration for use cases applications. *Transactions in GIS, 26*(7), 2874–2913. Advance online publication. DOI: 10.1111/tgis.12987

Nogare, D., Silveira, I. F., Cabral, P. P., Hauy, R. J., & Neves, V. (2024). Machine Learning Model: Perspectives for quality, observability, risk and continuous monitoring. *Congresso Latino-Americano de Software Livre e Tecnologias Abertas (Latinoware)*, 181–187. https://sol.sbc.org.br/index.php/latinoware/article/view/31526

Othman, K., Wang, X., Shalaby, A., & Abdulhai, B. (2025). Multimodal adaptive traffic signal control: A decentralized multiagent reinforcement learning approach. *Multimodal Transportation, 4*(1), 100190. DOI: 10.1016/j.multra.2025.100190

Ottun, A. R., Yin, Z., Liyanage, M., Boerger, M., Asadi, M., Hui, P., & Flores, H. (2023). Toward trustworthy and responsible autonomous drones in future smart cities. *Authorea Preprints*. DOI: 10.36227/techrxiv.21444102.v1

Ouyang, F., & Jiao, P. (2021). Artificial intelligence in education: The three paradigms. *Computers and Education: Artificial Intelligence, 2*, 100020. DOI: 10.1016/j.caeai.2021.100020

Pacheco-Cherrez, D. S., Mayo-Maldonado, J. C., Guillen, D., & Escobar, G. (2025). Cybersecurity by design for resilient voltage regulation in active distribution networks. *IEEE Transactions on Smart Grid, 16*(6), 5009–5023. DOI: 10.1109/TSG.2025.3601073

Pal, S., Umair, M., Tan, W.-H., & Foo, Y.-L. (2023). Practical evaluation of federated Learning in Edge AI for IoT. JOIV Int. J. Inform. Vis., 7(3–2), 2115.

Pang, G., Shen, C., Cao, L., & Van Den Hengel, A. (2021). Deep learning for anomaly detection: A review. *ACM Computing Surveys*, *54*(2), 1–38. DOI: 10.1145/3439950

Pang, G., Van Den Hengel, A., Shen, C., & Cao, L. (2021, August). Toward deep supervised anomaly detection: Reinforcement learning from partially labeled anomaly data. In *Proceedings of the 27th ACM SIGKDD Conference on Knowledge Discovery & Data Mining* (pp. 1298–1308). DOI: 10.1145/3447548.3467417

Paparrizos, J., Boniol, P., Palpanas, T., Tsay, R. S., Elmore, A., & Franklin, M. J. (2022). Volume under the surface: A new accuracy evaluation measure for time-series anomaly detection. *Proceedings of the VLDB Endowment International Conference on Very Large Data Bases*, *15*(11), 2774–2787. DOI: 10.14778/3551793.3551830

Parishwad, O., Jiang, S., & Gao, K. (2023). Investigating machine learning for simulating urban transport patterns: A comparison with traditional macro-models. *Multimodal Transportation*, *2*(3), 100085. DOI: 10.1016/j.multra.2023.100085

Pathirannahalage, I., Jayasooriya, V., Samarabandu, J., & Subasinghe, A. (2025). A comprehensive analysis of real-time video anomaly detection methods for human and vehicular movement. *Multimedia Tools and Applications*, *84*(10), 7519–7564. DOI: 10.1007/s11042-024-19204-w

Pawar, K., & Attar, V. (2019). Deep learning approaches for video-based anomalous activity detection. *World Wide Web (Bussum)*, *22*(2), 571–601. DOI: 10.1007/s11280-018-0582-1

Pooyandeh, M., & Sohn, I. (2021). Edge network optimization based on ai techniques: A survey. *Electronics (Basel)*, *10*(22), 2830. DOI: 10.3390/electronics10222830

Priyadarshini, I. (2024). Anomaly detection of IoT cyberattacks in smart cities using federated learning and split learning. *Big Data and Cognitive Computing*, *8*(3), 21. DOI: 10.3390/bdcc8030021

Raffin, T., Reichenstein, T., Klier, D., Kühl, A., & Franke, J. (2022). Qualitative assessment of the impact of manufacturing-specific influences on Machine Learning Operations. *Procedia CIRP*, *115*, 136–141. DOI: 10.1016/j.procir.2022.10.063

Rahimi, T., Ahmed, S., Cardenas-Barrera, J. L., & Diduch, C. (2025). Fairness-oriented Volt-Watt control methods of PV units for over-voltage suppression in PV-enriched smart cities. *Smart Cities*, *8*(3), 88. DOI: 10.3390/smartcities8030088

Rahman, M. A., Rashid, M. M., Hossain, M. S., Hassanain, E., Alhamid, M. F., & Guizani, M. (2019). Blockchain and IoT-based cognitive edge framework for sharing economy services in a smart city. *IEEE Access : Practical Innovations, Open Solutions*, *7*, 18611–18621. DOI: 10.1109/ACCESS.2019.2896065

Ranatunga, S., Ødegård, R. S., Jetlund, K., & Onstein, E. (2025). Use of Semantic Web Technologies to Enhance the Integration and Interoperability of Environmental Geospatial Data: A Framework Based on Ontology-Based Data Access. *ISPRS International Journal of Geo-Information*, *14*(2), 52. DOI: 10.3390/ijgi14020052

Rasheed, M. A., Uddin, S., Abdullah Tanweer, H., Ahmad Rasheed, M., Ahmed, M., & Murtaza, H. (2021). Data privacy issue in Federated Learning Resolution using Block Chain. *VFAST Trans. Softw. Eng.*, *9*(4), 51–61. DOI: 10.21015/vtse.v9i4.726

Rawat, R. (2023). Harnessing the Power of IoT and AI for Human Evolution. *International Journal of Research In Science & Engineering*, *33*(33), 58–68. DOI: 10.55529/ijrise.33.58.68

Rehman, M. H. ur, Salah, K., Damiani, E., & Svetinovic, D. (2020). Towards blockchain-based reputation-aware federated learning. In *Proceedings of the IEEE INFOCOM Conference on Computer Communications Workshops (INFOCOM WKSHPS)* (pp. 183–188). IEEE.

Rehman, A. U., Ali, M., Iqbal, S., Shafiq, A., Ullah, N., & Otaibi, S. A. (2022). Artificial intelligence-based control and coordination of multiple PV inverters for reactive power/voltage control of power distribution networks. *Energies*, *15*(17), 6297. DOI: 10.3390/en15176297

Reis, J., Santo, P. E., & Melão, N. (2019). Artificial intelligence in government services: A systematic literature review. In *Advances in intelligent systems and computing* (pp. 241–252). Springer., DOI: 10.1007/978-3-030-16181-1_23

Ren, X., Li, X., Ren, K., Song, J., Xu, Z., Deng, K., & Wang, X. (2021). Deep Learning-Based Weather Prediction: A Survey. *Big Data Research*, *23*, 100178. Advance online publication. DOI: 10.1016/j.bdr.2020.100178

Ruef, M., & Birkhead, C. (2024). *Learning from outliers and anomalies. Academy of Management Perspectives, (ja),* amp-2023.

Ruiz-Romero, S., Colmenar-Santos, A., Mur-Pérez, F., & López-Rey, Á. (2014). Integration of distributed generation in the power distribution network: The need for smart grid control systems, communication and equipment for a smart city—Use cases. *Renewable & Sustainable Energy Reviews*, *38*, 223–234. DOI: 10.1016/j.rser.2014.05.082

Rukat, T., Lange, D., Schelter, S., & Biessmann, F. (2020). *Towards automated ML model monitoring: Measure, improve and quantify data quality*. Amazon Science. https://www.amazon.science/publications/towards-automated-data-quality-management-for-machine-learning

S. Reis, J. C. (2025). Internet of Things and Artificial Intelligence for Secure and Sustainable Green Mobility: A Multimodal Data Fusion Approach to Enhance Efficiency and Security. *Multimodal Technologies and Interaction*, *9*(5). Advance online publication. DOI: 10.3390/mti9050039

Saini, K., & Sharma, S. (2025). Smart road traffic monitoring: Unveiling the synergy of IoT and AI for enhanced urban mobility. *ACM Computing Surveys*, *57*(11), 276. Advance online publication. DOI: 10.1145/3729217

Sallard, A., & Balać, M. (2023). Travel demand generation using Bayesian Networks: An application to Switzerland. *Procedia Computer Science*, *220*, 267–274. DOI: 10.1016/j.procs.2023.03.035

Samayamantri, L. S., Singhal, S., Krishnamurthy, O., & Regin, R. (2024). AI-driven multimodal approaches to human behavior analysis. In *Advancing Intelligent Networks Through Distributed Optimization* (pp. 485–506). IGI Global. DOI: 10.4018/979-8-3693-3739-4.ch025

San Emeterio de la Parte, M., Martínez-Ortega, J. F., Hernández Díaz, V., & Martínez, N. L. (2023). Big Data and precision agriculture: A novel spatio-temporal semantic IoT data management framework for improved interoperability. *Journal of Big Data*, *10*(1), 52. Advance online publication. DOI: 10.1186/s40537-023-00729-0

Santana, E. F. Z., Chaves, A. P., Gerosa, M. A., Kon, F., & Milojicic, D. S. (2018). Software platforms for smart cities: Concepts, requirements, challenges, and a unified reference architecture. *ACM Computing Surveys, 50*(6), 78:1–78:37.

Santhoshkumar, T., & Vanila, S. (2024). Exploratory data analysis and energy predictions with advanced AI and ML techniques. In *AI approaches to smart and sustainable power systems* (pp. 336–370). IGI Global. DOI: 10.4018/979-8-3693-1586-6.ch018

Sappa, A. (2023). Adaptive Machine Learning Algorithms for Anomaly Detection in Enterprise IT Infrastructure. *Research Briefs on Information and Communication Technology Evolution*, *9*, 206–227. DOI: 10.64799/rebicte.V9.14

Savithramma, R. M., Sumathi, R., & Sudhira, H. S. (2023). Reinforcement learning based traffic signal controller with state reduction. *Journal of Engineering Research*, *11*(1), 100017. DOI: 10.1016/j.jer.2023.100017

Scholapurapu, P. K. (2025). Power electronics for IoT-enabled smart grids and industrial automation. *SSRN*, *10*, 9789349552111–08. DOI: 10.2139/ssrn.5382142

Seenivasan, D. (2024). AI-driven enhancement of ETL workflows for scalable and efficient cloud data engineering. *International Journal of Engineering and Computer Science, 13*(6), 26837–26848. DOI: 10.18535/ijecs.v13i06.4824

Sehgal, A., Perelman, V., Kuryla, S., & Schonwalder, J. (2012). Management of resource constrained devices in the Internet of Things. *IEEE Communications Magazine, 50*(12), 144–149. DOI: 10.1109/MCOM.2012.6384464

Selmy, S. A. H., Kucher, D. E., Mozgeris, G., Moursy, A. R. A., Jimenez-Ballesta, R., Kucher, O. D., Fadl, M. E., & Mustafa, A. (2023). Detecting, Analyzing, and Predicting Land Use/Land Cover (LULC) Changes in Arid Regions Using Landsat Images, CA-Markov Hybrid Model, and GIS Techniques. *Remote Sensing (Basel), 15*(23), 5522. Advance online publication. DOI: 10.3390/rs15235522

Sezer, O. B., Dogdu, E., & Ozbayoglu, A. M. (2018). Context-aware computing, learning, and big data in Internet of Things: A survey. *IEEE Internet of Things Journal, 5*(1), 1–27. DOI: 10.1109/JIOT.2017.2773600

Shafique, K., Khawaja, B. A., Sabir, F., Qazi, S., & Mustaqim, M. (2020). Internet of things (IoT) for next-generation smart systems: A review of current challenges, future trends and prospects for emerging 5G-IoT scenarios. *IEEE Access : Practical Innovations, Open Solutions, 8*, 23022–23040. DOI: 10.1109/ACCESS.2020.2970118

Shah, D., & Crow, M. L. (2016). Online volt–var control for distribution systems with solid-state transformers. *IEEE Transactions on Power Delivery, 31*(1), 343–350. DOI: 10.1109/TPWRD.2015.2457442

Shaheed, S. H. (2023). Intelligent traffic congestion control using Black Widow Optimization with Hybrid Deep Learning on Smart City Environment. *Journal of Smart Internet of Things, 2023*(2), 44–61. DOI: 10.2478/jsiot-2023-0010

Shaheen, S., & Cohen, A. (2019). Shared ride services in North America: Definitions, impacts, and the future of pooling. *Transport Reviews, 39*(4), 427–442. DOI: 10.1080/01441647.2018.1497728

Shankar, S., Garcia, R., Hellerstein, J. M., & Parameswaran, A. G. (2024). "We Have No Idea How Models will Behave in Production until Production": How Engineers Operationalize Machine Learning. *Proceedings of the ACM on Human-Computer Interaction, 8*(CSCW1), 1–34. DOI: 10.1145/3653697

Shao, S., Su, L., Zhang, Q., Wu, S., Guo, S., & Qi, F. (2023). Multi task dynamic edge–end computing collaboration for urban Internet of Vehicles. *Computer Networks, 227*, 109690. DOI: 10.1016/j.comnet.2023.109690

Sharma, S. N., & Dehalwar, K. (2025). Examining the Inclusivity of India's National Urban Transport Policy for Senior Citizens. In D. S.-K. Ting & J. A. Stagner, *Transforming Healthcare Infrastructure* (1st ed., pp. 115–134). CRC Press. https://doi.org/DOI: 10.1201/9781003513834-5

Sharma, S. N., Singh, D., & Dehalwar, K. (2024). Surrogate Safety Analysis- Leveraging Advanced Technologies for Safer Roads. *Suranaree Journal of Science and Technology, 31*(4), 010320(1-14). https://doi.org/DOI: 10.55766/sujst-2024-04-e03837

Sharma, P., Kumawat, R., Jain, G., Keswani, G., Goyal, N. K., & Sheikh, M. F. (2026). Machine Learning-Based Intrusion Detection Systems for Healthcare Internet of Things (HIoT). In Shieh, C., Goel, P., Horng, M., & Pandey, H. (Eds.), *Resilient Healthcare Internet of Things Systems: Architectures, Security, and Trust* (pp. 185–214). IGI Global Scientific Publishing., DOI: 10.4018/979-8-3373-7610-3.ch007

Sharma, S. N., & Dehalwar, K. (2025). A systematic literature review of pedestrian safety in urban transport systems. *Journal of Road Safety*, *36*(4). Advance online publication. DOI: 10.33492/JRS-D-25-4-2707507

Sharma, S. N., & Dehalwar, K. (2025). A Systematic Literature Review of Transit-Oriented Development to Assess Its Role in Economic Development of City. *Transportation in Developing Economies*, *11*(2), 23. DOI: 10.1007/s40890-025-00245-1

Sharma, S. N., Kumar, A., & Dehalwar, K. (2024). The Precursors of Transit-oriented Development. *Economic and Political Weekly*, *59*(14), 16–20. DOI: 10.5281/ZENODO.10939448

Sharma, S., Goyal, N. K., Dandotiya, M., & Sharma, V. P. (2026). An Intelligent Data Encryption Approach with the Entangled Mermin-Based Qkd Protocol for Secure Quantum Communications. In Nayak, R., Mittal, N., Khunteta, A., & Kumar, M. (Eds.), *Recent Advancements in Artificial Intelligence. ICRAAI 2025. Lecture Notes in Networks and Systems* (Vol. 1468). Springer., DOI: 10.1007/978-981-96-7760-3_9

Sharma, S., Singh, V., Goyal, N. K., & Farooqi, A. (2026). Blockchain for Secure and Context-Aware Edge-Fog Computing. In Ramírez, J., Le, D., Kumar, A., Batta, P., & Rathore, P. (Eds.), *Semantic Annotation of Edge/Fog Services With Blockchain Integration* (pp. 263–292). IGI Global Scientific Publishing., DOI: 10.4018/979-8-3373-4387-7.ch009

Sheikh, N. (2024). AI-Driven Observability: Enhancing System Reliability and Performance. *Journal of Artificial Intelligence General Science (JAIGS) ISSN: 3006-4023, 7*(01), 229–239.

Sheikh, M. F. A., Kapoor, S. K., Goyal, N. K., Tuba, E., & Rao, A. (2026). Enhancing healthcare security with fog computing and blockchain technology. In Nayak, R. (Eds.), *Recent advancements in artificial intelligence* (pp. 467–478). Springer., DOI: 10.1007/978-981-96-7760-3_38

Shetty, G. S., & N, R. (2024). Strategies for Achieving energy efficiency and data Security through data aggregation in IoT Healthcare Applications: A Comprehensive study. *International Journal of Computer Networks and Applications, 11*(2), 191. DOI: 10.22247/ijcna/2024/224440

Sicari, S., Rizzardi, A., Grieco, L. A., & Coen-Porisini, A. (2015). Security, privacy and trust in Internet of Things: The road ahead. *Computer Networks, 76*, 146–164. DOI: 10.1016/j.comnet.2014.11.008

Silva, S., Altmann, A., Gutman, B., & Lorenzi, M. (2020). Fed-BioMed: A general open-source frontend framework for federated learning in healthcare. In Domain Adaptation and Representation Transfer, and Distributed and Collaborative Learning (pp. 201–210). Springer International Publishing. DOI: 10.1007/978-3-030-60548-3_20

Singh, I., Zhou, H., Yang, K., Ding, M., Lin, B., & Xie, P. (2020). Differentially-private federated neural architecture search.

Singh, K. P., & Gupta, S. (2012). Artificial intelligence based modeling for predicting disinfection by-products in water. *Chemometrics and Intelligent Laboratory Systems, 114*, 122–131. DOI: 10.1016/j.chemolab.2012.03.014

Singh, S., Rathore, S., & Park, J. (2020). BlockIoTIntelligence: A Blockchain-enabled Intelligent IoT Architecture with Artificial Intelligence. *Future Generation Computer Systems, 110*, 721–743. DOI: 10.1016/j.future.2019.09.002

Sivaraman, H. (2024). Machine Learning-Augmented Unified Testing and Monitoring Framework Reducing Costs and Ensuring Compliance, Quality, and Reliability with Shift-Left and Shift-Right Synergy for Cybersecurity Products. *Journal of Artificial Intelligence. Machine Learning and Data Science, 2*(2), 1645–1652. DOI: 10.51219/JAIMLD/hariprasad-sivaraman/367

SKB, S., Mathivanan, S. K., Rajadurai, H., Cho, J., & Easwaramoorthy, S. V. (. (2024). A multi-modal geospatial–temporal LSTM based deep learning framework for predictive modeling of urban mobility patterns. *Scientific Reports, 14*(1), 31579. DOI: 10.1038/s41598-024-74237-3 PMID: 39738120

Song, K., Yan, F., Ding, T., Gao, L., & Lu, S. (2019). A steel property optimization model based on the XGBoost algorithm and improved PSO. *Computational Materials Science, 174*, 109472. DOI: 10.1016/j.commatsci.2019.109472

Stephanie, V., Khalil, I., Atiquzzaman, M., & Yi, X. (2023). Trustworthy privacy-preserving hierarchical ensemble and federated learning in healthcare 4.0 with blockchain. *IEEE Transactions on Industrial Informatics, 19*(7), 7936–7945. DOI: 10.1109/TII.2022.3214998

Štuikys, V., & Burbaitė, R. (2024). Introduction to artificial neural networks and machine learning. In *Evolution of STEM-driven computer science education: The perspective of big concepts* (pp. 311–346). Springer Nature Switzerland. DOI: 10.1007/978-3-031-48235-9_11

Sugureddy, A. R. (2023). AI-driven solutions for robust data governance: A focus on generative AI applications. *International Journal of Data Analytics, 3*(1), 79–89.

Sumalee, A., & Ho, H. W. (2018). Smarter and more connected: Future intelligent transportation system. *IATSS Research, 42*(2), 67–71. DOI: 10.1016/j.iatssr.2018.05.005

Tahir, Z., Haseeb, M., Mahmood, S. A., Batool, S., Abdullah-Al-Wadud, M., Ullah, S., & Tariq, A. (2025). Predicting land use and land cover changes for sustainable land management using CA-Markov modelling and GIS techniques. *Scientific Reports, 15*(1), 3271. DOI: 10.1038/s41598-025-87796-w PMID: 39863826

Tang, B., Chen, Z., Hefferman, G., Wei, T., He, H., & Yang, Q. (2015). A hierarchical distributed fog computing architecture for big data analysis in smart cities. In *Proceedings of ASE BigData Social Informatics*. DOI: 10.1145/2818869.2818898

Tan, S.-Y., Taeihagh, A., & Sha, K. (2021). How transboundary learning occurs: Case study of the ASEAN Smart Cities Network (ASCN). *Sustainability (Basel), 13*(11), 6502. DOI: 10.3390/su13116502

Tanwar, S., Verma, A., & Bhattacharya, P. (2023). *Federated learning for internet of medical things*. CRC Press. DOI: 10.1201/9781003303374

Tonkoski, R., Lopes, L. A. C., & El-Fouly, T. H. M. (2011). Coordinated active power curtailment of grid-connected PV inverters for overvoltage prevention. *IEEE Transactions on Sustainable Energy, 2*(2), 139–147. DOI: 10.1109/TSTE.2010.2098483

Townsend, A. M. (2013). *Smart cities: Big data, civic hackers, and the quest for a new utopia*. W. W. Norton.

Truong, H.-L., & Nguyen, T.-M. (2021). QoA4ML - A Framework for Supporting Contracts in Machine Learning Services. *2021 IEEE International Conference on Web Services (ICWS)*, 465–475. DOI: 10.1109/ICWS53863.2021.00066

Truong, T. X., Nhu, V. H., Phuong, D. T. N., Nghi, L. T., Hung, N. N., Hoa, P. V., & Bui, D. T. (2023). A New Approach Based on TensorFlow Deep Neural Networks with ADAM Optimizer and GIS for Spatial Prediction of Forest Fire Danger in Tropical Areas. *Remote Sensing (Basel), 15*(14), 3458. Advance online publication. DOI: 10.3390/rs15143458

Uddin, M. Y., & Ahmad, S. (2020). A Review on Edge to Cloud: Paradigm Shift from Large Data Centers to Small Centers of Data Everywhere. *Proceedings of the 5th International Conference on Inventive Computation Technologies, ICICT 2020.* DOI: 10.1109/ICICT48043.2020.9112457

Ullah, W., Hussain, T., Ullah, F. U. M., Lee, M. Y., & Baik, S. W. (2023). TransCNN: Hybrid CNN and transformer mechanism for surveillance anomaly detection. *Engineering Applications of Artificial Intelligence, 123*, 106173. DOI: 10.1016/j.engappai.2023.106173

V. M. S. S. S., Krishnamurthy, T., Chaithra, R. N., & S. G. (2025). Advanced encryption strategies for protecting patient data in IoMT applications. In *2025 6th International Conference for Emerging Technology (INCET)* (pp. 1–6). IEEE. DOI: 10.1109/INCET64471.2025.11139937

Vansteenwegen, P., Melis, L., Aktaş, D., Montenegro, B. D. G., Sartori Vieira, F., & Sörensen, K. (2022). A survey on demand-responsive public bus systems. *Transportation Research Part C, Emerging Technologies, 137*, 103573. DOI: 10.1016/j.trc.2022.103573

Vartika, V., & William, P. (2023). Machine Learning and Cloud Computing Case Study for Online App Usage Tracking. *SciWaveBulletin, 1*(4), 32–37. DOI: 10.61925/SWB.2023.1405

Vasireddy, I., Bindu, G. H., & B, R. (2023). Transformative fusion: Vision transformers and GPT-2 – Unleashing new frontiers in image captioning within image processing. *International Journal of Innovative Research in Engineering and Management, 10*(6), 55–59. DOI: 10.55524/ijirem.2023.10.6.8

Veena, V., Manjunath, R., Asha, V., Gangaraju, V., Swetha, G., & Raghu, N. (2025). Smart Homes with IoT Technology Provide Safety and Security for Elderly Individuals. In Mallik, S., Mathivanan, S., Sangeetha, S., & Soufiene, B. (Eds.), *Real-World Applications of AI Innovation* (pp. 425–450). IGI Global Scientific Publishing., DOI: 10.4018/979-8-3693-4252-7.ch020

Verma, K. K., Singh, B. M., & Dixit, A. (2022). A review of supervised and unsupervised machine learning techniques for suspicious behavior recognition in intelligent surveillance system. *International Journal of Information Technology : an Official Journal of Bharati Vidyapeeth's Institute of Computer Applications and Management, 14*(1), 397–410. DOI: 10.1007/s41870-019-00364-0

Verma, R. K. (2025). Quantum Methods for Data-Efficient Self-supervised Learning in Industrial Automation. In Ghosh, A., Dutta, S., Das, A. K., Shukla, V. K., & Moreira, F. (Eds.), *Quantum Machine Learning in Industrial Automation. Information Systems Engineering and Management* (Vol. 65). Springer., DOI: 10.1007/978-3-031-99786-0_1

Verma, R. K., & Elngar, A. A. (2025). Smart city development in 6G era. In *Driving innovation through AI and digital twin for 6G powered sustainable ultra smart cities* (pp. 27–54). Wiley., DOI: 10.1002/9781394411320.ch2

Verma, R. K., Kumar, I., & Kumar, S. (2025). In Chand Kumain, S., Singh, M., Awasthi, L. K., & Singh, R. (Eds.), *"AI-driven insights: Revolutionizing satellite imagery applications," in *AI and ML Techniques in Image Processing and Object Detection* (pp. 89–104). Springer., DOI: 10.1007/978-981-96-7445-9_5

Vetri Selvi, M., Upadhyay, P., Revathi, A., & Punitha, N. S., R., & Saravanakumar, M. (2025). Enhancing Urban AI-Powered Transportation Systems With Machine Learning: In K. Bellam, K. Nagappan, N. Sankaran, S. Chowdhury, & D. Perakovic (Eds.), *Urban Mobility and Challenges of Intelligent Transportation Systems* (pp. 363–382). IGI Global. DOI: 10.4018/979-8-3693-7984-4.ch020

Von Ziegler, L., Sturman, O., & Bohacek, J. (2021). Big behavior: Challenges and opportunities in a new era of deep behavior profiling. *Neuropsychopharmacology : Official Publication of the American College of Neuropsychopharmacology, 46*(1), 33–44. DOI: 10.1038/s41386-020-0751-7 PMID: 32599604

Wagner, M., & Henzen, C. (2022). Quality Assurance for Spatial Research Data. *ISPRS International Journal of Geo-Information, 11*(6), 334. Advance online publication. DOI: 10.3390/ijgi11060334

Wali, B., Santi, P., & Ratti, C. (2023). A joint demand modeling framework for ride-sourcing and dynamic ridesharing services: A geo-additive Markov random field based heterogeneous copula framework. *Transportation, 50*(5), 1809–1845. *Scopus.* Advance online publication. DOI: 10.1007/s11116-022-10294-9

Wang, Z., Zhou, L., Das, A., Dave, V., Jin, Z., & Zou, J. (2020). Survive the schema changes: Integration of unmanaged data using deep learning. *arXiv Preprint arXiv:2010.07586*. DOI: 10.48550/2010.07586

Wang, B., Dabbaghjamanesh, M., Kavousi-Fard, A., & Yue, Y. (2024). AI-enhanced multi-stage learning-to-learning approach for secure smart cities load management in IoT networks. *Ad Hoc Networks, 164*, 103628. DOI: 10.1016/j.adhoc.2024.103628

Wang, B., Du, J., Cong, Y., & Yu, Y. (2025). Blockchain-based federated learning algorithm and incentive mechanism enhancing data privacy and training accuracy. *J. Comput. Methods Sci. Eng., 25*(1), 84–93. DOI: 10.1177/14727978241299490

Wang, C., Sun, Y., Lv, S., Wang, C., Liu, H., & Wang, B. (2023). Intrusion detection system based on one-class support vector machine and Gaussian mixture model. *Electronics (Basel), 12*(4), 930. DOI: 10.3390/electronics12040930

Wang, H., Zhang, R., Cheng, X., & Yang, L. (2022). Hierarchical Traffic Flow Prediction Based on Spatial-Temporal Graph Convolutional Network. *IEEE Transactions on Intelligent Transportation Systems, 23*(9), 16137–16147. Advance online publication. DOI: 10.1109/TITS.2022.3148105

Wang, S., Duan, J., Shi, D., Xu, C., Li, H., Diao, R., & Wang, Z. (2020). A data-driven multi-agent autonomous voltage control framework using deep reinforcement learning. *IEEE Transactions on Power Systems, 35*(6), 4644–4654. DOI: 10.1109/TPWRS.2020.2990179

Wang, Z., Qiu, J., Zhou, Y., Shi, Y., Fu, L., Chen, W., & Letaief, K. B. (2021). Federated learning via intelligent reflecting surface. *IEEE Transactions on Wireless Communications, 21*(2), 808–822. DOI: 10.1109/TWC.2021.3099505

Wilson, C. (2022). Public engagement and AI: A values analysis of national strategies. *Government Information Quarterly, 39*(1), 101652. DOI: 10.1016/j.giq.2021.101652

Wu, J., Wang, X., Dang, Y., & Lv, Z. (2022). Digital twins and artificial intelligence in transportation infrastructure: Classification, application, and future research directions. *Computers & Electrical Engineering, 101*, 107983. DOI: 10.1016/j.compeleceng.2022.107983

Wu, W., Li, P., Liu, R., Jin, W., Yao, B., Xie, Y., & Ma, C. (2020). Predicting peak load of bus routes with supply optimization and scaled Shepard interpolation: A newsvendor model. *Transportation Research Part E, Logistics and Transportation Review, 142*, 102041. DOI: 10.1016/j.tre.2020.102041

Wu, Y., Tan, H., Qin, L., Ran, B., & Jiang, Z. (2018). A hybrid deep learning based traffic flow prediction method and its understanding. *Transportation Research Part C, Emerging Technologies, 90*, 166–180. DOI: 10.1016/j.trc.2018.03.001

Yang, L., Elisa, N., & Eliot, N. (2018). Privacy and security aspects of e-government in smart cities. In *E-government, e-services and smart cities* (pp. 89–102). Elsevier., DOI: 10.1016/B978-0-12-815032-0.00007-X

Yang, Q., Liu, Y., Cheng, Y., Kang, Y., Chen, T., & Yu, H. (2019). Federated learning. *Synth. Lect. Artif. Intell. Mach. Learn.*, *13*(3), 1–207. PMID: 41336466

Yaqoob, I., Hashem, I. A. T., Mehmood, Y., Gani, A., Mokhtar, S., & Guizani, M. (2017). Internet of Things architecture: Recent advances, taxonomy, requirements, and open challenges. *IEEE Wireless Communications*, *24*(3), 10–16. DOI: 10.1109/MWC.2017.1600421

Yarlagadda, R. T. (2018). Internet of things & artificial intelligence in modern society. *International journal of creative research thoughts (IJCRT), ISSN*, 2320-2882.

Yi, C., Zhao, X., Sun, Q., & Wang, Z. (2024). Assessing the accuracy of remote sensing data products: A multi-granular spatial sampling method. *Future Generation Computer Systems*, *159*, 151–160. DOI: 10.1016/j.future.2024.04.062

Yoon, J., & Choi, M. K. (2023). Exploring video frame redundancies for efficient data sampling and annotation in instance segmentation. In *Proceedings of the IEEE/CVF Conference on Computer Vision and Pattern Recognition* (pp. 3308–3317). DOI: 10.1109/CVPRW59228.2023.00333

Yousefi, M. H. N., Degeler, V., & Lazovik, A. (2023). Empowering Machine Learning Development with Service-Oriented Computing Principles. In Aiello, M., Barzen, J., Dustdar, S., & Leymann, F. (Eds.), *Service-Oriented Computing* (Vol. 1847, pp. 24–44). Springer Nature Switzerland., DOI: 10.1007/978-3-031-45728-9_2

Yu, W., Liu, Y., Dillon, T., Rahayu, W., & Mostafa, F. (2022). An integrated framework for health state monitoring in a smart factory employing IoT and big data techniques. *IEEE Internet of Things Journal*, *9*(3), 2443–2454. DOI: 10.1109/JIOT.2021.3096637

Yu, X., She, X., Zhou, X., & Huang, A. Q. (2014). Power management for DC microgrid enabled by solid-state transformer. *IEEE Transactions on Smart Grid*, *5*(2), 954–965. DOI: 10.1109/TSG.2013.2277977

Zanella, A., Bui, N., Castellani, A., Vangelista, L., & Zorzi, M. (2014). Internet of Things for smart cities. *IEEE Internet of Things Journal*, *1*(1), 22–32. DOI: 10.1109/JIOT.2014.2306328

Zhang, H., Hou, N., Zhang, J., Li, X., & Huang, Y. (2021). Evaluating the Safety Impact of Connected and Autonomous Vehicles with Lane Management on Freeway Crash Hotspots Using the Surrogate Safety Assessment Model. *Journal of Advanced Transportation, 2021. Journal of Advanced Transportation, 2021*, 1–14. Advance online publication. DOI: 10.1155/2021/5565343

Zhang, H., Jia, X., & Chen, C. (2025). Deep learning-based real-time data quality assessment and anomaly detection for large-scale distributed data streams. *International Journal of Medical and All Body Health Research, 6*(1), 1–01. DOI: 10.54660/IJMBHR.2025.6.1.01-11

Zhang, J., & Tao, D. (2020). Empowering Things With Intelligence: A Survey of the Progress, Challenges, and Opportunities in Artificial Intelligence of Things. *IEEE Internet of Things Journal, 8*(10), 7789–7817. DOI: 10.1109/JIOT.2020.3039359

Zhang, K., Ni, J., Yang, K., Liang, X., Ren, J., & Shen, X. S. (2017). Security and privacy in smart city applications: Challenges and solutions. *IEEE Communications Magazine, 55*(1), 122–129. DOI: 10.1109/MCOM.2017.1600267CM

Zhou, H., Laval, J., Zhou, A., Wang, Y., Wu, W., Qing, Z., & Peeta, S. (2022). Review of Learning-Based Longitudinal Motion Planning for Autonomous Vehicles: Research Gaps Between Self-Driving and Traffic Congestion. *Transportation Research Record: Journal of the Transportation Research Board, 2676*(1), 324–341. DOI: 10.1177/03611981211035764

Zhou, Y., Liu, Y., Ning, N., Wang, L., Zhang, Z., Gao, X., & Lu, N. (2024). Integrating knowledge representation into traffic prediction: A spatial–temporal graph neural network with adaptive fusion features. *Complex & Intelligent Systems, 10*(2), 2883–2900. Advance online publication. DOI: 10.1007/s40747-023-01299-7

Zimelewicz, E., Kalinowski, M., Mendez, D., Giray, G., Alves, A. P. S., Lavesson, N., Azevedo, K., Villamizar, H., Escovedo, T., Lopes, H., Biffl, S., Musil, J., Felderer, M., Wagner, S., Baldassarre, T., & Gorschek, T. (2024). *ML-Enabled Systems Model Deployment and Monitoring: Status Quo and Problems* (No. arXiv:2402.05333). arXiv. /arXiv.2402.05333DOI: 10.17771/PUCRio.acad.68155

Zong, Z., Feng, T., Wang, J., Xia, T., & Li, Y. (2025). Deep Reinforcement Learning for Demand-Driven Services in Logistics and Transportation Systems: A Survey. *ACM Transactions on Knowledge Discovery from Data, 19*(4), 1–42. DOI: 10.1145/3708325

Zott, C., & Amit, R. (2012). Creating value through business model innovation. *MIT Sloan Management Review, 53*(3), 41–49.

Zuo, Z., Shuai, B., Wang, G., Liu, X., Wang, X., Wang, B., & Chen, Y. (2015). Convolutional recurrent neural networks: Learning spatial dependencies for image representation. In *Proceedings of the IEEE Conference on Computer Vision and Pattern Recognition Workshops* (pp. 18–26). DOI: 10.1109/CVPRW.2015.7301268

About the Contributors

Sultan Ahmad is associated with the Department of Computer Science at the College of Computer Engineering and Sciences, Prince Sattam Bin Abdulaziz University, Alkharj, Saudi Arabia, and also serves as an Adjunct Professor at Chandigarh University, Punjab, India. He earned his Master of Computer Science and Applications from Aligarh Muslim University in 2006 and completed his PhD in Computer Science and Engineering from Glocal University. With more than twenty years of teaching and research experience, he has established himself as a prolific academic and a strong contributor to global research. He has published around one hundred and eighty research papers and book chapters in respected SCI, SCIE, ESCI, and SCOPUS indexed journals and conferences. His work spans a broad range of emerging technologies, and he also serves as Guest Editor for several international journals in these indexing categories. Dr. Ahmad holds multiple design patents from Australia, China, India, and the United Kingdom and has authored and edited ten books covering advanced topics in computing. His research contributions have received support from national and international funding agencies, and he has delivered numerous talks as a resource person while serving on technical panels and committees of global conferences. With experience as principal and co-principal investigator on several funded research projects, he has also played a vital role in university accreditation processes such as ABET and NCAAA. His research expertise includes intelligent computing, data science, machine learning, and the Internet of Things. He is an active member of professional bodies including IEEE, IACSIT, and the Computer Society of India.

Sudan Jha is a Senior member of IEEE and a Professor in the Department of Computer Science & Engineering at Kathmandu University, Nepal with over 23+ years of combined teaching, research, and industrial experience. He is a lead researcher of the IoT R&D lab at Kathmandu University. His previous affiliations include KIIT University, Chandigarh University, Christ University, etc. He was

'Technical Director' at Nepal Television, 'Principal' at Nepal College of IT, and 'Individual Consultant' at Nepal Telecom Authority. He is dedicated to advancing higher education quality and actively works on smart platforms. His extensive research portfolio comprises 85+ SCI, SCIE indexed research papers and book chapters in international peer-reviewed journals and conferences. He serves as a Co-Editor-in-Chief in an international journal. He also serves as Guest Editor for SCIE/ESCI/SCOPUS indexed journals. With three patents to his name, he has authored and edited 7 books on cutting-edge topics in IoT, 5G, and AI, published by Elsevier, CRC, and AAP. His research has also secured funding for two international projects. Additionally, he contributes as a Keynote at more than 40 international conferences. In addition, he has also delivered faculty development programs, short-term training programs, and workshops in national and international conferences, and universities. He holds certifications in Microservices Architecture, Data Science, and Foundations of Artificial Intelligence. His primary research interests encompass Quality of Services in IoT enabled devices, Neutrosophic theory, and Neutrosophic Soft Set Systems.

Md. Alimul Haque is currently working as Assistant Professor in the Department of Computer Science, Veer Kunwar Singh University, Ara, India. He received his Bachelor and Master degree in Computer Application in 2004 and 2008 respectively and the Ph.D. Degree on "Security and Privacy aspects on Wireless Communication Networks" from V. K. S. University, Ara, India in 2017. His research interests include cyber security, wireless communication networks, IoT and Artificial Intelligence. He has served as a reviewer in many international journals such as the International Journal of Web-Based Learning and Teaching Technologies (IJWLTT) - IGI Global, Elsevier and IEEE Sensors Journals. He has published multi-focused research articles, and his research has led to numerous papers in peer-reviewed journals (Elsevier, Springer, IEEE and IGI) of the world. He has authored book on Blockchain that are available on Amazon. He has been a Life member of ISCA, Kolkata, India and the International Association of Engineers (IAENG), Hong Kong.

***

Munir Ahmad is a seasoned professional in the realm of Spatial Data Infrastructure (SDI), Geo-Information Productions, Information Systems, and Information Governance, boasting over 25 years of dedicated experience in the field. With a PhD in Computer Science, Dr. Ahmad's expertise spans Spatial Data Production, Management, Processing, Analysis, Visualization, and Quality Control. Throughout his career, Dr. Ahmad has been deeply involved in the development and deployment of SDI systems specially in the context of Pakistan, leveraging his proficiency in Spatial Database Design, Web, Mobile & Desktop GIS, and Geo Web Services Architecture. His contributions to Volunteered Geographic Information (VGI) and Open Source Geoportal & Metadata Portal have significantly enriched the geospatial community. As a trainer and researcher, Dr. Ahmad has authored over 50 publications, advancing the industry's knowledge base and fostering innovation in Geo-Tech, Data Governance, and Information Infrastructure, and Emerging Technologies. His commitment to Research and Development (R&D) is evident in his role as a dedicated educator and mentor in the field.

Archana Chaluvadi is a seasoned engineer with over 16 years of experience in Java development, cloud technologies, DevOps, and SecOps. Currently working as part of the SRE DevSecOps team at a leading IT company, Archana specializes in building secure, scalable, and resilient systems. Holding a Ph.D. in Computer Science and Engineering, they have contributed to the academic and professional community through published research papers in international conferences and authored book chapters on cutting-edge technology topics. Passionate about automation, security, and performance optimization, Archana continues to drive innovation in the ever-evolving tech landscape.

Bivash Ranjan Chowdhury holds an MBA in Management Information Systems from the International American University, Los Angeles, and a BSc and MSc in Physics from the National University, Gazipur, Bangladesh. His research interests encompass Artificial Intelligence, Data Management Systems, Machine Learning, and aspects of material science. He is currently engaged in interdisciplinary research focused on the integration of AI technologies within data-driven systems.

Rajeshwari Dembala holds a B.E., M.Tech, and Ph.D. in computer science and engineering from Visvesvaraya Technological University, Belagavi, India. With over 20 years of teaching experience, she is currently an associate professor in the Department of Information Science and Engineering at The National Institute of Engineering, Mysuru. Professor Rajeshwari D has published 18 journal papers in reputed journals, in addition to more than 5 conference publications, book chapters, and patents. Her research interests include data mining, artificial intelligence, the internet of things, blockchain technology, and data analytics. She currently supervises research scholars. She can be contacted at email: drrajeshwari@nie.ac.in.

Hossain Mohammad Dalim is a graduate student at International American University, Los Angeles, pursuing an MBA in Business Analytics. He holds a BBS in Marketing from National University, Gazipur, Bangladesh. His research interests in artificial intelligence (AI), data analysis, business analysis, and business administration. His work focuses on applying AI and machine-learning techniques to financial forecasting, cybersecurity, and risk assessment.

Kavita Dehalwar is an Assistant Professor in the Department of Architecture and Planning at Maulana Azad National Institute of Technology (MANIT), Bhopal, where she has been contributing to teaching and research since 2005. With over 15 years of academic experience, she has actively participated in numerous conferences and has a keen interest in scholarly writing. Dr. Dehalwar holds a PhD from MANIT Bhopal, earned in 2017. She is a life member of the Indian Society of Remote Sensing and the Indian Society of Geomatics, and a permanent member of the National Science Council and the Institute of Town Planners, India.

Mohammed Ezhan is a Software Engineer Intern at eBay and an MS in Computer Science student at Northeastern University, Boston. He has over two years of experience building secure and scalable backend systems, previously working at JPMorgan Chase. He has contributed to high-impact platforms, including a real-time payments dashboard, an OAuth 2.0-based IAM system that improved authentication efficiency, and a Kubernetes-based OCR platform that reduced infrastructure costs. His technical expertise includes Java, Spring Boot, Python, microservices, distributed systems, cloud technologies, and CI/CD. He is currently seeking Fall 2026 Software Engineering internship opportunities in backend, infrastructure, and platform engineering.

KarthiyayiniG. Assistant Professor, CSE(AIML) Mangayarkarasi College of Engineering Mrs G.Karthiyayini has been serving as an Assistant Professor in the Department of CSE(AIML) at Mangayarkarasi College of Engineering since Jan20, 2025.She holds a degree M.E(CSE) from Anna University. With over 11 years of teaching experience, Mrs G.Karthiyayini is committed to academic excellence and student development. Her areas of interest include Computer Networks, Cloud Computing, Artificial Intelligence and Machine Learning she has published papers around 12 in reputed national and international conferences and Published 1 IEEE Scopus Indexed journal and designed 2 Indian patents also. In addition to teaching Mrs G.Karthiyayini actively participates in curriculum development, mentoring students, and attending academic events such as workshops, seminars, FDP's. She believes in innovative teaching methodologies and integrates real-world applications into classroom instruction.

Deepak Gupta is a distinguished academician with over 20 years of teaching experience and expertise in Computer Science and Engineering. He holds a Ph.D. in Computer Engineering and has an extensive research portfolio, including many international journal papers, multiple conference proceedings, and several books.

His research interests span machine learning, artificial intelligence, and data mining. Currently, Dr. Gupta serves as an Associate Professor and Head of Department of CSE (Emerging Branches) at ITM Gwalior, MP, India. He has also contributed significantly to the field through patents and copyrights and actively engages as a reviewer, conference chair and editor.

Anorgul I. Ashirova is a faculty member in the Department of General Professional Sciences at Mamun University, Khiva, Uzbekistan. She is involved in teaching and academic development across general professional disciplines, contributing to foundational skill building and interdisciplinary education. Her academic interests include professional education, curriculum development, and the application of innovative instructional strategies to enhance student engagement and learning outcomes in higher education.

Jayakumar K has born in Tamil Nadu, India on 09th December 2005. He pursuing B.E. degree in Computer Science and Engineering, He is currently Studying in Computer Science and Engineeering at Chennai Institute of Technology, Chennai, India. He is a research scholar with good experience. Driven by a passion for innovation and a dedication to pushing the boundaries of design and technology. My portfolio showcases a commitment to impactful solutions, evidenced by the granting of 2 design patents, reflecting a keen eye for aesthetics and functionality. Beyond design, I hold 10+ published utility patents, demonstrating a deep understanding of practical problem-solving. My expertise is further validated by contributions to the academic community, including 2 publications at prestigious IEEE conferences. This blend of practical invention and scholarly contribution positions me as a versatile and accomplished professional ready to tackle complex challenges and create meaningful advancements.

Nosir Khurramov is a faculty member in the Department of Information Technology and Exact Sciences at Termez University of Economics and Service, Termez, Uzbekistan. He is involved in teaching and academic activities related to information technology and exact sciences, with academic interests in computing fundamentals, quantitative analysis, and technology-supported le

Sanjeev Kumar Is a Professor in Lovely Professional University Punjab India. He received Undergraduate in Hotel Management from Osmania University, Postgraduate in Tourism Management from Madurai Kamaraj University. He completed Doctor of Philosophy in Hospitality Management in 2019 from Amity University Jaipur India. His research areas lies in food Service Industry, his research focuses on Alcoholic beverages, Event management and Sustainable Management Practices, Metaverse, AI Machine Learning and Artificial Intelligence. He published more than 80 research papers, articles and chapters in Scopus Indexed, UGC Approved and peer reviewed Journals and books. Dr. Sanjeev Kumar participated and acted as resource person in various National and International conferences, seminars, session chair, research workshops and industry talks and his work has been widely cited. Dr. Sanjeev also edited more than 20 books with various International Publishers such as IGI Global, Cambridge Scholars, Emerald, CABI, Bentham Science, Taylor & Francis. Dr Sanjeev is an active researcher and has published numerous research articles in prestigious National and International journals and books.

Rampelli Manojkumar received his Ph.D. in Power Engineering from the Indian Institute of Technology Guwahati in 2023. He obtained his Master of Technology in Power and Energy Systems from the National Institute of Technology Karnataka, Surathkal, India, in 2015. He is currently an Assistant Professor in the Department of Electrical and Electronics Engineering at BVRIT Hyderabad College of Engineering for Women. He is currently the Early Career Editor for the International Journal of Ambient Energy and the Guest Editor for PLOS ONE. He has been a speaker at the IEEE Smart Grid webinar series and has contributed to multiple funded research projects on smart transformer-based microgrid systems and distributed generation management. He has published 9 SCI journal papers, including articles in IEEE Transactions on Power Systems, IEEE Transactions on Industrial Informatics, and Sustainable Cities and Society. His research expertise spans energy management, hybrid AC/DC distribution systems, energy storage, and power electronics applications in modern distribution systems. Additionally, he is an active reviewer for around 20 SCI journals, including IEEE Transactions on Industrial Informatics, IEEE Systems Journal, Journal of Energy Storage, Applied Energy, etc.

G. Muthupandi was born in Tamilnadu, India on 20th March 1990. He received B.E. degree in Computer Science and Engineering and M.E. degree in the discipline of Embedded and Real Time System, He is currently working as an Assistant Professor in Computer Science and Engineeering at Chennai Institute of Technology, Chennai, India. He is a researcher with more than Ten years of teaching experience and expertise in the field of the Internet of Things (IoT) and Embedded Systems. Driven by a passion for innovation and a dedication to pushing the boundaries of design and technology, I am a highly accomplished inventor and researcher. My portfolio showcases a commitment to impactful solutions, evidenced by the granting of 10 design patents, reflecting a keen eye for aesthetics and functionality. Beyond design, I hold 20 published utility patents, demonstrating a deep understanding of practical problem-solving. My expertise is further validated by contributions to the academic community, including 8 publications at prestigious IEEE conferences and 2 peer-reviewed journal papers. This blend of practical invention and scholarly contribution positions me as a versatile and accomplished professional ready to tackle complex challenges and create meaningful advancements.

Raghu N has Ten years of experience in Academic and Research. He received the Ph.D. (2020) in Electronics Engineering from Jain University, and B.E. in Electronics and Telecommunications Engineering from VTU University, Karnataka. His research interests were in RF Communication and Image Processing. He has published his research findings in various Scopus indexed Journals and International conferences. He is a member of various professional bodies.

Pramod Begur Nagaraj is a dynamic DevOps Engineer with over 3 years of experience specializing in cloud architecture, CI/CD pipeline management, container orchestration, and infrastructure automation. He holds a Master of Science in Information Systems from Northeastern University, Boston, and a Bachelor's degree in Computer Science from VTU. With a strong foundation in Infrastructure as Code (IaC) and cloud-native technologies, Pramod has a proven track record of building resilient, scalable systems and streamlining software delivery pipelines.

Winner Pulakhandam is an experienced Systems Administrator with a strong academic background, holding Masters degrees in Computer and Information Science. With a focus on managing and optimizing secure IT infrastructures, Winner brings expertise in network security, cloud technologies, and system performance. His work ensures reliable, scalable, and efficient IT operations that support organizational goals

Ravikumar R. Natarajan is an experienced academician with over 12+ years of teaching and research expertise across institutions in India, the UAE, and the Maldives. He currently serves as an Assistant Professor at Marwadi University, Rajkot, specializing in Machine Learning, Recommender Systems, and AI-driven applications in healthcare and security. He holds a Ph.D. in Machine Learning and Recommender Systems from Amity University, Jaipur, a Master of Engineering in CSE from VMRF University, Salem, and a Bachelor of Engineering in CSE from Anna University. He also has a MiniMBA from the Swiss eLearning Institute. Dr. Ravikumar has authored more than 8+ multiple SCIE/Scopus-indexed journal articles, 25+ IEEE/Springer conference papers, and co-authored a book on Artificial Intelligence. He has also published 9 patents and contributed 10+ book chapters in emerging domains such as Federated Learning, Blockchain, and Digital Twins for IGI Global, Wiley and Springer publisher. He actively mentors students through initiatives like the Research Activities Club (RAC) and Student Research Internship Training (SRIT), promoting innovation and research culture.

Arifur Rahman is a Business Analyst and researcher specializing in Artificial Intelligence (AI) and Machine Learning (ML). He uses data, AI, and ML to improve business decisions, enhance efficiency, and support sustainable growth. His work combines business understanding with technical expertise, bridging the gap between data science and real-world strategy. He earned a Master of Business Administration in Management from the National University, Bangladesh, one of the country's largest public universities known for its strong foundation in business and management education. He also completed an MBA in Business Analytics from the International American University (IAU), USA, which emphasizes practical learning in data analytics, technology, and leadership. His coursework included Business Analytics and Data Processing, Applied Decision Making, Strategic Management, and Management Information Systems, reflecting his focus on data-driven strategy and AI-based business transformation. Arifur has designed and applied AI and ML models for customer segmentation, predictive forecasting, fraud detection, energy management, and supply chain analysis. His research explores how organizations can use data responsibly to make informed decisions. He focuses on data-driven approaches that promote efficient and responsible resource use. His goal is to create simple, reliable analytical solutions that help organizations and communities make informed and sustainable choices. His work reflects a deep commitment to advancing responsible AI innovation that supports data-driven growth and sustainability in the digital economy.

Visrutatma Rao Vallu is an accomplished Clinical Data Management (CDM) professional with over 8 years of experience managing and leading complex clinical trials across therapeutic areas, including Oncology, Rare Diseases, Cardiovascular, Nephrology, and Renal Medicine. With advanced degrees in Information Technology and Business Administration, Vish combines technical expertise with strategic insights to deliver high-quality data management solutions.

Aarthi S is a dedicated Assistant Professor in Computer Science with over six years of teaching experience. She is currently pursuing her Ph.D. at Marwadi University, Gujarat, and holds an M.Phil. and MCA from Bharathiar University, Tamil Nadu. Throughout her academic career, Aarthi has specialized in programming languages such as C++, Java, and C, and contributed to research in recommender systems and AI-related data privacy. She has published journal articles, contributed to book chapters, and actively participated in technical workshops. Known for her student mentorship and engagement in academic events, Aarthi remains committed to fostering innovation and excellence in computer science education.

SindhujaS. is an Assistant Professor in the Department of Computer Technology at Kongu Engineering College, Erode. She is dedicated to advancing academic excellence through effective teaching, research, and student mentorship. With expertise in core areas of computer science and emerging technologies, she actively contributes to curriculum development and innovative learning practices. Her academic interests include data analytics, artificial intelligence, and software development methodologies. She is committed to fostering analytical thinking and technical competence among students while encouraging research-oriented learning. Through academic guidance and professional engagement, she strives to bridge the gap between theoretical knowledge and industry requirements.

Vishwanatha S, currently working as an Assistant Professor at JAIN (Deemed-to-Be University), obtained his Master's degree from Bangalore University, Bengaluru. He is pursuing a Ph.D. at Visvesvaraya Technological University (VTU), Belgaum. With over 13 years of experience in teaching and imparting knowledge in Mathematics. He has made significant contributions to the field through the publication of numerous research papers in various international journals. Additionally, he holds several patents, showcasing his innovative thinking and commitment to advancing knowledge. His research interests primarily focus on Fluid Mechanics and Magneto Hydrodynamics. He has presented at various national and international conferences. He is a member of the Indian Mathematical Society (IMS) and the International Association of Engineering (IAENG).

Shashikant Nishant Sharma is a prominent voice in the field of development discourse, known for his contributions as the author of 12 books on planning, management, and development. He currently serves as Research Head at Track2Training, an NGO he co-founded to promote training and skill development. A PhD Research Scholar at MANIT Bhopal and served as Assistant Professor at NIT Calicut, he is also a passionate speaker and writer who believes in fostering vibrant intellectual engagement to advance the goals of inclusive and sustainable development. Open for collaboration on research. *Drop a mail to urp2025@gmail .com

Ayesha Taranum, Associate Professor and Deputy HOD in the Department of Computer Science & Engineering at Vidyavardhaka college of Engineering, Mysuru. She is having 17+ years of Teaching. She completed Ph.D. from Visvesvaraya Technological University, Belagavi, Master of Technology, Software Engineering, from Visvesvaraya Technological University, Belagavi, Bachelor of Engineering in

Information Science & Engineering, Visvesvaraya Technological University, Belagavi, She is Examiner/valuator for Visvesvaraya Technological University, Belagavi. She has published many Research papers in International, National Journals/conferences. She is a Member of CSI, IBM Academic Initiative, and LM ISTE. Dr. Ayesha Taranum area of interest includes Networking, Performance enhancement algorithms & Internet of things. Has also received huge amount of funding from various funding agencies. Recipient of a ICTIEE 2024 Leadership of Teaching & Learning Award at ICTIEE 2024.IGIP International Engineering Educator Award under IIEECP (IUCEE International Engineering Educators Certification Program)

Suvarna Kisan Thakur is an Assistant Professor in the Department of Information Technology at Dr. Babasaheb Ambedkar Technological University (DBATU), Lonere, Maharashtra, India. She is currently pursuing her Ph.D. in the domain of Privacy-Preserving Data Mining. She has published research papers and book chapters in reputed journals and edited volumes indexed in Scopus and other recognized databases. Her research interests include Artificial Intelligence, Data Mining, Smart Healthcare Systems, Intelligent Retail Systems, and Privacy-Preserving Computing. She has actively contributed to academic development through organizing workshops, FDPs, and intellectual property awareness programs. Her current research explores AI-driven intelligent frameworks for healthcare, education, and retail automation systems.

C Venkata Subba Reddy working as an assistant professor in the Electrical and Electronics Engineering department, BVRIT HYDERABAD College of Engineering for Women. He has completed Ph. D from National Institute of Technology, warangal. He has extensive knowledge in power electronic converters and electrical drives.

Raj Kishor Verma is a seasoned academician and researcher with over 12 years of teaching experience in the field of Computer Science and Engineering (CSE). He is currently pursuing a Ph.D. in CSE from Galgotias University, focusing on developing innovative solutions to address contemporary challenges in technology. Raj Kishor Verma holds an M.Tech. in Mobile Computing from the Central University of Jharkhand, where he gained expertise in Automatic Text Summarization by the using Python, mobile systems, and pervasive computing. He completed his B.Tech. in Computer Science and Engineering from RVS College of Engineering & Technology, Jamshedpur, building a strong academic foundation. Throughout his career, Raj Kishor Verma has demonstrated excellence in teaching and research. He has published four research papers in Scopus-indexed journals, contributing significantly to fields such as mobile computing, artificial intelligence, wireless networks, and emerging technologies. He is actively working on additional research projects to advance knowledge in his domain. Currently, he serves as an Assistant Professor at ABES Institute of Technology (ABESIT),

Index

A

B

C

D

E

F

G

I

K

L

M

N

O

P

www.ingramcontent.com/pod-product-compliance
Lightning Source LLC
LaVergne TN
LVHW062334300926
849161LV00007B/1686